Student Solutions Manual

to accompany

Beginning Algebra

Sherri Messersmith
College of DuPage

Prepared by

Kris Mudunuri
Long Beach City College

The McGraw-Hill Companies

Student Solutions Manual to accompany
BEGINNING ALGEBRA
SHERRI MESSERSMITH

Published by McGraw-Hill Higher Education, an imprint of The McGraw-Hill Companies, Inc., 1221 Avenue of the Americas, New York, NY 10020.

This book is printed on acid-free paper.
Printed in the United States of America.

3 4 5 6 7 8 9 0 QVS/QVS 19 18 17 16 15 14

ISBN: 978-0-07-329716-3
MHID: 0-07-329716-X

www.mhhe.com

Student Solutions Manual

Beginning Algebra

Table of Contents

Chapter 1: The Real Number System and Geometry

Section 1.1: Exercises

1) a) $\frac{2}{5}$ b) $\frac{4}{6}=\frac{2}{3}$

c) $\frac{4}{4}=1$

3) $\frac{1}{2}$

5) a) $18=1\cdot 18$
$18=2\cdot 9$
$18=3\cdot 6$
The factors of 18 are 1, 2, 3, 6, 9, and 18.

b) $40=1\cdot 40$
$40=2\cdot 20$
$40=4\cdot 10$
$40=5\cdot 8$
The factors of 40 are 1, 2, 4, 5, 8, 10, 20, and 40.

c) $23=1\cdot 23$
The factors of 23 are 1 and 23.

$60=2\cdot 30$
$60=3\cdot 20$
$60=4\cdot 15$
$60=5\cdot 12$
$60=6\cdot 10$
The factors of 60 are 1, 2, 3, 4, 5, 6, 10, 12, 15, 20, 30, and 60.

7) a) Composite b) Composite

c) Prime

9) Composite. It is divisible by 2 and has other factors as well.

11) a) $18=2\cdot 3\cdot 3$ b) $54=2\cdot 3\cdot 3\cdot 3$

c) $42=2\cdot 3\cdot 7$ d) $150=2\cdot 3\cdot 5\cdot 5$

13) a) $\frac{9}{12}=\frac{3\cdot\cancel{3}}{2\cdot 2\cdot\cancel{3}}=\frac{3}{4}$

b) $\frac{54}{72}=\frac{\cancel{2}\cdot\cancel{3}\cdot\cancel{3}\cdot 3}{\cancel{2}\cdot 2\cdot 2\cdot\cancel{3}\cdot\cancel{3}}=\frac{3}{4}$

c) $\frac{84}{35}=\frac{2\cdot 2\cdot 3\cdot\cancel{7}}{5\cdot\cancel{7}}=\frac{12}{5}$ or $2\frac{2}{5}$

d) $\frac{120}{280}=\frac{\cancel{2}\cdot\cancel{2}\cdot\cancel{2}\cdot 3\cdot\cancel{5}}{\cancel{2}\cdot\cancel{2}\cdot\cancel{2}\cdot\cancel{5}\cdot 7}=\frac{3}{7}$

c) $\frac{125}{500}=\frac{\cancel{5}\cdot\cancel{5}\cdot\cancel{5}}{2\cdot 2\cdot\cancel{5}\cdot\cancel{5}\cdot\cancel{5}}=\frac{1}{4}$

d) $\frac{900}{450}=\frac{2\cdot\cancel{2}\cdot\cancel{3}\cdot\cancel{3}\cdot\cancel{5}\cdot\cancel{5}}{\cancel{2}\cdot\cancel{3}\cdot\cancel{3}\cdot\cancel{5}\cdot\cancel{5}}=2$

15) a) $\frac{2}{7}\cdot\frac{3}{5}=\frac{2\cdot 3}{7\cdot 5}=\frac{6}{35}$

b) $\frac{\overset{5}{\cancel{15}}}{\underset{13}{\cancel{26}}}\cdot\frac{\overset{2}{\cancel{4}}}{\underset{3}{\cancel{9}}}=\frac{5\cdot 2}{13\cdot 3}=\frac{10}{39}$

c) $\frac{1}{\underset{1}{\cancel{2}}}\cdot\frac{\overset{7}{\cancel{14}}}{15}=\frac{1\cdot 7}{1\cdot 15}=\frac{7}{15}$

d) $\frac{\overset{6}{\cancel{42}}}{\underset{5}{\cancel{55}}}\cdot\frac{\overset{2}{\cancel{22}}}{\underset{5}{\cancel{35}}}=\frac{6\cdot 2}{5\cdot 5}=\frac{12}{25}$

e) $4\cdot\frac{1}{8}=\frac{\overset{1}{\cancel{4}}}{1}\cdot\frac{1}{\underset{2}{\cancel{8}}}=\frac{1\cdot 1}{2}=\frac{1}{2}$

f) $6\frac{1}{8}\cdot\frac{2}{7}=\frac{\overset{7}{\cancel{49}}}{\underset{4}{\cancel{8}}}\cdot\frac{\overset{1}{\cancel{2}}}{\underset{1}{\cancel{7}}}=\frac{7\cdot 1}{4\cdot 1}=\frac{7}{4}$ or $1\frac{3}{4}$

17) She multiplied the whole numbers and multiplied the fractions. She should have converted the mixed numbers to improper fractions before multiplying.
$5\frac{1}{2}\cdot 2\frac{1}{3}=\frac{11}{2}\cdot\frac{7}{3}=\frac{11\cdot 7}{2\cdot 3}=\frac{77}{6}$ or $12\frac{5}{6}$

19) a) $\frac{1}{42}\div\frac{2}{7}=\frac{1}{\cancel{42}_{6}}\cdot\frac{\cancel{7}^{1}}{2}=\frac{1\cdot 1}{6\cdot 2}=\frac{1}{12}$

b) $\frac{3}{11}\div\frac{4}{5}=\frac{3}{11}\cdot\frac{5}{4}=\frac{3\cdot 5}{11\cdot 4}=\frac{15}{44}$

c) $\frac{18}{35}\div\frac{9}{10}=\frac{\cancel{18}^{2}}{\cancel{35}_{7}}\cdot\frac{\cancel{10}^{2}}{\cancel{9}_{1}}=\frac{2\cdot 2}{7\cdot 1}=\frac{4}{7}$

d) $\frac{14}{15}\div\frac{2}{15}=\frac{\cancel{14}^{7}}{\cancel{15}_{1}}\cdot\frac{\cancel{15}^{1}}{\cancel{2}_{1}}=\frac{7\cdot 1}{1\cdot 1}=7$

21) $10=2\cdot 5,\ 15=3\cdot 5$
LCM of 10 and 15 is $2\cdot 3\cdot 5=30$

23) a) $10=2\cdot 5,\ 30=2\cdot 3\cdot 5$
LCD of $\frac{9}{10}$ and $\frac{11}{30}$ is $2\cdot 3\cdot 5=30$

b) $8=2\cdot 2\cdot 2,\ 12=2\cdot 2\cdot 3$
LCD of $\frac{7}{8}$ and $\frac{5}{12}$ is $2\cdot 2\cdot 2\cdot 3=24$

c) $9=3\cdot 3,\ 6=2\cdot 3,\ 4=2\cdot 2$
LCD of $\frac{4}{9},\frac{1}{6},$ and $\frac{3}{4}$ is $2\cdot 2\cdot 3\cdot 3=36$

25) a) $\frac{6}{11}+\frac{2}{11}=\frac{6+2}{11}=\frac{8}{11}$

b) $\frac{19}{20}-\frac{7}{20}=\frac{19-7}{20}=\frac{12}{20}=\frac{3}{5}$

c) $\frac{4}{25}+\frac{2}{25}+\frac{9}{25}=\frac{4+2+9}{25}$
$=\frac{15}{25}=\frac{3}{5}$

d) $\frac{2}{9}+\frac{1}{6}=\frac{4}{18}+\frac{3}{18}=\frac{4+3}{18}=\frac{7}{18}$

e) $\frac{3}{5}+\frac{11}{30}=\frac{18}{30}+\frac{11}{30}=\frac{18+11}{30}=\frac{29}{30}$

f) $\frac{13}{18}-\frac{2}{3}=\frac{13}{18}-\frac{12}{18}=\frac{13-12}{18}=\frac{1}{18}$

g)
$\frac{4}{7}+\frac{5}{9}=\frac{36}{63}+\frac{35}{63}=\frac{36+35}{63}=\frac{71}{63}$ or $1\frac{8}{63}$

h) $\frac{5}{6}-\frac{1}{4}=\frac{10}{12}-\frac{3}{12}=\frac{10-3}{12}=\frac{7}{12}$

i) $\frac{3}{10}+\frac{7}{20}+\frac{3}{4}=\frac{6}{20}+\frac{7}{20}+\frac{15}{20}$
$=\frac{6+7+15}{20}$
$=\frac{28}{20}=\frac{7}{5}$ or $1\frac{2}{5}$

j) $\frac{1}{6}+\frac{2}{9}+\frac{10}{27}=\frac{9}{54}+\frac{12}{54}+\frac{20}{54}$
$=\frac{9+12+20}{54}$
$=\frac{41}{54}$

27) a) $8\frac{5}{11}+6\frac{2}{11}=14\frac{5+2}{11}=14\frac{7}{11}$

b) $2\frac{1}{10}+9\frac{3}{10}=11\frac{1+3}{10}$

$=11\frac{4}{10}=11\frac{2}{5}$

c) $7\frac{11}{12}-1\frac{5}{12}=6\frac{11-5}{12}=6\frac{6}{12}=6\frac{1}{2}$

d) $3\frac{1}{5}+2\frac{1}{4}=3\frac{4}{20}+2\frac{5}{20}$

$=5\frac{4+5}{20}=5\frac{9}{20}$

e) $5\frac{2}{3}-4\frac{4}{15}=5\frac{10}{15}-4\frac{4}{15}$

$=1\frac{10-4}{15}$

$=1\frac{6}{15}=1\frac{2}{5}$

f) $9\frac{5}{8}-5\frac{3}{10}=9\frac{25}{40}-5\frac{12}{40}$

$=4\frac{25-12}{40}=4\frac{13}{40}$

g) $4\frac{3}{7}+6\frac{3}{4}=4\frac{12}{28}+6\frac{21}{28}$

$=10\frac{12+21}{28}$

$=10\frac{33}{28}=11\frac{5}{28}$

h) $7\frac{13}{20}+\frac{4}{5}=7\frac{13}{20}+\frac{16}{20}$

$=7\frac{13+16}{20}$

$=7\frac{29}{20}=8\frac{9}{20}$

29) $7\div1\frac{2}{3}=7\div\frac{5}{3}=\frac{7}{1}\cdot\frac{3}{5}=\frac{21}{5}=4\frac{1}{5}$

Alex can make 4 whole bears.
Determine amount of fabric used to make 4 bears:

Amount of fabric remaining:

$= 7 \text{ yds} - \frac{20}{3} \text{ yds} = \frac{21}{3} \text{ yds} - \frac{20}{3} \text{ yds}$

$= \frac{1}{3}$ yd left over

31) $175\cdot\frac{2}{7}=\frac{\overset{25}{\cancel{175}}}{1}\cdot\frac{2}{\underset{1}{\cancel{7}}}=50$ hits

33) Add the width of the frame twice to each dimension since it is being added to both sides of the picture.

$18\frac{3}{8}+2\frac{1}{8}+2\frac{1}{8}=22\frac{3+1+1}{8}$

$=22\frac{5}{8}$

$$12\frac{1}{4}+2\frac{1}{8}+2\frac{1}{8}=2\frac{2}{8}+2\frac{1}{8}+2\frac{1}{8}$$
$$=16\frac{2+1+1}{8}$$
$$=16\frac{4}{8}=16\frac{1}{2}$$
$$16\frac{1}{2}\text{ in. by }22\frac{5}{8}\text{ in.}$$

35) $$\frac{2}{3}+1\frac{1}{4}+1\frac{1}{2}=\frac{8}{12}+1\frac{3}{12}+1\frac{6}{12}$$
$$=2\frac{8+3+6}{12}$$
$$=2\frac{17}{12}$$
$$=3\frac{5}{12}\text{ cups}$$

37) $$16\frac{3}{4}-11\frac{3}{5}=16\frac{15}{20}-11\frac{12}{20}$$
$$=5\frac{15-12}{20}$$
$$=5\frac{3}{20}\text{ gallons}$$

39) $$42\cdot\frac{5}{6}=\frac{\overset{7}{\cancel{42}}}{1}\cdot\frac{5}{\underset{1}{\cancel{6}}}=35\text{ problems}$$

41) Add the total length welded so far:
$$14\frac{1}{6}+10\frac{3}{4}=14\frac{2}{12}+10\frac{9}{12}$$
$$=24\frac{2+9}{12}=24\frac{11}{12}$$
Subtract the total from the desired length:

$$32\frac{7}{8}-24\frac{11}{12}=32\frac{21}{24}-24\frac{22}{24}$$
$$=31\frac{45}{24}-24\frac{22}{24}$$
$$=7\frac{45-22}{24}$$
$$=7\frac{23}{24}\text{ in.}$$

43) $$400\cdot\frac{3}{5}=\frac{\overset{80}{\cancel{400}}}{1}\cdot\frac{3}{\underset{1}{\cancel{5}}}=240\text{ students}$$

Section 1.2: Exercises

1) a) base: 6; exponent: 4

b) base: 2; exponent: 3

c) base: $\frac{9}{8}$; exponent: 5

3) a) $9\cdot9\cdot9\cdot9=9^4$

b) $2\cdot2\cdot2\cdot2\cdot2\cdot2\cdot2\cdot2=2^8$

c) $\frac{1}{4}\cdot\frac{1}{4}\cdot\frac{1}{4}=\left(\frac{1}{4}\right)^3$

5) a) $8^2=8\cdot8=64$

b) $(11)^2=11\cdot11=121$

c) $2^4=2\cdot2\cdot2\cdot2=16$

d) $5^3=5\cdot5\cdot5=125$

e) $3^4=3\cdot3\cdot3\cdot3=81$

f) $12^2=12\cdot12=144$

g) $1^2=1\cdot1=1$

h) $\left(\frac{3}{10}\right)^2 = \frac{3}{10}\cdot\frac{3}{10} = \frac{9}{100}$

i) $\left(\frac{1}{2}\right)^6 = \frac{1}{2}\cdot\frac{1}{2}\cdot\frac{1}{2}\cdot\frac{1}{2}\cdot\frac{1}{2}\cdot\frac{1}{2} = \frac{1}{64}$

j) $(0.3)^2 = (0.3)\cdot(0.3) = 0.09$

7) $(0.5)^2 = 0.5\cdot 0.5 = 0.25$ or

$(0.5)^2 = \left(\frac{1}{2}\right)^2 = \frac{1}{4}$

9) Answers may vary.

11) $17-2+4=15+4=19$

13) $48\div 2+14=24+14=38$

15) $20-3\cdot 2+9=20-6+9$
$=14+9=23$

17) $8+12\cdot\frac{3}{4}=8+9=17$

19) $\frac{2}{5}\cdot\frac{1}{8}+\frac{2}{3}\cdot\frac{9}{10}=\frac{\cancel{2}^{1}}{5}\cdot\frac{1}{\cancel{8}_{4}}+\frac{\cancel{2}^{1}}{\cancel{3}_{1}}\cdot\frac{\cancel{9}^{3}}{\cancel{10}_{5}}$
$=\frac{1}{20}+\frac{3}{5}=\frac{1}{20}+\frac{12}{20}$
$=\frac{13}{20}$

21) $2\cdot\frac{3}{4}-\left(\frac{2}{3}\right)^2=2\cdot\frac{3}{4}-\frac{4}{9}$
$=\frac{3}{2}-\frac{4}{9}$
$=\frac{27}{18}-\frac{8}{18}$
$=\frac{19}{18}$ or $1\frac{1}{18}$

23) $25-11\cdot 2+1=25-22+1$
$=3+1=4$

25) $39-3(9-7)^3=39-3(2)^3$
$=39-3(8)$
$=39-24=15$

27) $60\div 15+5\cdot 3=4+15=19$

29) $7[45\div(19-10)]+2$
$=7[45\div(9)]+2$
$=7\cdot 5+2$
$=35+2=37$

31) $1+2\left[(3+2)^3\div(11-6)^2\right]$
$=1+2\left[(5)^3\div(5)^2\right]$
$=1+2[125\div 25]$
$=1+2[5]$
$=1+10=11$

33) $\frac{4(7-2)^2}{12^2-8\cdot 3}=\frac{4\cdot(5)^2}{144-24}=\frac{4\cdot 25}{120}$
$=\frac{100}{120}=\frac{10}{12}=\frac{5}{6}$

35) $\dfrac{4(9-6)^3}{(2)^2+3\cdot 8}=\dfrac{4(3)^3}{4+24}$

$$=\frac{\overset{1}{\cancel{4}}(27)}{\underset{7}{\cancel{28}}}$$

$$=\frac{27}{7} \text{ or } 3\frac{6}{7}$$

Section 1.3: Exercises

1) Acute

3) Straight

5) Supplementary; complementary

7) $90^\circ-59^\circ=31^\circ$

9) $90^\circ-12^\circ=78^\circ$

11) $180^\circ-143^\circ=37^\circ$

13) $180^\circ-38^\circ=142^\circ$

15)

Angle Measure	Reason
$m\angle A=180^\circ-31^\circ=149^\circ$	supplementary
$m\angle B=31^\circ$	vertical
$m\angle C=m\angle A=149^\circ$	vertical

17) 180

19) the sum of the two known angles

$119^\circ+22^\circ=141^\circ$

the measure of the unknown angle

$180^\circ-141^\circ=39^\circ$

The triangle is obtuse since it contains one obtuse angle.

21) the sum of the two known angles

$90^\circ+51^\circ=141^\circ$

the measure of the unknown angle

$180^\circ-141^\circ=39^\circ$

The triangle is right since it contains one right angle.

23) Equilateral

25) Isosceles

27) True

29) $A=l\cdot w$ $\quad$ $P=2l+2w$

$=10\cdot 8$ $\quad$ $=2\cdot 10+2\cdot 8$

$=80\text{ ft}^2$ $\quad$ $=20+16=36\text{ ft}$

31) $A=\dfrac{1}{2}bh$

$$=\frac{1}{2}(14\text{ cm})(6\text{ cm})$$

$$=7\cdot 6=42\text{ cm}^2$$

$P=a+b+c$

$$=8\text{ cm}+14\text{ cm}+7.25\text{ cm}=29.25\text{ cm}$$

33) $A=s^2$

$$=(6.5\text{ mi})(6.5\text{ mi})$$

$$=42.25\text{ mi}^2$$

$P=4s$

$$=4\cdot 6.5\text{ mi}=26\text{ miles}$$

35) $A=\dfrac{1}{2}h(b_1+b_2)$

$$=\frac{1}{2}\cdot 12(11+16)$$

$$=6\cdot 27=162\text{ in}^2$$

$P=a+b+c+d$

$$=13\text{ in}+11\text{ in}+12\text{ in}+16\text{ in}$$

$$=52\text{ in}$$

37) a) $A = \pi r^2$
$= \pi(5 \text{ in})^2$
$= 25\pi \text{ in}^2$
$\approx 78.5 \text{ in}^2$

b) $C = 2\pi r$
$= 2\pi(5 \text{ in})$
$= 10\pi \text{ in}$
$\approx 31.4 \text{ in}$

39) a) $A = \pi r^2$
$= \pi(2.5 \text{ m})^2$
$= 6.25\pi \text{ m}^2 \approx 19.625 \text{ m}^2$

b) $C = 2\pi r$
$= 2\pi(2.5 \text{ m})$
$= 5\pi \text{ m} \approx 15.7 \text{ m}$

41) $A = \pi r^2$
$= \pi\left(\frac{1}{2} \text{ m}\right)^2$
$= \frac{1}{4}\pi \text{ m}^2$

$C = 2\pi r$
$= 2\pi\left(\frac{1}{2} \text{ m}\right)$
$= \pi \text{ m}$

43) $A = \pi r^2$
$= \pi(7 \text{ ft})^2$
$= 49\pi \text{ ft}^2$

$C = 2\pi r$
$= 2\pi(7 \text{ ft})$
$= 14\pi \text{ ft}$

45) $A = (23 \text{ m})(13 \text{ m}) + (11 \text{ m})(7 \text{ m})$
$= 299 \text{ m}^2 + 77 \text{ m}^2$
$= 376 \text{ m}^2$

$P = 2(23 \text{ m}) + 2(20 \text{ m})$
$= 46 \text{ m} + 40 \text{ m} = 86 \text{ m}$

47) $A = (20.5 \text{ in})(4.8 \text{ in}) + (5.7 \text{ in})(8.4 \text{ in})$
$+ (11.2 \text{ in})(4.9 \text{ in})$
$= 98.40 \text{ in}^2 + 47.88 \text{ in}^2 + 54.88 \text{ in}^2$
$= 201.16 \text{ in}^2$

$P = 2(20.5 \text{ in}) + 2(13.2 \text{ in})$
$= 41 \text{ in} + 26.4 \text{ in} = 67.4 \text{ in}$

49) $A = (14 \text{ in})(12 \text{ in}) - (10 \text{ in})(8 \text{ in})$
$= 168 \text{ in}^2 - 80 \text{ in}^2 = 88 \text{ in}^2$

51) $A = (4 \text{ ft})(7 \text{ ft}) - (1.5 \text{ ft})(1.5 \text{ ft})$
$= 28 \text{ ft}^2 - 2.25 \text{ ft}^2 = 25.75 \text{ ft}^2$

53) $A = (16 \text{ cm})^2 - 3.14(5 \text{ cm})^2$
$= 256 \text{ cm}^2 - 78.5 \text{ cm}^2 = 177.5 \text{ cm}^2$

55) $V = lwh = (7 \text{ m})(5 \text{ m})(2 \text{ m}) = 70 \text{ m}^3$

57) a) $V = \frac{4}{3}\pi r^3$
$= \frac{4}{3}\pi(6 \text{ in})^3 = \frac{4}{3}\pi(216 \text{ in}^3)$
$= 4\pi \cdot 72 \text{ in}^3 = 288\pi \text{ in}^3$

59) $V = \frac{4}{3}\pi r^3$
$= \frac{4}{3}\pi(5 \text{ ft})^3 = \frac{4}{3}\pi(125 \text{ ft}^3) = \frac{500}{3}\pi \text{ ft}^3$

61) $V = \pi r^2 h$
$= \pi(4 \text{ cm})^2(8.5 \text{ cm})$
$= \pi(16 \text{ cm}^2)(8.5 \text{ cm})$
$= 136\pi \text{ cm}^3$

63) a) $A = (9\text{ ft})(6.5\text{ ft}) = 58.5\text{ ft}^2$

$58.5\ \cancel{\text{ft}^2} \cdot \left(\frac{\$20}{\cancel{\text{ft}^2}}\right) = \$1170$

b) No, it would cost $1170 to use this glass.

65) a) $V = \pi(3\text{ ft})^2(8\text{ ft})$

$\approx (3.14)(9\text{ ft}^2)(8\text{ ft})$

$\approx 226.08\text{ ft}^3$

b) $\text{Capacity} = (7.48)(226.08) \approx 1691\text{ gal}$

67) a) $C = 2\pi(10\text{ in}) \approx 20(3.14)\text{ in}$

$\approx 62.8\text{ in}$

b) $A = \pi(10\text{ in})^2 \approx 3.14(100\text{ in}^2)$

$\approx 314\text{ in}^2$

69) $V = (30\text{ ft})(19\text{ ft})(1.5\text{ ft})$

$= 855\text{ ft}^3$

$\text{Capacity} = (855)(7.48) = 6395.4\text{ gal}$

71) First, find the area of the countertop; it is made of three sections:

$$A = 2\frac{1}{4}\left(10\frac{1}{4} - 2\frac{3}{4}\right) + 2\frac{1}{4}\left(9\frac{1}{6} - 2\frac{1}{2} - 2\cdot 2\frac{1}{4}\right)$$

$$+ 2\frac{1}{4}\left(6\frac{5}{6}\right)$$

$$= 2\frac{1}{4}\left(7\frac{1}{2} + 6\frac{5}{6} + 2\frac{1}{6}\right) = 2\frac{1}{4}\left(16\frac{1}{2}\right) = 37\frac{1}{8}$$

The area of the countertop is $37\frac{1}{8}\text{ ft}^2$.

To find the total cost, multiply the area by the unit cost.

$\text{Cost} = 37\frac{1}{8}\text{ ft}^2 \cdot (\$80.00) = \$2970.00$

No, she cannot afford her first choice granite. The granite countertop would cost $2970.00

73) $C = 2\pi r$

$= 2\pi(4.6\text{ in})$

$\approx 2(3.14)(4.6\text{ in})$

$\approx 28.9\text{ in}$

75) *a*) $A = \frac{1}{2}h(b_1 + b_2)$

$= \frac{1}{2}(4\text{ft})(14\text{ft} + 8\text{ft})$

$= 2\text{ft}(22\text{ft}) = 44\text{ft}^2$

b) First, find the perimeter of the garden

$P = 14\text{ft} + 5\text{ft} + 8\text{ft} + 5\text{ft} = 32\text{ft}$

To find the total cost, multiply the perimeter by unit cost

$\text{Cost} = 32\text{ft} \cdot \$23.50/\text{ft} = \$752$

77) $A = \frac{1}{2}bh = \frac{1}{2}(15.6\text{ in.})(18\text{ in.})$

$= 140.4\text{ in.}^2$

79) Use 3.14 for π

$V = \frac{1}{3}\pi r^2 h \approx \frac{1}{3}(3.14)(12\text{ ft})^2(8\text{ ft})$

$\approx 1205.76\text{ ft}^3$

Section 1.4: Exercises

1) Answers may vary.

3) a) 17 b) 17,0 c) 17,0,−25

d) $17, 3.8, \frac{4}{5}, 0, -25, 6.\overline{7}, -2\frac{1}{8}$

e) $\sqrt{10}, 9.721983...$

f) all numbers in the set.

5) true

7) false

9) true

11)

13)

15) The distance of the number from zero

17) −8

19) 15

21) $\frac{3}{4}$

23) $|-10| = 10$

25) $\left|\frac{9}{4}\right| = \frac{9}{4}$

27) $-|-14| = -14$

29) $|17-4| = |13| = 13$

31) $-\left|-4\frac{1}{7}\right| = -4\frac{1}{7}$

33) $-10,\ -2,\ 0,\ \frac{9}{10},\ 3.8,\ 7$

35) $-6.51,\ -6.5,\ -5,\ 2, 7\frac{1}{3},\ 7\frac{5}{6}$

37) True

39) False

41) False

43) False

45) −53

47) 14 million

49) −419,000

Section 1.5: Exercises

1) Answers may vary.

3) Answers may vary.

5) $6-11=-5$

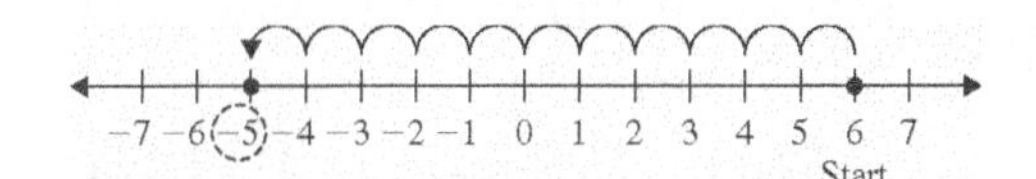

7) $-2+(-7)=-9$

9) $8+(-15)=-7$

11) $-3-11=-14$

13) $-31+54=23$

15) $-26-(-15)=-26+15=-11$

17) $-352-498=-850$

19) $-\frac{7}{12}+\frac{3}{4}=-\frac{7}{12}+\frac{9}{12}=\frac{2}{12}=\frac{1}{6}$

21) $-\frac{1}{6}-\frac{7}{8}=-\frac{4}{24}+\left(-\frac{21}{24}\right)=-\frac{25}{24}\text{ or }-1\frac{1}{24}$

23) $-\frac{4}{9}-\left(-\frac{4}{15}\right)=-\frac{20}{45}+\frac{12}{45}=-\frac{8}{45}$

25) $19.4+(-16.7)=2.7$

27) $-25.8-(-16.57)=-25.8+16.57=-9.23$

29) $9-(5-11)=9-(-6)=9+6=15$

31) $-1+(-6-4)=-1+(-10)=-11$

33) $(-3-1)-(-8+6)=(-4)-(-2)$
$=-4+2=-2$

35) $-16+4+3-10=-12+3-10$
$=-9-10=-19$

37) $5-(-30)-14+2=5+30-14+2$
$=35-14+2$
$=21+2=23$

39) $\frac{4}{9}-\left(\frac{2}{3}+\frac{5}{6}\right)=\frac{8}{18}-\left(\frac{12}{18}+\frac{15}{18}\right)$
$=\frac{8}{18}-\left(\frac{27}{18}\right)$
$=-\frac{19}{18}\text{ or }-1\frac{1}{18}$

41) $\left(\frac{1}{8}-\frac{1}{2}\right)+\left(\frac{3}{4}-\frac{1}{6}\right)$
$=\left(\frac{3}{24}-\frac{12}{24}\right)+\left(\frac{18}{24}-\frac{4}{24}\right)$
$=-\frac{9}{24}+\frac{14}{24}=\frac{5}{24}$

43) $(2.7+3.8)-(1.4-6.9)=6.5-(-5.5)$
$=6.5+5.5=12$

45) $|7-11|+|6+(-13)|=|-4|+|-7|$
$=4+7=11$

47) $-|-2-(-3)|-2|-5+8|$
$=-|-2+3|-2|3|$
$=-|1|-2(3)$
$=-1-6=-7$

49) false

51) false

53) true

55) $-18+6=-12$
His score in the 2005 Masters was -12.

57) $6{,}110{,}000-5{,}790{,}000=320{,}000$
The carbon emissions in China were 320 thousand metric tons more than those of the United States.

59) $881,566+45,407=926,973$

There were 926,973 flights at O'Hare in 2007.

61) a) $1745-1748=-3$

b) $1611-1745=-134$

c) $1480-1611=-131$

d) $1480-1763=-283$

63) a) $2.1-2.8=-0.7$

b) $2.6-2.5=0.1$

c) $2.5-2.1=0.4$

d) $2.6-2.8=-0.2$

65) $5+7=12$

67) $10-16=-6$

69) $9-(-8)=17$

71) $-21+13=-8$

73) $-20+30=10$

75) $23-19=4$

77) $(-5+11)-18=6-18=-12$

Section 1.6: Exercises

1) negative

3) $-8\cdot7=-56$

5) $-15\cdot(-3)=45$

7) $-4\cdot3\cdot(-7)=-12\cdot(-7)=84$

9) $\frac{4}{33}\cdot\left(-\frac{11}{10}\right)=\frac{\overset{2}{\cancel{4}}}{\underset{3}{\cancel{33}}}\cdot\left(-\frac{\overset{1}{\cancel{11}}}{\underset{5}{\cancel{10}}}\right)=-\frac{2}{15}$

11) $(-0.5)(-2.8)=1.4$

13) $-9\cdot(-5)\cdot(-1)\cdot(-3)=45\cdot3=135$

15) $\frac{3}{10}\cdot(-7)\cdot(8)\cdot(-1)\cdot(-5)=\frac{3}{\underset{1}{\cancel{10}}}\cdot\frac{\overset{-28}{\cancel{-280}}}{1}=-84$

17) when k is negative

19) when $k\neq0$

21) $(-6)^2=36$

23) $-5^3=-125$

25) $(-3)^2=9$

27) $-7^2=-49$

29) $-2^5=-32$

31) positive

33) $-50\div(-5)=10$

35) $\frac{64}{-16}=-4$

37) $\frac{-2.4}{0.3}=-8$

39) $-\frac{12}{13}\div\left(-\frac{6}{5}\right)=\frac{10}{13}$

41) $-\frac{0}{7}=0$

43) $\frac{270}{-180}=-\frac{3}{2}$ or $-1\frac{1}{2}$

45) $7+8(-5)=7+(-40)=-33$

47) $(9-14)^2-(-3)(6)=(-5)^2-(-(-18)$
$=25+18$
$=43$

49) $10-2(1-4)^3\div 9$
$=10-2(-3)^3\div 9$
$=10-2(-27)\div 9$
$=10-(-54)\div 9$
$=10+6=16$

51) $\left(-\frac{3}{4}\right)(8)-2\left[7-(-3)(-6)\right]$
$=-6-2[7-18]$
$=-6-2[-11]$
$=-6+22=16$

53) $\frac{-46-3(-12)}{(-5)(-2)(-4)}=\frac{-46+36}{-40}$
$=\frac{-10}{-40}=\frac{1}{4}$

55) $-12\cdot 6=-72$

57) $9+(-7)(-5)=9+35=44$

59) $\frac{63}{-9}+7=-7+7=0$

61) $(-4)(-8)-19=32-19=13$

63) $\frac{-100}{4}-(-7+2)=-25-(-7+2)$
$=-25-(-5)=-25+5=-20$

65) $2\left[18+(-31)\right]=2[-13]=-26$

67) $\frac{2}{3}(-27)=\frac{2}{\not{3}}\left(-\overset{9}{\not{27}}\right)=-18$

69) $12(-5)+\frac{1}{2}(36)=-60+18=-42$

Section 1.7: Exercises

1) The constant is 4.

Term	Coeff.
$7p^2$	7
$-6p$	-6
4	4

3) The constant is 11.

Term	Coeff.
x^2y^2	1
$2xy$	2
$-y$	-1
11	11

5) The constant is -1.

Term	Coeff.
$-2g^5$	-2
$\frac{g^4}{5}$	$\frac{1}{5}$
$3.8g^2$	3.8
g	1
-1	-1

7) a) $4c+3$ when $c=2$
$=4(2)+3=8+3=11$

b) $4c+3$ when $c=-5$
$=4(-5)+3=-20+3=-17$

9) $x+4y=3+4(-5)=3-20=-17$

11)
$$\begin{aligned} &z^2-xy-19 \\ &=(-2)^2-(3)(-5)-19 \\ &=4-(-15)-19=4+15-19 \\ &=19-19=0 \end{aligned}$$

13) $\dfrac{x^3}{2y+1}=\dfrac{(3)^3}{(2)(-5)+1}$
$=\dfrac{27}{-10+1}=\dfrac{27}{-9}=-3$

15) $\dfrac{z^2-y^2}{2y-4(x+z)}$
$=\dfrac{(-2)^2-(-5)^2}{2(-5)-4(3-2)}$
$=\dfrac{4-25}{-10-4(1)}$
$=\dfrac{-21}{-14}=\dfrac{\cancel{-21}^{3}}{\cancel{-14}_{2}}=\dfrac{3}{2}$

17) No, the exponents are different.

19) Yes. both are a^3b-terms

21) 1 22) 0 23) -5 24) $\dfrac{1}{8}$

25) distributive

27) identity

29) commutative

31) associative

33) $19+p$

35) $(8+1)+9$

37) $3k-21$

39) y

41) No. Subtraction is not commutative.

43) $2(1+9)=2\cdot1+2\cdot9=2+18=20$

45) $-2(5+7)=-2\cdot5+(-2)\cdot7$
$=-10+(-14)=-24$

47) $4(8-3)=4\cdot8+4\cdot(-3)=32+(-12)=20$

49) $-(10-4)=-10+4=-6$

51) $8(y+3)=8y+8\cdot3=8y+24$

53) $-10(z+6)=-10z+(-10)\cdot6=10z-60$

55) $-3(x-4y-6)=-3x+(-3)\cdot(-4y)+(-3)\cdot(-6)$
$=-3x+12y+18$

57) $-(-8c+9d-14)=8c-9d+14$

59) $10p+9+14p-2=10p+14p+9-2$
$=24p+7$

61) $-18y^2-2y^2+19+y^2-2+13$
$=-18y^2-2y^2+y^2+19-2+13$
$=-19y^2+30$

63) $\frac{4}{9}+3r-\frac{2}{3}+\frac{1}{5}r=3r+\frac{1}{5}r+\frac{4}{9}-\frac{2}{3}$
$=\frac{15}{5}r+\frac{1}{5}r+\frac{4}{9}-\frac{6}{9}$
$=\frac{16}{5}r-\frac{2}{9}$

65) $2(3w+5)+w=6w+10+w$
$=6w+w+10$
$=7w+10$

67) $9-4(3-x)-4x+3$
$=9-12+4x-4x+3$
$=4x-4x+9-12+3$
$=0x+0=0$

69) $3g-(8g+3)+5=3g-8g-3+5$
$=-5g+2$

71) $-5(t-2)-(10-2t)$
$=-5t+10-10+2t$
$=-5t+2t+10-10$
$=-3t$

73) $3[2(5x+7)-11]+4(7-x)$
$=3[10x+14-11]+28-4x$
$=3[10x+3]+28-4x$
$=30x+9+28-4x$
$=30x-4x+9+28$
$=26x+37$

75) $\frac{4}{5}(2z+10)-\frac{1}{2}(z+3)$
$=\frac{8}{5}z+8-\frac{1}{2}z-\frac{3}{2}$
$=\frac{8}{5}z-\frac{1}{2}z+8-\frac{3}{2}$
$=\frac{16}{10}z-\frac{5}{10}z+\frac{16}{2}-\frac{3}{2}$
$=\frac{11}{10}z+\frac{13}{2}$

77) $1+\frac{3}{4}(10t-3)+\frac{5}{8}\left(t+\frac{1}{10}\right)$
$=1+\frac{15}{2}t-\frac{9}{4}+\frac{5}{8}t+\frac{1}{16}$
$=\frac{15}{2}t+\frac{5}{8}t+1-\frac{9}{4}+\frac{1}{16}$
$=\frac{60}{8}t+\frac{5}{8}t+\frac{16}{16}-\frac{36}{16}+\frac{1}{16}$
$=\frac{65}{8}t-\frac{19}{16}$

79) $2.5(x-4)-1.2(3x+8)$
$=2.5x-10-3.6x-9.6$
$=2.5x-3.6x-10-9.6$
$=-1.1x-19.6$

81) $x+18$

83) $x-6$

85) $x-3$

87) $12+2x$

89) $(3+2x)-7=2x+3-7=2x-4$

91) $(x+15)-5=x+15-5=x+10$

Chapter 1 Review

1) a) $16 = 1 \cdot 16$
$16 = 2 \cdot 8$
$16 = 4 \cdot 4$
The factors of 16 are 1, 2, 4, 8, and 16.

b) $37 = 1 \cdot 37$
The factors of 37 are 1 and 37.

3) a) $\frac{12}{30} = \frac{12 \div 6}{30 \div 6} = \frac{2}{5}$

b) $\frac{414}{702} = \frac{414 \div 18}{702 \div 18} = \frac{23}{39}$

5) $\frac{45}{64} \cdot \frac{32}{75} = \frac{\overset{3}{\cancel{45}}}{\underset{2}{\cancel{64}}} \cdot \frac{\overset{1}{\cancel{32}}}{\underset{5}{\cancel{75}}} = \frac{3}{10}$

7) $35 \div \frac{7}{8} = 35 \cdot \frac{8}{7} = \overset{5}{\cancel{35}} \cdot \frac{8}{\underset{1}{\cancel{7}}} = 40$

9) $\frac{30}{49} \div 2\frac{6}{7} = \frac{30}{49} \div \frac{20}{7}$
$= \frac{30}{49} \cdot \frac{7}{20}$
$= \frac{\overset{3}{\cancel{30}}}{\underset{7}{\cancel{49}}} \cdot \frac{\overset{1}{\cancel{7}}}{\underset{2}{\cancel{20}}} = \frac{3}{14}$

11) $\frac{2}{3} + \frac{1}{4} = \frac{8}{12} + \frac{3}{12} = \frac{11}{12}$

13) $\frac{1}{5} + \frac{1}{3} + \frac{1}{6} = \frac{6}{30} + \frac{10}{30} + \frac{5}{30} = \frac{21}{30} = \frac{7}{10}$

15) $\frac{5}{8} - \frac{2}{7} = \frac{35}{56} - \frac{16}{56} = \frac{19}{56}$

17) $9\frac{3}{8} - 2\frac{5}{6} = 9\frac{9}{24} - 2\frac{20}{24}$
$= 8\frac{33}{24} - 2\frac{20}{24}$
$= 6\frac{33-20}{24} = 6\frac{13}{24}$

19) $3^4 = 3 \cdot 3 \cdot 3 \cdot 3 = 81$

21) $\left(\frac{3}{4}\right)^3 = \left(\frac{3}{4}\right) \cdot \left(\frac{3}{4}\right) \cdot \left(\frac{3}{4}\right) = \frac{27}{64}$

23) $13 - 7 + 4 = 6 + 4 = 10$

25) $\frac{12 - 56 \div 8}{(1+5)^2 - 2^4} = \frac{12-7}{(6)^2 - 2^4}$
$= \frac{5}{36-16}$
$= \frac{5}{20} = \frac{1}{4}$

27) $180^\circ - 78^\circ = 102^\circ$

29) $A = lw$
$= \left(3\frac{1}{2} \text{ mi}\right)\left(1\frac{7}{8} \text{ mi}\right)$
$= \left(\frac{7}{2} \text{ mi}\right)\left(\frac{15}{8} \text{ mi}\right)$
$= \frac{105}{16} \text{mi}^2 = 6\frac{9}{16} \text{ mi}^2$

$P = 2l + 2w$
$= 2\left(3\frac{1}{2} \text{ mi}\right) + 2\left(1\frac{7}{8} \text{ mi}\right)$
$= 7 \text{ mi} + \frac{15}{4} \text{ mi} = 7 \text{ mi} + 3\frac{3}{4} \text{ mi}$
$= 10\frac{3}{4} \text{ mi}$

31) $A=\frac{1}{2}bh+lw$
$=\frac{1}{2}(8\text{in})(3\text{in})+(11\text{in})(8\text{in})$
$=\frac{1}{2}(24\text{ in}^2)+(88\text{ in}^2)$
$=12\text{ in}^2+88\text{ in}^2$
$=100\text{ in}^2$

$P=11\text{ in}+8\text{ in}+11\text{ in}+5\text{ in}+5\text{ in}$
$=40\text{ in}$

33) a) $A=\pi r^2$
$=\pi(3\text{ in})^2$
$=9\pi\text{ in}^2\approx 28.26\text{ in}^2$

b) $C=2\pi r$
$=2\pi(3\text{ in})$
$=6\pi\text{ in}\approx 18.84\text{ in}$

35) $A=\pi(13\text{ cm})^2-\frac{1}{2}(20\text{ cm})(17\text{ cm})$
$=(3.14)169\text{ cm}^2-170\text{ cm}^2$
$=530.66\text{ cm}^2-170\text{ cm}^2$
$=360.66\text{ cm}^2$

37) $V=\pi r^2h$
$=\pi(1\text{ ft})^2(1.3\text{ ft})=1.3\pi\text{ ft}^3$

39) $V=s^3=\left(2\frac{1}{2}\text{ in}\right)^3$
$=\left(\frac{5}{2}\text{ in}\right)^3=\frac{125}{8}\text{ in}^3\text{ or }15\frac{5}{8}\text{ in}^3$

41) a) $-16, 0, 4$
b) $\frac{7}{15}, -16, 0, 3.\overline{2}, 8.5, 4$
c) 4
d) $0, 4$
e) $\sqrt{31}, 6.01832...$

43) a) $|-18|=18$ b) $-|7|=-7$

45) $-21-(-40)=-21+40=19$

47) $\frac{5}{12}-\frac{5}{8}=\frac{5}{12}\cdot\frac{2}{2}-\frac{5}{8}\cdot\frac{3}{3}=\frac{10-15}{24}=-\frac{5}{24}$

49) $\left(-\frac{3}{2}\right)(8)=\left(-\frac{3}{\cancel{2}_1}\right)\left(\cancel{8}^4\right)=-12$

51) $(-4)(3)(-2)(-1)(-3)=72$

53) $-108\div 9=-12$

55) $-3\frac{1}{8}\div\left(\frac{5}{6}\right)=-\frac{\cancel{25}^5}{\cancel{8}_4}\cdot\frac{\cancel{6}^3}{\cancel{5}_1}=\frac{15}{4}=3\frac{3}{4}$

57) $-6^2=-36$

59) $(-2)^6=64$

61) $3^3=27$

63) $56\div(-7)-1=-8-1=-9$

65) $-11+4\cdot 3+(-8+6)^5=-11+4\cdot 3+(-2)^5$
$=-11+12+(-32)$
$=1+(-32)$
$=-31$

67) $\frac{-120}{-3}=40$

69) $(-4)\cdot 7-15=-28-15=-43$

71)

Term	Coeff.
$5z^4$	5
$-8z^3$	-8
$\frac{3}{5}z^2$	$\frac{3}{5}$
$-z$	-1
14	14

73) $\frac{2a+b}{a^3-b^2}$ when $a=-3$ and $t=5$

$$=\frac{2(-3)+5}{(-3)^3-(5)^2}=\frac{-6+5}{-27-25}$$
$$=\frac{-1}{52}=\frac{1}{52}$$

75) inverse

77) distributive

79) $7(3-9)=7\cdot 3-7\cdot 9=21-63=-42$

81) $-(15-3)=-15+3=-12$

83) $9m-14+3m+4=12m-10$

85) $15y^2+8y-4+2y^2-11y+1$

$$=15y^2+2y^2+8y-11y-4+1$$
$$=17y^2-3y-3$$

87) $\frac{3}{2}(5n-4)+\frac{1}{4}(n+6)$

$$=\frac{15}{2}n-6+\frac{1}{4}n+\frac{3}{2}$$
$$=\frac{15}{2}n+\frac{1}{4}n-6+\frac{3}{2}$$
$$=\frac{30}{4}n+\frac{1}{4}n-\frac{12}{2}+\frac{3}{2}$$
$$=\frac{31}{4}n-\frac{9}{2}$$

Chapter 1 Test

1) $210=2\cdot 3\cdot 5\cdot 7$

3) $\frac{7}{16}\cdot\frac{10}{21}=\frac{\overset{1}{\cancel{7}}}{\underset{8}{\cancel{16}}}\cdot\frac{\overset{5}{\cancel{10}}}{\underset{3}{\cancel{21}}}=\frac{5}{24}$

5) $10\frac{2}{3}-3\frac{1}{4}=10\frac{8}{12}-3\frac{3}{12}$

$$=7\frac{8-3}{12}=7\frac{5}{12}$$

7) $\frac{3}{5}-\frac{17}{20}=\frac{12}{20}-\frac{17}{20}=-\frac{\overset{1}{\cancel{5}}}{\underset{4}{\cancel{20}}}=-\frac{1}{4}$

9) $16+8\div 2=16+4=20$

11) $-15\cdot(-4)=60$

13) $23-6\left[-4+(9-11)^4\right]$
$= 23-6\left[-4+(-2)^4\right]$
$= 23-6[-4+16] = 23-6[12]$
$= 23-72 = -49$

15) $14{,}693 \text{ ft} - (-518 \text{ ft})$
$= 14{,}693 \text{ ft} + 518 \text{ ft} = 15{,}211 \text{ ft}$

17) $180^\circ - 31^\circ = 149^\circ$

19) a) $A = \frac{1}{2}bh = \frac{1}{2}(6 \text{ mm})(3 \text{ mm})$
$= 9 \text{ mm}^2$

$P = a+b+c$
$= 5\text{mm} + 6\text{mm} + 3.6\text{mm}$
$= 14.6 \text{ mm}$

b) $A = lw = (15 \text{ cm})(7 \text{ cm})$
$= 105 \text{ cm}^2$

$P = 2l + 2w$
$= 2(15 \text{ cm}) + 2(7 \text{ cm})$
$= 30 \text{ cm} + 14 \text{ cm} = 44 \text{ cm}$

c) $A = l \cdot w + 2(l \cdot w)$
$(16\text{in})\left[(14-4)\text{in}\right] + 2(5\text{in})(4\text{in})$
$= 160 \text{ in}^2 + 40 \text{ in}^2 = 200 \text{ in}^2$

$P = (5+4+6+4+5+14+16+14) \text{ in}$
$= 68 \text{ in}$

21) a) $A = \pi r^2 = \pi(9 \text{ ft})^2 = 81\pi \text{ ft}^2$

b) $A = \pi r^2 \approx 3.14(81) \text{ ft}^3$
$= 254.34 \text{ ft}^3$

23)

$-5 \quad -3\frac{1}{2} \quad -\frac{5}{6} \quad \frac{2}{3} \quad 2.2 \quad 4$

−5 −4 −3 −2 −1 0 1 2 3 4 5

25)

Term	Coeff.
$4p^3$	4
$-p^2$	−1
$\frac{1}{3}p$	$\frac{1}{3}$
−10	−10

27) a) commutative

b) associative

c) inverse

d) distributive

29) a) $-8k^2 + 3k - 5 + 2k^2 + k - 9$
$= -8k^2 + 2k^2 + 3k + k - 5 - 9$
$= -6k^2 + 4k - 14$

b) $\frac{4}{3}(6c-5) - \frac{1}{2}(4c+3)$
$= 8c - \frac{20}{3} - 2c - \frac{3}{2}$
$= 8c - 2c - \frac{20}{3} - \frac{3}{2}$
$= 8c - 2c - \frac{40}{6} - \frac{9}{6} = 6c - \frac{49}{6}$

Chapter 2: The Rules of Exponents

Section 2.1A: Exercises

1) $9 \cdot 9 \cdot 9 \cdot 9 \cdot 9 \cdot 9 = 9^6$

3) $\left(\frac{1}{7}\right)\left(\frac{1}{7}\right)\left(\frac{1}{7}\right)\left(\frac{1}{7}\right) = \left(\frac{1}{7}\right)^4$

5) $(-5)(-5)(-5)(-5)(-5)(-5)(-5)$
$= (-5)^7$

7) $(-3y)(-3y)(-3y)(-3y)$
$\cdot(-3y)(-3y)(-3y)(-3y)$
$= (-3y)^8$

9) base: 6; exponent: 8

11) base: 0.05; exponent: 7

13) base: -8; exponent: 5

15) base: $9x$; exponent: 8

17) base: $-11a$; exponent: 2

19) base: p; exponent: 4

21) base: y; exponent: 2

23) $(3+4)^2 = 7^2 = 49,\ 3^2 + 4^2 = 9 + 16 = 25.$

"adding then squaring" and "squaring then adding" are not the same

25) Answers may vary.

27) No. $3t^4 = 3 \cdot t^4$
$(3t)^4 = 3^4 \cdot t^4 = 81t^4$

29) $2^5 = 32$ 30) $9^2 = 81$

31) $(11)^2 = 121$ 32) $4^3 = 64$

33) $(-2)^4 = 16$

35) $-3^4 = -81$

37) $-2^3 = -8$

39) $\left(\frac{1}{5}\right)^3 = \frac{1}{125}$

41) $2^2 \cdot 2^3 = 2^{2+3} = 2^5 = 32$

43) $3^2 \cdot 3^2 = 3^{2+2} = 3^4 = 81$

45) $5^2 \cdot 2^3 = 25 \cdot 8 = 200$

47) $\left(\frac{1}{2}\right)^4 \cdot \left(\frac{1}{2}\right)^2 = \left(\frac{1}{2}\right)^{4+2} = \left(\frac{1}{2}\right)^6 = \frac{1}{64}$

49) $8^3 \cdot 8^9 = 8^{3+9} = 8^{12}$

51) $5^2 \cdot 5^4 \cdot 5^5 = 5^{2+4+5} = 5^{11}$

53) $(-7)^2 \cdot (-7)^3 \cdot (-7)^3 = (-7)^{2+3+3}$
$= (-7)^8$

55) $b^2 \cdot b^4 = b^{2+4} = b^6$

57) $k \cdot k^2 \cdot k^3 = k^{1+2+3} = k^6$

59) $8y^3 \cdot y^2 = 8(y^3 \cdot y^2) = 8y^5$

61) $(9m^4)(6m^{11}) = (9 \cdot 6)(m^4 \cdot m^{11})$
$= 54m^{15}$

63) $(-6r)(7r^4) = (-6 \cdot 7)(r \cdot r^4) = -42r^5$

65) $(-7t^6)(t^3)(-4t^7)$
$=[-7\cdot 1\cdot(-4)](t^6\cdot t^3\cdot t^7)$
$=28t^{16}$

67) $\left(\frac{5}{3}x^2\right)(12x)(-2x^3)$
$=\left[\frac{5}{3}\cdot 12\cdot(-2)\right](x^2\cdot x\cdot x^3)$
$=-40x^6$

69) $\left(\frac{8}{21}b\right)(-6b^8)\left(-\frac{7}{2}b^6\right)$
$=\left[\frac{8}{21}\cdot(-6)\cdot\left(-\frac{7}{2}\right)\right](b\cdot b^8\cdot b^6)$
$=8b^{15}$

71) $(y^3)^4=y^{3\cdot 4}=y^{12}$

73) $(w^{11})^7=w^{11\cdot 7}=w^{77}$

75) $(3^3)^2=3^{3\cdot 2}=3^6=729$

77) $((-5)^3)^2=(-5)^{3\cdot 2}=(-5)^6$

79) $\left(\frac{1}{3}\right)^4=\frac{1}{3^4}=\frac{1}{81}$

81) $\left(\frac{6}{a}\right)^2=\frac{6^2}{a^2}=\frac{36}{a^2}$

83) $\left(\frac{m}{n}\right)^5=\frac{m^5}{n^5}$

85) $(10y)^4=10^4y^4=10{,}000y^4$

87) $(-3p)^4=(-3)^4\cdot p^4=81p^4$

89) $(-4ab)^3=(-4)^3\cdot a^3\cdot b^3=-64a^3b^3$

91) $6(xy)^3=6\cdot x^3\cdot y^3=6x^3y^3$

93) $-9(tu)^4=-9\cdot t^4\cdot u^4=-9t^4u^4$

95) a) $A=lw$
$=(3w)(w)$
$=3(w\cdot w)$
$=3w^2$ sq units

$P=2l+2w$
$=2(3w)+2(w)$
$=6w+2w$
$=8w$ units

b) $A=lw$
$=(5k^3)(k^2)$
$=5(k^3\cdot k^2)$
$=5k^5$ sq units

$P=2l+2w$
$=2(5k^3)+2(k^2)$
$=10k^3+2k^2$ units

97) $A=\frac{1}{2}bh$
$=\frac{1}{2}(x)\left(\frac{3}{4}x\right)$
$=\left(\frac{1}{2}\cdot\frac{3}{4}\right)(x\cdot x)$
$=\frac{3}{8}x^2$ sq units

Section 2.1B: Exercises

1) operations

3) $(k^9)^2(k^3)^2 = (k^{18})(k^6) = k^{24}$

5) $(5z^4)^2(2z^6)^3 = (5^2)(z^4)^2(2^3)(z^6)^3$
$= (25z^8)(8z^{18})$
$= 200z^{26}$

7) $6ab(-a^{10}b^2)^3$
$= 6ab\cdot(-1)^3(a^{10})^3(b^2)^3$
$= 6ab\cdot(-1a^{30}b^6)$
$= -6a^{31}b^7$

9) $(9+2)^2 = 11^2 = 121$

11) $(-4t^6u^2)^3(u^4)^5$
$= (-4)^3(t^6)^3(u^2)^3\cdot u^{20}$
$= (-64t^{18}u^6)\cdot u^{20}$
$= -64t^{18}u^{26}$

13) $8(6k^7l^2)^2 = 8(6^2)(k^7)^2(l^2)^2$
$= 8(36k^{14}l^4)$
$= 288k^{14}l^4$

15) $\left(\frac{3}{g^5}\right)^3\left(\frac{1}{6}\right)^2 = \frac{3^3}{(g^5)^3}\cdot\frac{1}{6^2} = \frac{27}{g^{15}}\cdot\frac{1}{36}$
$= \frac{27}{36g^{15}} = \frac{\overset{3}{\cancel{27}}}{\underset{4}{\cancel{36}}\,g^{15}}$
$= \frac{3}{4g^{15}}$

17) $\left(\frac{7}{8}n^2\right)^2(-4n^9)^2$
$= \left(\frac{7^2}{8^2}\right)(n^2)^2(-4)^2(n^9)^2$
$= \frac{49}{64}(n^4)(16)(n^{18})$
$= \frac{49}{\underset{4}{\cancel{64}}}(n^4)\left(\overset{1}{\cancel{16}}\right)(n^{18})$
$= \frac{49}{4}n^{22}$

19) $h^4(10h^3)^2(-3h^9)^2$
$= h^4(10^2)(h^3)^2(-3)^2(h^9)^2$
$= h^4(100)(h^6)(9)(h^{18})$
$= 900h^{28}$

21) $3w^{11}(7w^2)^2(-w^6)^5$
$= 3w^{11}(7^2)(w^2)^2(-1)^5(w^6)^5$
$= 3w^{11}(49)(w^4)(-1)(w^{30})$
$= -147w^{45}$

23) $\frac{(12x^3)^2}{(10y^5)^2} = \frac{(12^2)(x^3)^2}{(10^2)(y^5)^2}$
$= \frac{144x^6}{100y^{10}}$
$= \frac{36x^6}{25y^{10}}$

25) $\dfrac{(4d^9)^2}{(-2c^5)^6} = \dfrac{(4^2)(d^9)^2}{(-2)^6(c^5)^6}$

$= \dfrac{16d^{18}}{64c^{30}}$

$= \dfrac{d^{18}}{4c^{30}}$

27) $\dfrac{8(a^4b^7)^9}{(6c)^2} = \dfrac{8(a^4)^9(b^7)^9}{(6^2)(c^2)}$

$= \dfrac{8a^{36}b^{63}}{36c^2}$

$= \dfrac{2a^{36}b^{63}}{9c^2}$

29) $\dfrac{r^4(r^5)^7}{2t(11t^2)^2} = \dfrac{r^4(r^{35})}{2t(11^2)(t^2)^2}$

$= \dfrac{r^{39}}{2t(121t^4)}$

$= \dfrac{r^{39}}{242t^5}$

31) $\left(\dfrac{4}{9}x^3y\right)^2\left(\dfrac{3}{2}x^6y^4\right)^3$

$= \left(\dfrac{4^2}{9^2}\right)(x^3)^2(y^2)\left(\dfrac{3^3}{2^3}\right)(x^6)^3(y^4)^3$

$= \left(\dfrac{16}{81}x^6y^2\right)\left(\dfrac{27}{8}x^{18}y^{12}\right)$

$= \left(\dfrac{\overset{2}{\cancel{16}}}{\underset{3}{\cancel{81}}}x^6y^2\right)\left(\dfrac{\overset{1}{\cancel{27}}}{\underset{1}{\cancel{8}}}x^{18}y^{12}\right)$

$= \dfrac{2}{3}x^{24}y^{14}$

33) $\left(-\dfrac{2}{5}c^9d^2\right)^3\left(\dfrac{5}{4}cd^6\right)^2$

$= (-1)^3\left(\dfrac{2^3}{5^3}\right)(c^9)^3(d^2)^3\left(\dfrac{5^2}{4^2}\right)(c^2)(d^6)^2$

$= \left(-\dfrac{8}{125}c^{27}d^6\right)\left(\dfrac{25}{16}c^2d^{12}\right)$

$= \left(-\dfrac{\overset{1}{\cancel{8}}}{\underset{5}{\cancel{125}}}c^{27}d^6\right)\left(\dfrac{\overset{1}{\cancel{25}}}{\underset{2}{\cancel{16}}}c^2d^{12}\right)$

$= -\dfrac{1}{10}c^{29}d^{18}$

35) $\left(\dfrac{5x^5y^2}{z^4}\right)^3 = \dfrac{(5^3)(x^5)^3(y^2)^3}{(z^4)^3} = \dfrac{125x^{15}y^6}{z^{12}}$

37) $\left(-\dfrac{3t^4u^9}{2v^7}\right)^4 = (-1)^4\cdot\dfrac{(3^4)(t^4)^4(u^9)^4}{(2^4)(v^7)^4}$

$= \dfrac{81t^{16}u^{36}}{16v^{28}}$

39) $\left(\dfrac{12w^5}{4x^3y^6}\right)^2 = \left(\dfrac{\overset{3}{\cancel{12}}w^5}{\underset{1}{\cancel{4}}x^3y^6}\right)^2$

$= \left(\dfrac{3w^5}{x^3y^6}\right)^2 = \dfrac{(3)^2(w^5)^2}{(x^3)^2(y^6)^2}$

$= \dfrac{9w^{10}}{x^6y^{12}}$

41) a) $P = 4(5l^2 \text{ units}) = 20l^2 \text{ units}$

b) $A = (5l^2 \text{ units})^2$

$= (5^2)(l^2)^2 \text{ sq units}$

$= 25l^4 \text{ sq units}$

43) a) $A = (x \text{ units})\left(\frac{3}{8}x \text{ units}\right)$

$= \frac{3}{8}x^2 \text{ sq. units}$

b) $P = 2(x \text{ units}) + 2\left(\frac{3}{8}x \text{ units}\right)$

$= 2x \text{ units} + \frac{3}{4}x \text{ units}$

$= \frac{8}{4}x \text{ units} + \frac{3}{4}x \text{ units}$

$= \frac{11}{4}x \text{ units}$

Section 2.2A: Exercises

1) false

3) true

5) $2^0 = 1$

7) $-5^0 = -1 \cdot 5^0 = -1 \cdot 1 = -1$

9) $0^8 = 0$

11) $(5)^0 + (-5)^0 = 1 + 1 = 2$

13) $6^{-2} = \left(\frac{1}{6}\right)^2 = \frac{1^2}{6^2} = \frac{1}{36}$

15) $2^{-4} = \left(\frac{1}{2}\right)^4 = \frac{1^4}{2^4} = \frac{1}{16}$

17) $5^{-3} = \left(\frac{1}{5}\right)^3 = \frac{1^3}{5^3} = \frac{1}{125}$

19) $\left(\frac{1}{8}\right)^{-2} = 8^2 = 64$

21) $\left(\frac{1}{2}\right)^{-5} = 2^5 = 32$

23) $\left(\frac{4}{3}\right)^{-3} = \left(\frac{3}{4}\right)^3 = \frac{3^3}{4^3} = \frac{27}{64}$

25) $\left(\frac{9}{7}\right)^{-2} = \left(\frac{7}{9}\right)^2 = \frac{7^2}{9^2} = \frac{49}{81}$

27) $\left(-\frac{1}{4}\right)^{-3} = (-4)^3 = -64$

29) $\left(-\frac{3}{8}\right)^{-2} = \left(\frac{8}{-3}\right)^2 = \frac{8^2}{(-3)^2} = \frac{64}{9}$

31) $-2^{-6} = -\left(\frac{1}{2}\right)^6 = -\frac{1^6}{2^6} = -\frac{1}{64}$

33) $-1^{-5} = -(1)^5 = -1$

35) $2^{-3} - 4^{-2} = \left(\frac{1}{2}\right)^3 - \left(\frac{1}{4}\right)^2$

$= \frac{1^3}{2^3} - \frac{1^2}{4^2} = \frac{1}{8} - \frac{1}{16}$

$= \frac{2}{16} - \frac{1}{16} = \frac{1}{16}$

37) $2^{-2} + 3^{-2} = \left(\frac{1}{2}\right)^2 + \left(\frac{1}{3}\right)^2$

$= \frac{1^2}{2^2} + \frac{1^2}{3^2} = \frac{1}{4} + \frac{1}{9}$

$= \frac{9}{36} + \frac{4}{36} = \frac{13}{36}$

39) $-9^{-2}+3^{-3}+(-7)^{0}$
$=-\left(\frac{1}{9}\right)^{2}+\left(\frac{1}{3}\right)^{3}+1$
$=-\frac{1^2}{9^2}+\frac{1^3}{3^3}+1=-\frac{1}{81}+\frac{1}{27}+1$
$=-\frac{1}{81}+\frac{3}{81}+\frac{81}{81}=\frac{83}{81}$

Section 2.2B: Exercises

1) a) w b) n c) $2p$ d) c

3) $r^0=1$ 4) $(5m)^0=1$

5) $-2k^0=-2\cdot 1=-2$

7) $x^0+(2x)^0=1+1=2$

9) $d^{-3}=\left(\frac{1}{d}\right)^{3}=\frac{1^3}{d^3}=\frac{1}{d^3}$

11) $p^{-1}=\left(\frac{1}{p}\right)^{1}=\frac{1}{p}$

13) $\left(\frac{a^{-10}}{b^{-3}}\right)=\frac{b^3}{a^{10}}$

15) $\frac{y^{-8}}{x^{-5}}=\frac{x^5}{y^8}$

17) $\frac{t^5}{8u^{-3}}=\frac{t^5u^3}{8}$

19) $5m^6n^{-2}=\frac{5m^6}{n^2}$

21) $\frac{2}{t^{-11}u^{-5}}=2t^{11}u^5$

23) $\frac{8a^6b^{-1}}{5c^{-10}d}=\frac{8a^6c^{10}}{5bd}$

25) $\frac{2z^4}{x^{-7}y^{-6}}=2x^7y^6z^4$

27) $\left(\frac{a}{6}\right)^{-2}=\left(\frac{6}{a}\right)^{2}=\frac{6^2}{a^2}=\frac{36}{a^2}$

29) $\left(\frac{2n}{q}\right)^{-5}=\left(\frac{q}{2n}\right)^{5}=\frac{q^5}{2^5n^5}=\frac{q^5}{32n^5}$

31) $\left(\frac{12b}{cd}\right)^{-2}=\left(\frac{cd}{12b}\right)^{2}=\frac{c^2d^2}{12^2b^2}=\frac{c^2d^2}{144b^2}$

33) $-9k^{-2}=-9\cdot\frac{1}{k^2}=-\frac{9}{k^2}$

35) $3t^{-3}=3\cdot\frac{1}{t^3}=\frac{3}{t^3}$

37) $-m^{-9}=-1\cdot\frac{1}{m^9}=-\frac{1}{m^9}$

39) $\left(\frac{1}{z}\right)^{-10}=z^{10}$

41) $\left(\frac{1}{j}\right)^{-1}=j$

43) $5\left(\frac{1}{n}\right)^{-2}=5n^2$

45) $c\left(\frac{1}{d}\right)^{-3}=cd^3$

Section 2.3: Exercises

1) You must subtract the denominators exponent from the numerators exponent; $\frac{a^5}{a^3}$=a$^{5-3}$=a^2

3) $\frac{d^{10}}{d^5}=d^{10-5}=d^5$

5) $\frac{m^9}{m^5}=m^{9-5}=m^4$

7) $\frac{8t^{15}}{t^8}=8t^{15-8}=8t^7$

9) $\frac{6^{12}}{6^{10}}=6^{12-10}=6^2=36$

11) $\frac{3^{12}}{3^8}=3^{12-8}=3^4=81$

13) $\frac{2^5}{2^9}=2^{5-9}=2^{-4}=\left(\frac{1}{2}\right)^4=\frac{1}{16}$

15) $\frac{5^6}{5^9}=5^{6-9}=5^{-3}=\left(\frac{1}{5}\right)^3=\frac{1}{125}$

17) $\frac{10d^4}{d^2}=10d^{4-2}=10d^2$

19) $\frac{20c^{11}}{30c^6}=\frac{\overset{2}{\cancel{20}}}{\underset{3}{\cancel{30}}}c^{11-6}=\frac{2}{3}c^5$

21) $\frac{y^3}{y^8}=y^{3-8}=y^{-5}=\frac{1}{y^5}$

23) $\frac{x^{-3}}{x^6}=x^{-3-6}=x^{-9}=\frac{1}{x^9}$

25) $\frac{t^{-6}}{t^{-3}}=t^{-6+3}=t^{-3}=\frac{1}{t^3}$

27) $\frac{a^{-1}}{a^9}=a^{-1-9}=a^{-10}=\frac{1}{a^{10}}$

29) $\frac{t^4}{t}=t^{4-1}=t^3$

31) $\frac{15w^2}{w^{10}}=15w^{2-10}=15w^{-8}=\frac{15}{w^8}$

33) $\frac{-6k}{k^4}=-6k^{1-4}=-6k^{-3}=-\frac{6}{k^3}$

35) $\frac{a^4b^9}{ab^2}=a^{4-1}b^{9-2}=a^3b^7$

37) $\frac{10k^{-2}l^{-6}}{15k^{-5}l^2}=\frac{\overset{2}{\cancel{10}}k^{-2+5}l^{-6-2}}{\underset{3}{\cancel{15}}}$

$=\frac{2k^3l^{-8}}{3}=\frac{2k^3}{3l^8}$

39) $\frac{300x^7y^3}{30x^{12}y^8}=\frac{\overset{10}{\cancel{300}}x^{7-12}y^{3-8}}{\cancel{30}}$

$=10x^{-5}y^{-5}$

$=\frac{10}{x^5y^5}$

41) $\frac{6v^{-1}w}{54v^2w^{-5}}=\frac{\overset{1}{\cancel{6}}v^{-1}w}{\underset{9}{\cancel{54}}v^2w^{-5}}$

$=\frac{1}{9}v^{-1-2}w^{1+5}$

$=\frac{1}{9}v^{-3}w^6=\frac{w^6}{9v^3}$

43) $\dfrac{3c^5d^{-2}}{8cd^{-3}}=\dfrac{3}{8}c^{5-1}d^{-2+3}$
$=\dfrac{3}{8}c^4d$

45) $\dfrac{(x+y)^9}{(x+y)^2}=(x+y)^{9-2}=(x+y)^7$

47) $\dfrac{(c+d)^{-5}}{(c+d)^{-11}}=(c+d)^{-5+11}=(c+d)^6$

Putting It All Together

1) $\left(\dfrac{2}{3}\right)^4=\dfrac{2^4}{3^4}=\dfrac{16}{81}$

3) $\dfrac{3^9}{3^5\cdot 3^4}=\dfrac{3^9}{3^{5+4}}=\dfrac{3^9}{3^9}=3^{9-9}=3^0=1$

5) $\left(\dfrac{10}{3}\right)^{-2}=\left(\dfrac{3}{10}\right)^2=\dfrac{3^2}{10^2}=\dfrac{9}{100}$

7) $(9-6)^2=3^2=9$

9) $10^{-2}=\dfrac{1}{10^2}=\dfrac{1}{100}$

11) $\dfrac{2^7}{2^{12}}=2^{7-12}=2^{-5}=\dfrac{1}{2^5}=\dfrac{1}{32}$

13) $\left(-\dfrac{5}{3}\right)^{-7}\cdot\left(-\dfrac{5}{3}\right)^4=\left(-\dfrac{5}{3}\right)^{-7+4}$
$=\left(-\dfrac{5}{3}\right)^{-3}$
$=\left(-\dfrac{3}{5}\right)^3=-\dfrac{27}{125}$

15) $3^{-2}-12^{-1}=\left(\dfrac{1}{3}\right)^2-\dfrac{1}{12}$
$=\dfrac{1}{9}-\dfrac{1}{12}$
$=\dfrac{4}{36}-\dfrac{3}{36}=\dfrac{1}{36}$

17) $-10\left(-3g^4\right)^3=-10\cdot(-27)g^{12}$
$=270g^{12}$

19) $\dfrac{33s}{s^{12}}=33s^{1-12}=33s^{-11}=\dfrac{33}{s^{11}}$

21) $\left(\dfrac{2xy^4}{3x^{-9}y^{-2}}\right)^4=\left(\dfrac{2}{3}x^{1+9}y^{4+2}\right)^4$
$=\left(\dfrac{2}{3}x^{10}y^6\right)^4$
$=\dfrac{16}{81}x^{40}y^{24}$

23) $\left(\dfrac{9m^8}{n^3}\right)^{-2}=\left(\dfrac{n^3}{9m^8}\right)^2=\dfrac{n^6}{81m^{16}}$

25) $\left(-b^5\right)^3=-b^{15}$ 26) $\left(h^{11}\right)^8=h^{88}$

27) $\left(-3m^5n^2\right)^3=-27m^{15}n^6$

29) $\left(-\dfrac{9}{4}z^5\right)\left(\dfrac{8}{3}z^{-2}\right)=\left(-\dfrac{\overset{3}{\cancel{9}}}{\underset{1}{\cancel{4}}}z^5\right)\left(\dfrac{\overset{2}{\cancel{8}}}{\underset{1}{\cancel{3}}}z^{-2}\right)$
$=-6z^{5-2}=-6z^3$

31) $\left(\dfrac{s^7}{t^3}\right)^{-6}=\left(\dfrac{t^3}{s^7}\right)^6=\dfrac{t^{18}}{s^{42}}$

33) $\left(-ab^3c^5\right)^2\left(\dfrac{a^4}{bc}\right)^3=\left(a^2b^6c^{10}\right)\left(\dfrac{a^{12}}{b^3c^3}\right)$
$=a^{2+12}b^{6-3}c^{10-3}$
$=a^{14}b^3c^7$

35) $\left(\dfrac{48u^{-7}v^2}{36u^3v^{-5}}\right)^{-3}=\left(\dfrac{36u^3v^{-5}}{48u^{-7}v^2}\right)^3$
$=\left(\dfrac{\overset{3}{\cancel{36}}\,u^3v^{-5}}{\underset{4}{\cancel{48}}\,u^{-7}v^2}\right)^3$
$=\left(\dfrac{3}{4}u^{3-(-7)}v^{-5-2}\right)^3$
$=\left(\dfrac{3}{4}u^{10}v^{-7}\right)^3$
$=\dfrac{27}{64}u^{30}v^{-21}=\dfrac{27u^{30}}{64v^{21}}$

37) $\left(\dfrac{-3t^4u}{t^2u^{-4}}\right)^3=\left(-3t^{4-2}u^{1+4}\right)^3$
$=\left(-3t^2u^5\right)^3=-27t^6u^{15}$

39) $\left(h^{-3}\right)^6=h^{-18}=\dfrac{1}{h^{18}}$

41) $\left(\dfrac{h}{2}\right)^4=\dfrac{h^4}{16}$

43) $-7c^4\left(-2c^2\right)^3=-7c^4\cdot(-8)c^6=56c^{10}$

45) $\left(12a^7\right)^{-1}(6a)^2=\dfrac{(6a)^2}{12a^7}=\dfrac{36a^2}{12a^7}$
$=\dfrac{\overset{3}{\cancel{36}}\,a^2}{\underset{1}{\cancel{12}}\,a^7}=3a^{2-7}$
$=3a^{-5}=\dfrac{3}{a^5}$

47) $\left(\dfrac{9}{20}r^4\right)\left(4r^{-3}\right)\left(\dfrac{2}{33}r^9\right)$
$=\dfrac{9}{20}\cdot4\cdot\dfrac{2}{33}r^4r^{-3}r^9$
$=\dfrac{\overset{3}{\cancel{9}}\cdot\overset{1}{\cancel{4}}\cdot2}{\underset{5}{\cancel{20}}\cdot1\cdot\underset{11}{\cancel{33}}}r^{4-3+9}$
$=\dfrac{6}{55}r^{10}$

49) $\dfrac{(a^2b^{-5}c)^{-3}}{(a^4b^{-3}c)^{-2}}=\dfrac{a^{-6}b^{15}c^{-3}}{a^{-8}b^6c^{-2}}=a^{-6+8}b^{15-6}c^{-3+2}=\dfrac{a^2b^9}{c}$

51) $\dfrac{(2mn^{-2})^3(5m^2n^{-3})^{-1}}{(3m^{-3}n^3)^{-2}}$
$=\dfrac{(2mn^{-2})^3(3m^{-3}n^3)^2}{(5m^2n^{-3})^1}$
$=\dfrac{8m^3n^{-6}9m^{-6}n^6}{5m^2n^{-3}}$
$=\dfrac{72}{5}m^{3-6-2}n^{-6+6+3}=\dfrac{72}{5}m^{-5}n^3$
$=\dfrac{72n^3}{5m^5}$

53) $\left(\dfrac{4n^{-3}m}{n^8m^2}\right)^0=1$

55) $\left(\dfrac{49c^4d^8}{21c^4d^5}\right)^{-2}=\left(\dfrac{21c^4d^5}{49c^4d^8}\right)^2$
$=\left(\dfrac{\overset{3}{\cancel{21}}}{\underset{7}{\cancel{49}}}c^{4-4}d^{5-8}\right)^2$
$=\left(\dfrac{3}{7}c^0d^{-3}\right)^2=\left(\dfrac{3}{7d^3}\right)^2=\dfrac{9}{49d^6}$

57) $(p^{2c})^6=p^{12c}$

59) $y^{m} \cdot y^{3m} = y^{4m}$

61) $t^{5b} \cdot t^{-8b} = t^{-3b} = \dfrac{1}{t^{3b}}$

63) $\dfrac{\overset{5}{\cancel{25}}\, c^{2x}}{\underset{8}{\cancel{40}}\, c^{9x}} = \dfrac{5c^{2x-9x}}{8} = \dfrac{5c^{-7x}}{8} = \dfrac{5}{8c^{7x}}$

Section 2.4: Exercises

1) yes

3) no

5) no

7) yes

9) Answers may vary.

11) Answers may vary.

13) 71.765×10^{2} : $71.765 = 7176.5$

15) 40.6×10^{-3} : $0.0406 = 0.0406$

17) $1{,}200{,}006\times10^{-7}$: $0.1200006 = 0.1200006$

19) -6.8×10^{-5} : $-00006.8 = -0.000068$

21) -5.26×10^{4} : $-5.2600 = -52{,}600$

23) 8×10^{-6} : $000008. = 0.000008$

25) 6.021967×10^{5} : 6.021967
$= 602{,}196.7$

27) 3×10^{6} : $3.000000 = 3{,}000{,}000$

29) -7.44×10^{-4} : $-0007.44 = -0.000744$

31) 2.4428×10^{7} : $2.4428000 = 24{,}428{,}000$

33) 2.5×10^{-11} : 00000000025.0
$= 0.000000000025$ *meters*

35) $2110.5 = 2110.5 = 2.1105\times10^{3}$

37) $0.000096 = 0.000096 = 9.6\times10^{-5}$

39) $-7{,}000{,}000 = -7{,}000{,}000. = -7\times10^{6}$

41) $3400 = 3400. = 3.4\times10^{3}$

43) $0.0008 = 0.0008 = 8\times10^{-4}$

45) $-0.076 = -0.076 = -7.6\times10^{-2}$

47) $6000 = 6000. = 6\times10^{3}$

49) $380{,}800{,}000 \text{ kg} = 3.808\times10^{8} \text{ kg}$

51) $0.00000001 \text{ cm} = 1\times10^{-8} \text{ cm}$

53) $\dfrac{6\times10^{9}}{2\times10^{5}} = \dfrac{6}{2}\times\dfrac{10^{9}}{10^{5}} = 3\times10^{4} = 30{,}000$

55) $(2.3\times10^{3})(3\times10^{2})$
$= (2.3\times3)(10^{3}\times10^{2})$
$= 6.9\times10^{5}$
$= 690{,}000$

57) $\dfrac{8.4\times10^{12}}{-7\times10^{9}}=-\dfrac{8.4}{7}\times\dfrac{10^{12}}{10^{9}}$

$=-1.2\times10^{3}=-1200$

59) $\left(-1.5\times10^{-8}\right)\left(4\times10^{6}\right)$

$=\left(-1.5\times4\right)\left(10^{-8}\times10^{6}\right)$

$=-6.0\times10^{-2}$

$=-0.06$

61) $\dfrac{-3\times10^{5}}{6\times10^{8}}=-\dfrac{3}{6}\times\dfrac{10^{5}}{10^{8}}$

$=-0.5\times10^{-3}=-0.0005$

63) $\left(9.75\times10^{4}\right)+\left(6.25\times10^{4}\right)$

$=(9.75+6.25)10^{4}$

$=16\times10^{4}$

$=160,000$

65) $\left(3.19\times10^{-5}\right)+\left(9.2\times10^{-5}\right)$

$=(3.19+9.2)10^{-5}$

$=12.39\times10^{-5}$

$=0.0001239$

67) $365\left(1.44\times10^{7}\right)$

$=(365\cdot1.44)10^{7}$

$=525.6\times10^{7}$

$=5,256,000,000$ particles

69) $\dfrac{2.21\times10^{10}\text{ lb}}{1,300,000\text{ cow}}=\dfrac{2.21\times10^{10}\text{ lb}}{1.3\times10^{6}\text{ cow}}$

$=\dfrac{2.21}{1.3}\times\dfrac{10^{10}\text{ lb}}{10^{6}\text{ cow}}$

$=1.7\times10^{4}\dfrac{\text{lb}}{\text{cow}}$

$=17,000\dfrac{\text{lb}}{\text{cow}}$

71) First determine the area of the photo.

$A=4\text{ in}\cdot6\text{ in}=24\text{ in}^2$

Then multiply by the rate.

$24\text{ in}^2\cdot\dfrac{1.1\times10^{6}\text{ droplets}}{\text{in}^2}$

$=2.4\times10\ \cancel{\text{in}^2}\cdot\dfrac{1.1\times10^{6}\text{ droplets}}{\cancel{\text{in}^2}}$

$=(2.4\times1.1)\left(10\times10^{6}\right)$ droplets

$=2.64\times10^{7}$ droplets

$=26,400,000$ droplets

73) $\begin{pmatrix}\text{money spent by}\\ \text{average household}\\ \text{on food}\end{pmatrix}=\dfrac{\begin{pmatrix}\text{total \$ spent}\\ \text{on food}\end{pmatrix}}{\begin{pmatrix}\text{\# of house}\\ \text{holds}\end{pmatrix}}$

$=\dfrac{7.3\times10^{11}}{1.2\times10^{8}}=\$6083$

75) $(\text{distance traveled})=(\text{speed})(\text{time})$

$=(7800\text{ m/s})(2\text{ days})$

$=(7800)(2\cdot24\cdot60\cdot60\text{ seconds})$

$=\left(7.8\times10^{3}\right)\left(1.728\times10^{5}\right)$

$=13.4784\times10^{8}$

$=1.34784\times10^{9}\text{ m}$

77) $\begin{pmatrix}\text{\# of metric tons of}\\ \text{carbon emissions}\\ \text{per person}\end{pmatrix}=\dfrac{\begin{pmatrix}\text{total \# of metric tons}\\ \text{of carbon emissions}\end{pmatrix}}{\begin{pmatrix}\text{total US}\\ \text{population}\end{pmatrix}}$

$=\dfrac{6\times10^{9}}{300,000,000}=\dfrac{6\times10^{9}}{3\times10^{8}}=20\text{ metric tons.}$

Chapter 2: The Rules of Exponents

Chapter 2 Review

(2.1A)

1) a) $8 \cdot 8 \cdot 8 \cdot 8 \cdot 8 \cdot 8 = 8^6$

b) $(-7)(-7)(-7)(-7) = (-7)^4$

3) a) $2^3 \cdot 2^2 = 2^{3+2} = 2^5 = 32$

b) $\left(\frac{1}{3}\right)^2 \cdot \left(\frac{1}{3}\right) = \left(\frac{1}{3}\right)^{2+1} = \left(\frac{1}{3}\right)^3 = \frac{1}{27}$

c) $(7^3)^4 = 7^{3 \cdot 4} = 7^{12}$

d) $(k^5)^6 = k^{5 \cdot 6} = k^{30}$

5) a) $(5y)^3 = 5^3 y^3 = 125y^3$

b) $(-7m^4)(2m^{12}) = -14m^{4+12}$
$= -14m^{16}$

c) $\left(\frac{a}{b}\right)^6 = \frac{a^6}{b^6}$

d) $6(xy)^2 = 6x^2 y^2$

e) $\left(\frac{10}{9}c^4\right)(2c)\left(\frac{15}{4}c^3\right)$

$= \frac{10}{9} \cdot 2 \cdot \frac{15}{4} c^4 c c^3$

$= \frac{\cancel{10}^{5}}{\cancel{9}_{3}} \cdot \cancel{2}^{1} \cdot \frac{\cancel{15}^{5}}{\cancel{4}_{1}} c^{4+1+3}$

$= \frac{25}{3} c^8$

(2.1B)

7) a) $(z^5)^2 (z^3)^4 = z^{10} z^{12} = z^{22}$

b) $-2(3c^5 d^8)^2 = -2(9c^{10} d^{16})$
$= -18c^{10} d^{16}$

c) $(9-4)^3 = 5^3 = 125$

d) $\frac{(10t^3)^2}{(2u^7)^3} = \frac{100t^6}{8u^{21}} = \frac{25t^6}{2u^{21}}$

(2.2A)

9) a) $8^0 = 1$ b) $-3^0 = -1$

c) $9^{-1} = \frac{1}{9}$

d) $3^{-2} - 2^{-2} = \left(\frac{1}{3}\right)^2 - \left(\frac{1}{2}\right)^2$

$= \frac{1}{9} - \frac{1}{4}$

$= \frac{4}{36} - \frac{9}{36} = -\frac{5}{36}$

e) $\left(\frac{4}{5}\right)^{-3} = \left(\frac{5}{4}\right)^3 = \frac{125}{64}$

(2.2B)

11) a) $v^{-9} = \frac{1}{v^9}$ b) $\left(\frac{9}{c}\right)^{-2} = \left(\frac{c}{9}\right)^2 = \frac{c^2}{81}$

c) $\left(\frac{1}{y}\right)^{-8} = y^8$ d) $-7k^{-9} = -\frac{7}{k^9}$

e) $\frac{19z^{-4}}{a^{-1}} = \frac{19a}{z^4}$ f) $20m^{-6}n^5 = \frac{20n^5}{m^6}$

g) $\left(\frac{2j}{k}\right)^{-5} = \left(\frac{k}{2j}\right)^5 = \frac{k^5}{32j^5}$

(2.3)

13) a) $\frac{3^8}{3^6} = 3^{8-6} = 3^2 = 9$

b) $\frac{r^{11}}{r^3} = r^{11-3} = r^8$

c) $\frac{48t^{-2}}{32t^3} = \frac{3}{2}t^{-2-3} = \frac{3}{2}t^{-5} = \frac{3}{2t^5}$

d) $\frac{21xy^2}{35x^{-6}y^3} = \frac{3}{5}x^{1-(-6)}y^{2-3}$

$= \frac{3}{5}x^7y^{-1} = \frac{3x^7}{5y}$

15) a) $\left(-3s^4t^5\right)^4 = 81s^{16}t^{20}$

b) $\frac{\left(2a^6\right)^5}{\left(4a^7\right)^2} = \frac{32a^{30}}{16a^{14}} = 2a^{30-14} = 2a^{16}$

c) $\left(\frac{z^4}{y^3}\right)^{-6} = \left(\frac{y^3}{z^4}\right)^6 = \frac{y^{18}}{z^{24}}$

d) $\left(-x^3y\right)^5\left(6x^{-2}y^3\right)^2$

$= \left(-x^{15}y^5\right)\left(36x^{-4}y^6\right)$

$= -36x^{11}y^{11}$

e) $\left(\frac{cd^{-4}}{c^8d^{-9}}\right)^5 = \left(c^{1-8}d^{-4-(-9)}\right)^5$

$= \left(c^{-7}d^5\right)^5$

$= c^{-35}d^{25}$

$= \frac{d^{25}}{c^{35}}$

f) $\left(\frac{14m^5n^5}{7m^4n}\right)^3 = \left(2m^{5-4}n^{5-1}\right)^3$

$= \left(2mn^4\right)^3$

$= 8m^3n^{12}$

g) $\left(\frac{3k^{-1}t}{5k^{-7}t^4}\right)^{-3} = \left(\frac{3}{5}k^{-1-(-7)}t^{1-4}\right)^{-3}$

$= \left(\frac{3}{5}k^6t^{-3}\right)^{-3}$

$= \left(\frac{3k^6}{5t^3}\right)^{-3}$

$= \left(\frac{5t^3}{3k^6}\right)^3 = \frac{125t^9}{27k^{18}}$

h) $\left(\frac{40}{21}x^{10}\right)\left(3x^{-12}\right)\left(\frac{49}{20}x^{2}\right)$

$=\frac{40}{21}\cdot 3\cdot\frac{49}{20}x^{10}x^{-12}x^{2}$

$=\frac{\overset{2}{\cancel{40}}\cdot\overset{1}{\cancel{3}}\cdot\overset{7}{\cancel{49}}}{\underset{1}{\cancel{21}}\cdot 1\cdot\underset{1}{\cancel{20}}}x^{10-12+2}=14x^{0}=14$

17) a) $y^{3k}\cdot y^{7k}=y^{3k+7k}=y^{10k}$

b) $\left(x^{5p}\right)^{2}=x^{2\cdot 5p}=x^{10p}$

c) $\frac{z^{12c}}{z^{5c}}=z^{12c-5c}=z^{7c}$

d) $\frac{t^{6d}}{t^{11d}}=t^{6d-11d}=t^{-5d}=\frac{1}{t^{5d}}$

(2.4)

19) $-4.185\times 10^{2}=-418.5$

21) $6.7\times 10^{-4}=0.00067$

23) $2\times 10^{4}=20,000$

25) $0.0000575=5.75\times 10^{-5}$

27) $32,000,000=3.2\times 10^{7}$

29) $178,000=1.78\times 10^{5}$

31) $0.0009315=9.315\times 10^{-4}$

33) $\frac{8\times 10^{6}}{2\times 10^{13}}=\frac{8}{2}\times\frac{10^{6}}{10^{13}}$
$=4\times 10^{-7}$
$=0.0000004$

35) $\left(9\times 10^{-8}\right)\left(4\times 10^{7}\right)$
$=\left(9\times 4\right)\left(10^{-8}\times 10^{7}\right)$
$=36\times 10^{-1}$
$=3.6$

37) $\frac{-3\times 10^{10}}{-4\times 10^{6}}=\frac{3}{4}\times\frac{10^{10}}{10^{6}}$
$=0.75\times 10^{4}$
$=7500$

39) $\frac{2.4\times 10^{5}}{0.8\times 10^{1}}=\frac{2.4}{0.8}\times\frac{10^{5}}{10^{1}}$
$=0.3\times 104$
$=30,000$ quills

41) $\frac{2.99\times 10^{-23}\text{ g}}{\cancel{\text{molecule}}}\cdot 100,000,000\ \cancel{\text{molecules}}$

$=2.99\times 10^{-23}\cdot 1\times 10^{8}\text{ g}$
$=2.99\left(10^{-23}\times 10^{8}\right)\text{ g}$
$=2.99\times 10^{-15}\text{ g}$
$=0.00000000000000299\text{ g}$

43) $\left(\frac{143,000\text{ visits}}{\cancel{\text{second}}}\right)\text{x}\left(\frac{60\ \cancel{\text{seconds}}}{\cancel{\text{minute}}}\right)\text{x}\left(3\ \cancel{\text{minutes}}\right)$
$=\left(1.43\text{x}10^{5}\right)\left(6\times 10^{1}\right)\left(3\right)\text{visits}$
$=\left(1.43\cdot 6\cdot 3\right)\left(10^{5}\cdot 10^{1}\right)\text{visits}$
$=25.74\times 10^{6}\text{ visits}$
$=25,740,000\text{ visits}$

Chapter 2 Test

1) $(-3)(-3)(-3)=(-3)^3$

3) $5^2\cdot 5=5^3=125$

5) $\left(8^3\right)^{12}=8^{36}$

7) $3^4=81$

9) $2^{-5}=\left(\frac{1}{2}\right)^5=\frac{1}{32}$

11) $\left(-\frac{3}{4}\right)^3=-\frac{27}{64}$

13) $\left(5n^6\right)^3=125n^{18}$

15) $\frac{m^{10}}{m^4}=m^{10-4}=m^6$

17) $$\left(\frac{-12t^{-6}u^8}{4t^5u^{-1}}\right)^{-3}=\left(-3t^{-6-5}u^{8-(-1)}\right)^{-3}$$
$$=\left(-3t^{-11}u^9\right)^{-3}$$
$$=\left(\frac{-3u^9}{t^{11}}\right)^{-3}$$
$$=\left(-\frac{t^{11}}{3u^9}\right)^3=-\frac{t^{33}}{27u^{27}}$$

19) $\left(\frac{(9x^2y^{-2})^3}{4xy}\right)^0=1$

21) $\frac{12a^4b^{-3}}{20c^{-2}d^3}=\frac{\overset{3}{\cancel{12}}\,a^4b^{-3}}{\underset{5}{\cancel{20}}\,c^{-2}d^3}=\frac{3a^4c^2}{5b^3d^3}$

23) $t^{10k}\cdot t^{3k}=t^{10k+3k}=t^{13k}$

25) $0.000165=1.65\times 10^{-4}$

27) $2.18\times 10^7=21{,}800{,}000$

Cumulative Review: Chapters 1-2

1) $\frac{90}{150}=\frac{90\div 30}{150\div 30}=\frac{3}{5}$

3) $\frac{4}{15}\div\frac{20}{21}=\frac{4}{15}\cdot\frac{21}{20}=\frac{\overset{1}{\cancel{4}}}{\underset{5}{\cancel{15}}}\cdot\frac{\overset{7}{\cancel{21}}}{\underset{5}{\cancel{20}}}=\frac{7}{25}$

5) $-26+5-7=-21-7=-28$

7) $(-1)^5=-1$

9) a) $$P=2\left(53\frac{1}{3}\text{ yd}\right)+2(120\text{ yd})$$
$$=2\left(\frac{160}{3}\text{ yd}\right)+240\text{ yd}$$
$$=\frac{320}{3}\text{ yd}+240\text{ yd}$$
$$=\frac{320}{3}\text{ yd}+\frac{720}{3}\text{ yd}$$
$$=\frac{1040}{3}\text{ yd}=346\frac{2}{3}\text{ yd}$$

b) First, find the area and then multiply by the cost per yd^2.
$$A=\left(53\frac{1}{3}\text{ yd}\right)(120\text{ yd})$$
$$=\left(\frac{160}{\underset{1}{\cancel{3}}}\text{ yd}\right)\left(\overset{40}{\cancel{120}}\text{ yd}\right)$$
$$=6{,}400\text{ yd}^2$$

$$6{,}400\text{ yd}^2\cdot\frac{\$1.80}{\text{yd}^2}=\$11{,}520$$

11) $V = \frac{4}{3}\pi r^3$ where r is the radius of the sphere

13) $4x^3 + 2x - 3$, when $x = 4$

$$= 4(4)^3 + 2(4) - 3$$
$$= 4(64) + 8 - 3$$
$$= 256 + 8 - 3$$
$$= 261$$

15) $5(t^2 + 7t - 3) - 2(4t^2 - t + 5)$

$$= 5t^2 + 35t - 15 - 8t^2 + 2t - 10$$
$$= 5t^2 - 8t^2 + 35t + 2t - 15 - 10$$
$$= -3t^2 + 37t - 25$$

17) $4^3 \cdot 4^7 = 4^{10}$

19) $\left(\frac{32x^3}{8x^{-2}}\right)^{-1} = \left(\frac{32}{8}x^{3-(-2)}\right)^{-1} = \left(4x^5\right)^{-1}$

$$= \frac{1}{4x^5}$$

21) $(4z^3)(-7z^5) = -28z^8$

23) $\left(-2a^{-6}b\right)^5 = (-2)^5(a^{-6})^5(b^5)$

$$= -32a^{-30}b^5 = -\frac{32b^5}{a^{30}}$$

25) $\left(6.2\times10^5\right)\left(9.4\times10^{-2}\right)$

$$= (6.2\times9.4)\left(10^5\times10^{-2}\right)$$
$$= 58.28\times10^3$$
$$= 58{,}280$$

Section 3.1 Exercises

1) Expression

3) Equation

5) No, it is an expression.

7) b, c

9) No.
$$a-4=-9$$
$$5-4=-9$$
$$1\neq -9$$

11) Yes.
$$-12y=8$$
$$-12\left(-\frac{2}{3}\right)=8$$
$$\overset{4}{\cancel{12}}\left(\frac{2}{\cancel{3}}\right)=8$$
$$8=8$$

13) Yes.
$$1.3=2p-1.7$$
$$1.3=2(1.5)-1.7$$
$$1.3=3.0-1.7$$
$$1.3=1.3$$

15)
$$n-5=12$$
$$n-5+5=12+5$$
$$n=17$$
The solution set is $\{17\}$.

17)
$$b+10=4$$
$$b+10-10=4-10$$
$$b=-6$$
The solution set is $\{-6\}$.

19)
$$-16=k-12$$
$$-16+12=k-12+12$$
$$-4=k$$
The solution set is $\{-4\}$.

21)
$$6=6+y$$
$$6-6=6+y-6$$
$$0=y$$
The solution set is $\{0\}$.

23)
$$a-2.9=-3.6$$
$$a-2.9+2.9=-3.6+2.9$$
$$a=-0.7$$
The solution set is $\{-0.7\}$.

25)
$$12=x+7.2$$
$$12-7.2=x+7.2-7.2$$
$$4.8=x$$
The solution set is $\{4.8\}$.

27)
$$h+\frac{5}{6}=\frac{1}{3}$$
$$h+\frac{5}{6}-\frac{5}{6}=\frac{1}{3}-\frac{5}{6}$$
$$h=\frac{2}{6}-\frac{5}{6}$$
$$h=-\frac{3}{6}=-\frac{1}{2}$$
The solution set is $\left\{-\frac{1}{2}\right\}$.

29)
$$-\frac{2}{5}=-\frac{5}{4}+c$$
$$-\frac{2}{5}+\frac{5}{4}=-\frac{5}{4}+\frac{5}{4}+c$$
$$-\frac{8}{20}+\frac{25}{20}=c$$
$$\frac{17}{20}=c$$
The solution set is $\left\{\frac{17}{20}\right\}$.

31) Answers may vary.

33) $2n=8$
$$\frac{\cancel{2}n}{\cancel{2}}=\frac{8}{2}$$
$$n=4$$
The solution set is $\{4\}$.

35) $-5z=35$
$$\frac{\cancel{-5}z}{\cancel{-5}}=\frac{35}{-5}$$
$$z=-7$$
The solution set is $\{-7\}$.

37) $-48=-4r$
$$\frac{\overset{12}{\cancel{-48}}}{\cancel{-4}}=\frac{\cancel{-4}r}{\cancel{-4}}$$
$$12=r$$
The solution set is $\{12\}$.

39) $63=-28y$
$$\frac{\overset{9}{\cancel{63}}}{\underset{-4}{\cancel{-28}}}=\frac{\cancel{-28}y}{\cancel{-28}}$$
$$-\frac{9}{4}=y$$
The solution set is $\left\{-\frac{9}{4}\right\}$.

41) $10n=2.3$
$$\frac{\cancel{10}n}{\cancel{10}}=\frac{2.3}{10}$$
$$n=0.23$$
The solution set is $\{0.23\}$.

43) $-7=-0.5d$
$$\frac{-7}{-0.5}=\frac{-0.5d}{-0.5}$$
$$14=d$$
The solution set is $\{14\}$.

45) $-x=1$
$$\frac{-x}{-1}=\frac{1}{-1}$$
$$x=-1$$
The solution set is $\{-1\}$.

47) $-6.5=-v$
$$\frac{-6.5}{-1}=\frac{-v}{-1}$$
$$6.5=v$$
The solution set is $\{6.5\}$.

49)
$$\frac{a}{4}=12$$
$$4\cdot\frac{a}{4}=4\cdot 12$$
$$1a=48$$
$$a=48$$
The solution set is $\{48\}$.

51)
$$-\frac{m}{3}=13$$
$$-\frac{m}{3}(-3)=13(-3)$$
$$1m=-39$$
$$m=-39$$

53)
$$\frac{w}{6}=-\frac{3}{4}$$
$$6\cdot\frac{w}{6}=-\frac{3}{4}\cdot 6$$
$$1w=-\frac{3}{\cancel{4}_2}\cdot\cancel{6}^3$$
$$w=-\frac{9}{2}$$
The solution set is $\left\{-\frac{9}{2}\right\}$.

55)
$$\frac{1}{5}q=-9$$
$$5\cdot\frac{1}{5}q=-9\cdot(5)$$
$$1q=-45$$
$$q=-45$$
The solution set is $\{-45\}$.

57)
$$-\frac{1}{6}m=-14$$
$$(-6)\cdot\left(-\frac{1}{6}m\right)=(-14)(-6)$$
$$1m=84$$
$$m=84$$
The solution set is $\{84\}$.

59)
$$\frac{5}{12}=\frac{1}{4}c$$
$$\frac{4}{1}\cdot\left(\frac{5}{12}\right)=\frac{4}{1}\cdot\left(\frac{1}{4}c\right)$$
$$\frac{\cancel{4}^1}{1}\cdot\left(\frac{5}{\cancel{12}_3}\right)=1c$$
$$\frac{5}{3}=c$$
The solution set is $\left\{\frac{5}{3}\right\}$.

61)
$$\frac{1}{3}y=-\frac{11}{15}$$
$$\frac{3}{1}\cdot\left(\frac{1}{3}y\right)=\frac{3}{1}\cdot\left(-\frac{11}{15}\right)$$
$$1y=\frac{\cancel{3}^1}{1}\cdot\left(-\frac{11}{\cancel{15}_5}\right)$$
$$y=-\frac{11}{5}$$
The solution set is $\left\{-\frac{11}{5}\right\}$.

63)
$$-\frac{5}{3}d=-30$$
$$-\frac{3}{5}\cdot\left(-\frac{5}{3}d\right)=-\frac{3}{5}\cdot\left(-\frac{30}{1}\right)$$
$$1d=-\frac{3}{\cancel{5}_1}\cdot\left(-\frac{\cancel{30}^6}{1}\right)$$
$$d=18$$
The solution set is $\{18\}$.

65) $-21 = \frac{3}{2}d$

$\frac{2}{3}\cdot(-21) = \frac{2}{3}\cdot\left(\frac{3}{2}d\right)$

$\frac{2}{\cancel{3}_1}\cdot\left(-\cancel{21}^{7}\right) = 1d$

$-14 = d$

The solution set is $\{-14\}$.

67) $5z + 8 = 43$

$5z + 8 - 8 = 43 - 8$

$5z = 35$

$\frac{\cancel{5}z}{\cancel{5}} = \frac{\cancel{35}^{7}}{\cancel{5}_1}$

$z = 7$

The solution set is $\{7\}$.

69) $4n - 15 = -19$

$4n = -19 + 15$

$\frac{4n}{4} = \frac{-4}{4}$

$n = -1$

The solution set is $\{4\}$.

71) $8d - 15 = -15$

$8d - 15 + 15 = -15 + 15$

$8d = 0$

$\frac{\cancel{8}d}{\cancel{8}} = \frac{0}{8}$

$d = 0$

The solution set is $\{0\}$.

73) $-11 = 5t - 9$

$-11 + 9 = 5t - 9 + 9$

$-2 = 5t$

$\frac{-2}{5} = \frac{\cancel{5}t}{\cancel{5}}$

$-\frac{2}{5} = t$

The solution set is $\left\{-\frac{2}{5}\right\}$.

75) $-6h + 19 = 3$

$-6h + 19 - 19 = 3 - 19$

$-6h = -16$

$\frac{-6h}{-6} = \frac{\cancel{-16}^{8}}{\cancel{-6}_3}$

$h = \frac{8}{3}$

The solution set is $\left\{\frac{8}{3}\right\}$.

77) $10 = 3 - 7y$

$10 - 3 = 3 - 3 - 7y$

$7 = -7y$

$\frac{7}{-7} = \frac{\cancel{-7}y}{\cancel{-7}}$

$-1 = y$

The solution set is $\{-1\}$.

79) $\frac{1}{2}d+7=12$

$$\frac{1}{2}d+7-7=12-7$$

$$\frac{1}{2}d=5$$

$$\frac{2}{1}\cdot\frac{1}{2}d=5\cdot\frac{2}{1}$$

$$d=10$$

The solution set is $\{10\}$.

81) $\frac{4}{5}b-9=-13$

$$\frac{4}{5}b-9+9=-13+9$$

$$\frac{4}{5}b=-4$$

$$\frac{5}{4}\cdot\frac{4}{5}b=\frac{5}{4}\cdot(-4)$$

$$1b=\frac{5}{\cancel{4}}\cdot\left(-\overset{-1}{\cancel{4}}\right)$$

$$b=-5$$

The solution set is $\{-5\}$.

83) $-1=\frac{10}{11}c+5$

$$-1-5=\frac{10}{11}c+5-5$$

$$-6=\frac{10}{11}c$$

$$\frac{11}{10}\cdot\frac{10}{11}c=\frac{11}{10}\cdot(-6)$$

$$x=\frac{11}{\underset{5}{\cancel{10}}}\cdot\left(-\overset{3}{\cancel{6}}\right)$$

$$x=-\frac{33}{5}$$

The solution set is $\left\{-\frac{33}{5}\right\}$.

85) $2-\frac{5}{6}t=-2$

$$2-\frac{5}{6}t-2=-2-2$$

$$-\frac{5}{6}t=-4$$

$$\left(-\frac{6}{5}\right)\left(-\frac{5}{6}\right)t=-4\left(-\frac{6}{5}\right)$$

$$t=\frac{24}{5}$$

The solution set is $\left\{\frac{24}{5}\right\}$.

87)
$$\frac{3}{4}=\frac{1}{2}-\frac{1}{6}z$$
$$\frac{3}{4}-\frac{1}{2}=\frac{1}{2}-\frac{1}{6}z-\frac{1}{2}$$
$$\frac{3}{4}-\frac{2}{4}=-\frac{1}{6}z$$
$$\frac{1}{4}=-\frac{1}{6}z$$
$$-6\cdot\frac{1}{4}=-6\cdot\left(-\frac{1}{6}z\right)$$
$$-\overset{3}{\cancel{6}}\cdot\frac{1}{\underset{2}{\cancel{4}}}=1z$$
$$-\frac{3}{2}=z$$
The solution set is $\left\{-\frac{3}{2}\right\}$.

89)
$$0.2p+9.3=5.7$$
$$0.2p+9.3-9.3=5.7-9.3$$
$$0.2p=-3.6$$
$$\frac{\cancel{0.2}p}{\cancel{0.2}}=\frac{-3.6}{0.2}$$
$$p=-18$$
$$p=-18$$
The solution set is $\{18\}$.

91)
$$3.8c-7.62=2.64$$
$$3.8c-7.62+7.62=2.64+7.62$$
$$3.8c=10.26$$
$$\frac{\cancel{3.8}c}{\cancel{3.8}}=\frac{10.26}{3.8}$$
$$x=2.7$$
The solution set is $\{2.7\}$.

93)
$$14.74=-20.6-5.7u$$
$$14.74+20.6=20.6-20.6-5.7u$$
$$35.34=-5.7u$$
$$\frac{35.34}{-5.7}=\frac{-5.7u}{-5.7}$$
$$-6.2=u$$
The solution set is $\{-6.2\}$.

95) $-9z=6$
$$\frac{-9z}{-9}=\frac{\overset{2}{\cancel{6}}}{-\underset{3}{\cancel{9}}}$$
$$1z=-\frac{2}{3}$$
$$z=-\frac{2}{3}$$
The solution set is $\left\{-\frac{2}{3}\right\}$.

97)
$$3a-11=16$$
$$3a-11+11=16+11$$
$$3a=27$$
$$\frac{3a}{3}=\frac{\overset{9}{\cancel{27}}}{\cancel{3}}$$
$$a=9$$
The solution set is $\{9\}$.

99)
$$-\frac{c}{6}=-9$$
$$\left(-\overset{1}{\cancel{6}}\right)-\frac{c}{\cancel{6}}=-9\cdot(-6)$$
$$c=54$$
The solution set is $\{54\}$.

101)
$$-34=n+15$$
$$-34-15=n+15-15$$
$$-49=n$$
The solution set is $\{-49\}$.

103)
$$-\frac{1}{7}p=-8$$
$$(-7)\left(-\frac{1}{7}p\right)=(-7)(-8)$$
$$p=56$$
The solution set is $\{56\}$.

105)
$$8.33-6.35d=17.22$$
$$8.33-6.35d-8.33=17.22-8.33$$
$$-6.35d=8.89$$
$$\frac{-6.35d}{-6.35}=\frac{8.89}{-6.35}$$
$$1d=-1.4$$
$$d=-1.4$$
The solution set is $\{-1.4\}$.

107)
$$-15=9+\frac{4}{5}c$$
$$-15-9=9+\frac{4}{5}c-9$$
$$-24=\frac{4}{5}c$$
$$\left(\frac{5}{\cancel{4}_1}\right)\left(-\cancel{24}^6\right)=\left(\frac{5}{4}\right)\frac{4}{5}c$$
$$-30=c$$
The solution set is $\{-30\}$.

109)
$$-\frac{3}{4}k+\frac{2}{5}=-2$$
$$-\frac{3}{4}k+\frac{2}{5}=-2$$
$$-\frac{3}{4}k+\frac{2}{5}-\frac{2}{5}=-2-\frac{2}{5}$$
$$-\frac{3}{4}k=-\frac{12}{5}$$
$$\left(-\frac{4}{3}\right)\left(-\frac{3}{4}k\right)=\left(-\frac{4}{\cancel{3}_1}\right)\left(-\frac{\cancel{12}^4}{5}\right)$$
$$k=\frac{16}{5}$$
The solution set is $\left\{\frac{16}{5}\right\}$.

Section 3.2 Exercises

1) Step1: Clear parentheses and combine like terms on each side of the equation.
Step2: Isolate the variable.
Step3: Solve for the variable.
Step4: Check the solution.

3) $3x+7+5x+4=27$

$8x+11=27$	*Combine like terms.*
$8x+11-11=27-11$	Subtraction Property of Equality.
$8x=16$	*Combine like terms.*
$\frac{\cancel{8}x}{\cancel{8}}=\frac{16}{8}$	Division Property of Equality.
$x=2$	Simplify.

The solution set is $\{2\}$.

5)
$$6a-10+4a+9=39$$
$$10a-1=39$$
$$10a-1+1=39+1$$
$$10a=40$$
$$\frac{10a}{10}=\frac{40}{10}$$
$$a=4$$
The solution set is $\{4\}$.

7)
$$-15+8y-10y+1=8$$
$$-14-2y=8$$
$$-14-2y+14=8+14$$
$$-2y=22$$
$$\frac{-2y}{-2}=\frac{22}{-2}$$
$$y=-11$$
The solution set is $\{-11\}$.

9)
$$30=5c+14-11c+1$$
$$30=-6c+15$$
$$30-15=-6c+15-15$$
$$15=-6c$$
$$\frac{\overset{5}{\cancel{15}}}{-\underset{2}{\cancel{6}}}=\frac{-6c}{-6}$$
$$-\frac{5}{2}=c$$
The solution set is $\left\{-\frac{5}{2}\right\}$.

11)
$$5-3m+9m+10-7m=-4$$
$$15-m=-4$$
$$15-m-15=-4-15$$
$$-m=-19$$
$$\frac{-m}{-1}=\frac{-19}{-1}$$
$$m=19$$
The solution set is $\{19\}$.

13)
$$5=-12p+7+4p-12$$
$$5=-8p-5$$
$$5+5=-8p-5+5$$
$$10=-8p$$
$$\frac{\overset{5}{\cancel{10}}}{-\underset{4}{\cancel{8}}}=\frac{-8p}{-8}$$
$$-\frac{5}{4}=p$$
The solution set is $\left\{-\frac{5}{4}\right\}$.

15)
$$\frac{1}{4}n+2+\frac{1}{2}n-\frac{3}{2}=\frac{11}{4}$$
$$\frac{1}{4}n+\frac{4}{2}+\frac{2}{4}n-\frac{3}{2}=\frac{11}{4}$$
$$\frac{3}{4}n+\frac{1}{2}=\frac{11}{4}$$
$$\frac{3}{4}n+\frac{2}{4}=\frac{11}{4}$$
$$\frac{3}{4}n+\frac{2}{4}-\frac{2}{4}=\frac{11}{4}-\frac{2}{4}$$
$$\frac{3}{4}n=\frac{9}{4}$$
$$\frac{\cancel{4}}{\cancel{3}}\cdot\frac{\cancel{3}}{\cancel{4}}n=\frac{\cancel{4}}{\cancel{3}}\cdot\frac{\overset{3}{\cancel{9}}}{\cancel{4}}$$
$$n=3$$
The solution set is $\{3\}$.

17)
$$4.2d-1.7-2.2d+4.3=-1.4$$
$$2.0d+2.6=-1.4$$
$$2.0d+2.6-2.6=-1.4-2.6$$
$$2d=-4$$
$$\frac{2d}{2}=\frac{-4}{2}$$
$$d=-2$$
The solution set is $\{-2\}$.

19) $2(5x+3)-3x+4=11$

$$10x+6-3x+4=-11$$
$$7x+10=-11$$
$$7x+10-10=-11-10$$
$$7x=-21$$
$$\frac{7x}{7}=\frac{-21}{7}$$
$$x=-3$$

The solution set is $\{-3\}$.

21) $7(b-5)+5(b+4)=45$

$$7b-35+5b+20=45$$
$$12b-15=45$$
$$12b-15+15=45+15$$
$$12b=60$$
$$\frac{12b}{12}=\frac{60}{12}$$
$$b=5$$

The solution set is $\{5\}$.

23) $8-3(2k+9)+2(7+k)=2$

$$8-6k-27+14+2k=2$$
$$-4k-5=2$$
$$-4k-5+5=2+5$$
$$-4k=7$$
$$\frac{-4k}{-4}=\frac{7}{-4}$$
$$k=-\frac{7}{4}$$

The solution set is $\left\{-\frac{7}{4}\right\}$.

25) $-23=4(3x-7)-(8x-5)$

$$-23=12x-28-8x+5$$
$$-23=4x-23$$
$$-23+23=-4x-23+23$$
$$0=-4x$$
$$\frac{0}{-4}=\frac{-4x}{-4}$$
$$0=x$$

The solution set is $\{0\}$.

27) $8=5(4n+3)-3(2n-7)-20$

$$8=20n+15-6n+21-20$$
$$8=14n+16$$
$$8-16=14n+16-16$$
$$-8=14n$$
$$\frac{-\overset{4}{\cancel{8}}}{\underset{7}{\cancel{14}}}=\frac{14n}{14}$$
$$-\frac{4}{7}=n$$

The solution set is $\left\{-\frac{4}{7}\right\}$.

29) $2(7u-3)-(u+9)-3(2u+1)=24$

$$14u-6-u-9-6u-3=24$$
$$7u-18=24$$
$$7u-18+18=24+18$$
$$7u=42$$
$$\frac{7u}{7}=\frac{42}{7}$$
$$u=6$$

The solution set is $\{6\}$.

31) $\frac{1}{3}(3w+4)-\frac{2}{3}=-\frac{1}{3}$

$$w+\frac{4}{3}-\frac{2}{3}=-\frac{1}{3}$$
$$w+\frac{2}{3}=-\frac{1}{3}$$
$$w+\frac{2}{3}-\frac{2}{3}=-\frac{1}{3}-\frac{2}{3}$$
$$w=-\frac{3}{3}$$
$$w=-1$$

The solution set is $\{-1\}$.

33) $\frac{1}{2}(c-2)+\frac{1}{4}(2c+1)=\frac{5}{4}$

$$\frac{1}{2}c-1+\frac{1}{2}c+\frac{1}{4}=\frac{5}{4}$$
$$c-\frac{4}{4}+\frac{1}{4}=\frac{5}{4}$$
$$c-\frac{3}{4}=\frac{5}{4}$$
$$c-\frac{3}{4}+\frac{3}{4}=\frac{5}{4}+\frac{3}{4}$$
$$c=\frac{8}{4}$$
$$c=\frac{\overset{2}{\cancel{8}}}{\underset{1}{\cancel{4}}}$$
$$c=2$$

The solution set is $\{2\}$.

35) $\frac{4}{3}(t+1)-\frac{1}{6}(4t-3)=2$

$$\frac{4}{3}t+\frac{4}{3}-\frac{4}{6}t+\frac{3}{6}=2$$
$$\frac{8}{6}t+\frac{8}{6}-\frac{4}{6}t+\frac{3}{6}=2$$
$$\frac{4}{6}t+\frac{11}{6}=\frac{12}{6}$$
$$\frac{4}{6}t+\frac{11}{6}-\frac{11}{6}=\frac{12}{6}-\frac{11}{6}$$
$$\frac{4}{6}t=\frac{1}{6}$$
$$\left(\frac{6}{4}\right)\left(\frac{4}{6}t\right)=\left(\frac{\overset{1}{\cancel{6}}}{4}\right)\left(\frac{1}{\underset{1}{\cancel{6}}}\right)$$
$$t=\frac{1}{4}$$

The solution set is $\left\{\frac{1}{4}\right\}$.

37) $9y+8=4y-17$

$$9y+8-4y=4y-17-4y$$
$$5y+8=-17$$
$$5y+8-8=-17-8$$
$$5y=-25$$
$$\frac{5y}{5}=-\frac{25}{5}$$
$$y=-5$$

The solution set is $\{-5\}$.

39) $$\begin{aligned} 5k-6&=7k-8 \\ 5k-6-7k&=7k-8-7k \\ -2k-6&=-8 \\ -2k-6+6&=-8+6 \\ -2k&=-2 \\ \frac{-2k}{-2}&=\frac{-2}{-2} \\ k&=1 \end{aligned}$$
The solution set is $\{1\}$.

41) $$\begin{aligned} -15w+4&=24-7w \\ -15w+4+7w&=24-7w+7w \\ -8w+4&=24 \\ -8w+4-4&=24-4 \\ -8w&=20 \\ \frac{-8w}{-8}&=\frac{\overset{5}{\cancel{20}}}{\underset{-2}{\cancel{-8}}} \\ w&=-\frac{5}{2} \end{aligned}$$
The solution set is $\left\{-\frac{5}{2}\right\}$.

43) $$\begin{aligned} 1.8z-1.1&=1.4z+1.7 \\ 1.8z-1.1-1.4z&=1.4z+1.7-1.4z \\ 0.4z-1.1&=1.7 \\ 0.4z-1.1+1.1&=1.7+1.1 \\ 0.4z&=2.8 \\ \frac{0.4z}{0.4}&=\frac{2.8}{0.4} \\ z&=7 \end{aligned}$$
The solution set is $\{7\}$.

45) $$\begin{aligned} 18-h+5h-11&=9h+19-3h \\ 4h+7&=6h+19 \\ 4h+7-6h&=6h+19-6h \\ -2h+7&=19 \\ -2h+7-7&=19-7 \\ -2h&=12 \\ \frac{-2h}{-2}&=\frac{12}{-2} \\ h&=-6 \end{aligned}$$
The solution set is $\{-6\}$.

47) $$\begin{aligned} 2t+7-6t+12&=4t+5-7t-1 \\ -4t+19&=-3t+4 \\ -4t+19+3t&=-3t+4+3t \\ -t+19&=4 \\ -t+19-19&=4-19 \\ -t&=-15 \\ \frac{-t}{-1}&=\frac{-15}{-1} \\ t&=15 \end{aligned}$$
The solution set is $\{15\}$.

49) $$\begin{aligned} 6.1r+1.6-3.7r-0.3&=r-1.7-0.2r-0.6 \\ 2.4r+1.3&=1.2r-2.3 \\ 2.4r+1.3-1.2r&=1.2r-2.3-1.2r \\ 1.2r+1.3&=-2.3 \\ 1.2r+1.3-1.3&=-2.3-1.3 \\ 1.2r&=-3.6 \\ \frac{1.2r}{1.2}&=\frac{-3.6}{1.2} \\ r&=-3 \end{aligned}$$
The solution set is $\{-3\}$.

51) $1+5(4n-7)=4(7n-3)-30$

$$1+20n-35=28n-12-30$$
$$20n-34=28n-42$$
$$20n-34-28n=28n-42-28n$$
$$-8n-34=-42$$
$$-8n-34+34=-42+34$$
$$-8n=-8$$
$$\frac{-8n}{-8}=\frac{-8}{-8}$$
$$n=1$$

The solution set is $\{1\}$.

53) $2(1-8c)=5-3(6c+1)+4c$

$$2-16c=5-18c-3+4c$$
$$2-16c=2-14c$$
$$2-16c+14c=2-14c+14c$$
$$2-2c=2$$
$$2-2c-2=2-2$$
$$-2c=0$$
$$\frac{-2c}{-2}=\frac{0}{-2}$$
$$c=0$$

The solution set is $\{0\}$.

55) $9-(8p-5)+4p=6(2p+1)$

$$9-8p+5+4p=12p+6$$
$$14-4p=12p+6$$
$$14-4p-12p=12p+6-12p$$
$$14-16p=6$$
$$14-16p-14=6-14$$
$$-16p=-8$$
$$\frac{-16p}{-16}=\frac{-8}{-16}$$
$$p=\frac{1}{2}$$

The solution set is $\left\{\frac{1}{2}\right\}$.

57) $-3(4r+9)+2(3r+8)=r-(9r-5)$

$$-12r-27+6r+16=r-9r+5$$
$$-6r-11=-8r+5$$
$$-6r-11+8r=-8r+5+8r$$
$$2r-11=5$$
$$2r-11+11=5+11$$
$$2r=16$$
$$r=\frac{16}{2}$$
$$r=8$$

The solution set is $\{8\}$.

59) $m+\frac{1}{2}(3m+4)-5=\frac{2}{3}(2m-1)+\frac{5}{6}$

$$m+\frac{3}{2}m+2-5=\frac{4}{3}m-\frac{2}{3}+\frac{5}{6}$$
$$\frac{2}{2}m+\frac{3}{2}m-3=\frac{4}{3}m-\frac{4}{6}+\frac{5}{6}$$
$$\frac{5}{2}m-3=\frac{4}{3}m+\frac{1}{6}$$
$$\frac{5}{2}m-\frac{4}{3}m-3=\frac{4}{3}m+\frac{1}{6}-\frac{4}{3}m$$
$$\frac{15}{6}m-\frac{8}{6}m-\frac{18}{6}=\frac{1}{6}$$
$$\frac{7}{6}m-\frac{18}{6}=\frac{1}{6}$$
$$\frac{7}{6}m-\frac{18}{6}+\frac{18}{6}=\frac{1}{6}+\frac{18}{6}$$
$$\frac{7}{6}m=\frac{19}{6}$$
$$\frac{6}{7}\cdot\frac{7}{6}m=\frac{\not{6}}{7}\cdot\frac{19}{\not{6}}$$
$$m=\frac{19}{7}$$

The solution set is $\left\{\frac{19}{7}\right\}$.

Section 3.3: Exercises

1) Eliminate fractions by multiplying both sides of the equation by the LCD of all the fractions in the equation.

3) Multiply both sides of the equation by 8.

5) $\frac{3}{8}x-\frac{1}{2}=\frac{1}{8}x+\frac{3}{4}$

$$8\left(\frac{3}{8}x-\frac{1}{2}\right)=8\left(\frac{1}{8}x+\frac{3}{4}\right)$$
$$3x-4=x+6$$
$$3x-x-4=x-x+6$$
$$2x-4=6$$
$$2x-4+4=6+4$$
$$2x=10$$
$$\frac{2x}{2}=\frac{10}{2}$$
$$x=5$$

The solution set is $\{5\}$.

7) $\frac{4}{7}t+\frac{1}{14}=\frac{3}{14}t+\frac{3}{2}$

$$14\left(\frac{4}{7}t+\frac{1}{14}\right)=14\left(\frac{3}{14}t+\frac{3}{2}\right)$$
$$8t+1=3t+21$$
$$8t-3t+1=3t-3t+21$$
$$5t+1=21$$
$$5t+1-1=21-1$$
$$5t=20$$
$$\frac{5t}{5}=\frac{20}{5}$$
$$t=4$$

The solution set is $\{4\}$

9) $\frac{1}{3}-\frac{1}{2}m=\frac{1}{6}m+\frac{7}{9}$

$$18\left(\frac{1}{3}-\frac{1}{2}m\right)=18\left(\frac{1}{6}m+\frac{7}{9}\right)$$
$$6-9m=3m+14$$
$$6-9m-3m=3m+14-3m$$
$$6-12m=14$$
$$6-12m-6=14-6$$
$$-12m=8$$
$$\frac{-12m}{-12}=\frac{8}{-12}$$
$$m=-\frac{2}{3}$$

The solution set is $\left\{-\frac{2}{3}\right\}$.

11) $\frac{1}{3}+\frac{1}{9}(k+5)-\frac{k}{4}=2$

$$36\left(\frac{1}{3}+\frac{1}{9}(k+5)-\frac{k}{4}\right)=36(2)$$
$$12+4(k+5)-9k=72$$
$$12+4k+20-9k=72$$
$$32-5k=72$$
$$32-32-5k=72-32$$
$$-5k=40$$
$$\frac{-5k}{-5}=\frac{40}{-5}$$
$$k=-8$$

The solution set is $\{-8\}$.

13) $\frac{3}{4}(y+7)+\frac{1}{2}(3y-5)=\frac{9}{4}(2y-1)$

$$4\cdot\frac{3}{4}(y+7)+4\cdot\frac{1}{2}(3y-5)=4\cdot\frac{9}{4}(2y-1)$$
$$3(y+7)+2(3y-5)=9(2y-1)$$
$$3y+21+6y-10=18y-9$$
$$9y+11=18y-9$$
$$9y+11-18y=18y-9-18y$$
$$-9y+11=-9$$
$$-9y+11-11=-9-11$$
$$-9y=-20$$
$$y=\frac{-20}{-9}$$

The solution set is $\left\{\frac{20}{9}\right\}$.

15) $\frac{1}{2}(4r+1)-r=\frac{2}{5}(2r-3)+\frac{3}{2}$

$$10\cdot\frac{1}{2}(4r+1)-10\cdot r=10\cdot\frac{2}{5}(2r-3)+10\cdot\frac{3}{2}$$
$$5(4r+1)-10r=4(2r-3)+15$$
$$20r+5-10r=8r-12+15$$
$$10r+5=8r+3$$
$$10r+5-8r=8r+3-8r$$
$$2r+5=3$$
$$2r+5-5=3-5$$
$$2r=-2$$
$$r=\frac{-2}{2}$$
$$r=-1$$

The solution set is $\{-1\}$.

17)
$$0.06d+0.13=0.31$$
$$100(0.06d+0.13)=100(0.31)$$
$$6d+13=31$$
$$6d+13-13=31-13$$
$$6d=18$$
$$\frac{6d}{6}=\frac{18}{6}$$
$$d=3$$

The solution set is $\{3\}$.

19)
$$0.04n-0.05(n+2)=0.1$$
$$100(0.04n-0.05(n+2))=100(0.1)$$
$$4n-5(n+2)=10$$
$$4n-5n-10=10$$
$$-n=20$$
$$\frac{-n}{-1}=\frac{20}{-1}$$
$$n=-20$$

The solution set is $\{-20\}$.

21)
$$0.2(c-4)+1=0.15(c+2)$$
$$100(0.2(c-4)+1)=100(0.15(c+2))$$
$$20(c-4)+100=15(c+2)$$
$$20c-80+100=15c+30$$
$$20c+20=15c+30$$
$$20c+20-15c=15c+30-15c$$
$$5c+20=30$$
$$5c+20-20=30-20$$
$$5c=10$$
$$\frac{5c}{5}=\frac{10}{5}$$
$$c=2$$

The solution set is $\{2\}$.

23) $0.35a - a = 0.03(5a + 4)$

$$100(0.35a - a) = 100(0.03)(5a + 4)$$
$$35a - 100a = 3(5a + 4)$$
$$-65a = 15a + 12$$
$$-65a - 15a = 12$$
$$-80a = 12$$
$$a = \frac{12}{-80}$$
$$a = -0.15$$

The solution set is $\{-0.15\}$.

25) $0.06w + 0.1(20 - w) = 0.08(20)$

$$100(0.06w + 0.1(20 - w)) = 100(0.08(20))$$
$$6w + 10(20 - w) = 8(20)$$
$$6w + 200 - 10w = 160$$
$$200 - 4w = 160$$
$$200 - 4w - 200 = 160 - 200$$
$$-4w = -40$$
$$\frac{-4w}{-4} = \frac{-40}{-4}$$
$$w = 10$$

The solution set is $\{10\}$.

27) $0.07k + 0.15(200) = 0.09(k + 200)$

$$100(0.07k + 0.15(200)) = 100(0.09(k + 200))$$
$$7k + 15(200) = 9(k + 200)$$
$$7k + 3000 = 9k + 1800$$
$$7k + 3000 - 9k = 9k + 1800 - 9k$$
$$-2k + 3000 = 1800$$
$$-2k + 3000 - 3000 = 1800 - 3000$$
$$-2k = -1200$$
$$\frac{-2k}{-2} = \frac{-1200}{-2}$$
$$k = 600$$

The solution set is $\{600\}$.

29) The variable is eliminated, and you get a false statement like $5 = -12$

31) $9(c + 6) - 2c = 4c + 1 + 3c$

$$9c + 54 - 2c = 4c + 1 + 3c$$
$$7c + 54 = 7c + 1$$
$$7c - 7c + 54 = 7c + 1 - 7c$$
$$54 \neq 1$$

The solution set is $\varnothing$.

33) $5t + 2(t + 3) - 4t = 4(t + 1) - (t - 2)$

$$5t + 2t + 6 - 4t = 4t + 4 - t + 2$$
$$3t + 6 = 3t + 6$$
$$3t + 6 - 3t = 3t + 6 - 3t$$
$$6 = 6$$

The solution set is $\{\text{all real numbers}\}$.

35) $\frac{5}{6}k-\frac{2}{3}=\frac{1}{6}(5k-4)+\frac{1}{2}$

$$18\left(\frac{5}{6}k-\frac{2}{3}\right)=18\cdot\frac{1}{6}(5k-4)+18\cdot\frac{1}{2}$$

$$18\cdot\frac{5}{6}k-18\cdot\frac{2}{3}=3(5k-3)+9$$

$$3\cdot 5k-6\cdot 2=15k-12+9$$

$$15k-12=15k-3$$

$$15k-12-15k=15k-3-15k$$

$$-12\neq-3$$

The solution set is $\varnothing$.

37) $\frac{n}{5}=20$

$$5\cdot\frac{n}{5}=5\cdot 20$$

$$1n=100$$

$$n=100$$

The solution set is $\{100\}$.

39) $-19=6-p$

$$-19-6=6-p-6$$

$$-25=-p$$

$$\frac{-25}{-1}=\frac{-p}{-1}$$

$$25=p$$

The solution set is $\{25\}$.

41) $-5.4=-0.9m$

$$\frac{-5.4}{-0.9}=-\frac{-0.9m}{-0.9}$$

$$6=m$$

The solution set is $\{6\}$.

43) $51=4y-13$

$$51+13=4y-13+13$$

$$64=4y$$

$$\frac{\cancel{64}^{16}}{\cancel{4}_{1}}=\frac{\cancel{4}^{1}y}{\cancel{4}_{1}}$$

$$16=y$$

The solution set is $\{16\}$.

45) $9-(7k-2)+2k=4(k+3)+5$

$$9-7k+2+2k=4k+12+5$$

$$11-5k=4k+17$$

$$11-5k+5k=4k+17+5k$$

$$11=17+9k$$

$$11-17=17+9k-17$$

$$-6=9k$$

$$\frac{-6}{9}=\frac{9k}{9}$$

$$-\frac{2}{3}=k$$

The solution set is $\left\{-\frac{2}{3}\right\}$.

47) $-\frac{5}{4}r+17=7$

$$-\frac{5}{4}r+17-17=7-17$$

$$-\frac{5}{4}r=-10$$

$$\left(-\frac{\cancel{4}}{\cancel{5}}\right)\left(-\frac{\cancel{5}}{\cancel{4}}r\right)=\left(-\frac{4}{\cancel{5}_{1}}\right)\left(-\cancel{10}^{2}\right)$$

$$r=8$$

The solution set is $\{8\}$.

49) $8(3t+4)=10t-3+7(2t+5)$

$$24t+32=10t-3+14t+35$$
$$24t+32=24t+32$$
$$24t-24t+32=24t-24t+32$$
$$32=32$$

The solution set is $\{\text{all real numbers}\}$.

51) $\frac{5}{3}w+\frac{2}{5}=w-\frac{7}{3}$

$$15\left(\frac{5}{3}w+\frac{2}{5}\right)=15\cdot w-15\cdot\frac{7}{3}$$
$$15\cdot\frac{5}{3}w+15\cdot\frac{2}{5}=15w-15\cdot\frac{7}{3}$$
$$5\cdot 5w+3\cdot 2=15w-5\cdot 7$$
$$25w+6=15w-35$$
$$25w+6-15w=15w-35-15w$$
$$10w+6=-35$$
$$10w-6+6=-35-6$$
$$10w=-41$$
$$w=-\frac{41}{10}$$

The solution set is $\left\{-\frac{41}{10}\right\}$.

53) $7(2q+3)-3(q+5)=6$

$$14q+21-3q-15=6$$
$$11q+6=6$$
$$11q+6-6=6-6$$
$$11q=0$$
$$q=0$$

The solution set is $\{0\}$.

55) $016h+0.4(2000)=0.22(2000+h)$

$$100\cdot 0.16h+100\cdot 0.4(2000)=100\cdot(0.22(2000+h))$$
$$16h+40(2000)=22(2000+h)$$
$$16h+80,000=44,000+22h$$
$$16h-22h+80,000=44,000$$
$$-6h+80,000-80,000=44,000-80,000$$
$$-6h=-36,000$$
$$\frac{-6h}{-6}=\frac{-36,000}{-6}$$
$$h=6000$$

The solution set is $\{6000\}$.

57) $-9r+4r-11+2=3r+7-8r+9$

$$-5r-9=-5r+16$$
$$-5r+5r-9=-5r+5r+16$$
$$-9\neq 16$$

The solution set is $\varnothing$.

59) $\frac{1}{2}(2r+9)=1+\frac{1}{3}(r+12)$

$$6\cdot\frac{1}{2}(2r+9)=6\cdot 1+6\cdot\frac{1}{3}(r+12)$$
$$3(2r+9)=6+2(r+12)$$
$$6r+27=6+2r+2.12$$
$$6r+27=6+2r+24$$
$$6r+27=2r+30$$
$$6r+27-2r=2r+30-2r$$
$$4r+27=30$$
$$4r+27-27=30-27$$
$$4r=3$$
$$r=\frac{3}{4}$$

The solution set is $\left\{\frac{3}{4}\right\}$.

61) 1) Read the problem until you understand it
2) Choose a variable to represent an unknown quantity.
3) Translate the problem from English into an equation.
4) Solve the equation.
5) Check the answer in the original problem and interpret the meaning of the solution as it relates to the problem.

63) Let $x =$ a number

$$x+12=5$$
$$x+12-12=5-12$$
$$x=-7$$

The number is -7.

65) Let $x =$ a number

$$x-9=12$$
$$x-9+9=12+9$$
$$x=21$$

The number is 21.

67) Let $x =$ a number

$$2x+5=17$$
$$2x+5-5=17-5$$
$$2x=12$$
$$\frac{2x}{2}=\frac{12}{2}$$
$$x=6$$

The number is 6.

69) Let $x =$ a number

$$2x+11=13$$
$$2x+11-11=13-11$$
$$2x=2$$
$$\frac{2x}{2}=\frac{2}{2}$$
$$x=1$$

The number is 1.

71) Let $x =$ a number

$$3x-8=40$$
$$3x-8+8=40+8$$
$$3x=48$$
$$\frac{3x}{3}=\frac{48}{3}$$
$$x=16$$

The number is 16.

73) Let $x =$ a number

$$\frac{3}{4}x=33$$
$$\frac{4}{3}\cdot\frac{3}{4}x=\frac{4}{3}\cdot 33$$
$$1x=4\cdot 11$$
$$x=44$$

The number is 44.

75) Let $x =$ a number

$$\frac{1}{2}x-9=3$$
$$\frac{1}{2}x-9+9=3+9$$
$$\frac{1}{2}x=12$$
$$2\cdot\frac{1}{2}x=2\cdot(12)$$
$$x=24$$

The number is 24.

77) Let x = a number

$$2x-3=x+8$$
$$2x-3-x=x+8-x$$
$$x-3=8$$
$$x-3+3=8+3$$
$$x=11$$

The number is 11.

79) Let x = a number

$$\frac{1}{3}x+10=x-2$$
$$\frac{1}{3}x+10+2=x-2+2$$
$$\frac{1}{3}x+12=x$$
$$3\cdot\frac{1}{3}x+3\cdot 12=3\cdot x$$
$$x+36=3x$$
$$x+36-x=3x-x$$
$$36=2x$$
$$\frac{36}{2}=\frac{2x}{2}$$
$$18=x$$

The number is 18.

81) Let x be the number.

$$x-24=\frac{x}{9}$$
$$9(x-24)=9\left(\frac{x}{9}\right)$$
$$9x-216=x$$
$$9x-216-9x=x-9x$$
$$-216=-8x$$
$$-216=-8x$$
$$\frac{-216}{-8}=\frac{-8x}{-8}$$
$$27=x$$

The number is 27.

83) Let x be the number.

$$x+\frac{2}{3}x=25$$
$$3\left(x+\frac{2}{3}x\right)=3(25)$$
$$3x+2x=75$$
$$5x=75$$
$$x=\frac{75}{5}$$
$$x=15$$

The number is 15.

85) Let x be the number.

$$x-2x=13$$
$$-x=13$$
$$\frac{-x}{-1}=\frac{13}{-1}$$
$$x=-13$$

The number is -13.

Section 3.4: Exercises

1) $c+14$ 3) $c-37$ 5) $\frac{1}{2}s$ 7) $14-x$

9) The number of children must be whole number.

11) It is an even number.

13) Let x=the amount of rain in 1905. Then the amount of rain in $2004 = x-1.2$

$$\begin{pmatrix}\text{amount in}\\ 1905\end{pmatrix}+\begin{pmatrix}\text{amount in}\\ 2004\end{pmatrix}=7.2$$

$$x + x-1.2 = 7.2$$

$$2x-1.2=7.2$$

$$2x=8.4$$

$$x=4.2$$

$$\text{amount in } 1905 = 4.2 \text{ inches}$$

$$\text{amount in } 2004 = 4.2-1.2$$

$$=3.0 \text{ inches}$$

1905: 4.2 inches. 2004: 3.0 inches.

15) Let x be the number of titles won by Lance Armstrong.
Then the number of titles won by Miguel Indurain is $x-2$

$$\begin{pmatrix}Number\ by\\ Lance\end{pmatrix}+\begin{pmatrix}Number\ by\\ Miguel\end{pmatrix}=12$$

$$x + x-2 = 12$$

$$2x-2=12$$

$$2x=14$$

$$x=7$$

$$Number\ by\ Lance = 7$$

$$Number\ by\ Miguel = x-2$$

$$=5$$

$$=8$$

Lance: 7 Miguel: 5

17) Let x be the amount of caffeine in decaffeinated coffee.
Then the amount of caffeine regular coffee is $13x$

$$\begin{pmatrix} Amount\ of \\ caffeine\ in \\ decaffeinated \\ coffee \end{pmatrix} + \begin{pmatrix} Amount\ of \\ caffeine\ in \\ regular \\ coffee \end{pmatrix} = 280$$

$$x + 13x = 280$$
$$14x = 280$$
$$x = 20$$

$$Amount\ of\ caffeine\ in\ decaffeinated\ coffee = 20\,mg$$
$$Amount\ of\ caffeine\ in\ regular\ coffee = 13x = 13(20) = 260\,mg$$

Decaffeinated coffee: 20 mg.; Regular coffee: 260 mg.

19) Let $x = number\ of\ students\ taking\ the\ spanish\ class.$

Then the number of students taking the $French\ class = \frac{2}{3}x.$

$$\begin{pmatrix} Number\ of\ students \\ \text{in Spanish class} \end{pmatrix} + \begin{pmatrix} Number\ of\ students \\ \text{in French class} \end{pmatrix} = 310$$

$$x + \frac{2}{3}x = 310$$
$$\frac{5}{3}x = 310$$
$$x = \frac{3}{5} \cdot 310$$
$$x = 186$$

$$number\ of\ students\ in\ Spanish\ class = 186$$
$$number\ of\ students\ in\ French\ class = \frac{2}{3} \cdot x = \frac{2}{3} \cdot 186 = 124$$

Spanish: 186; French: 124

21) Let $x =$ the length of the shorter piece.
Then the length of the longer piece $= x + 14$.
$(\text{shorter piece}) + (\text{longer piece}) = 36$

$$x + x + 14 = 36$$
$$2x = 22$$
$$x = 11$$
$$\textit{shorter piece} = 11\,\textit{inches}$$
$$\textit{longer piece} = x + 14$$
$$= 11 + 14$$
$$= 25\,\textit{inches}$$

The pipe will be cut into an 11 in. piece and a 25 in. piece.

23) Let $x =$ the length of the shorter piece.
Then the length of the longer piece $= 2x$.
$(\text{longer piece}) + (\text{shorter piece}) = 28.5$

$$2x + x = 28.5$$
$$3x = 28.5$$
$$x = 9.5$$
$$\textit{Shorter piece} = 9.5\,\textit{inches}$$
$$\textit{Longer piece} = 2x$$
$$= 2(9.5)$$
$$= 19\,\textit{inches}$$

The wire will be cut into a 9.5 in. piece for the bracelet
and a 19 in. piece for the necklace.

25) Let x = the length Derek's sandwich.

Then the length of Tamara's sandwich $= \frac{1}{2}x$

Then the length of Cory's sandwich $= \frac{1}{2}x + 2$

$$\begin{pmatrix}\text{length of} \\ \text{Derk's sandwich}\end{pmatrix} + \begin{pmatrix}\text{length of} \\ \text{Cory's sandwich}\end{pmatrix} + \begin{pmatrix}\text{length of} \\ \text{Tamara's sandwich}\end{pmatrix} = 6$$

$$x + \frac{1}{2}x + \frac{1}{2}x + 2 = 6$$

$$2x + 2 = 6$$

$$2x = 6 - 2$$

$$2x = 4$$

$$x = 2\ ft$$

$$Derek's\ sandwich = 2\ ft$$

$$Tamara's\ sandwich = \frac{1}{2}x = 1\ ft$$

$$Cory's\ sandwich = \frac{1}{2}x + 2 = 3\ ft$$

Derek: 2 ft. Cory: 3 ft. Tamara: 1 ft.

27) x = first integer

$x+1$ = second integer

$x+2$ = third integer

$$\begin{pmatrix}\text{first} \\ \text{integer}\end{pmatrix} + \begin{pmatrix}\text{second} \\ \text{integer}\end{pmatrix} + \begin{pmatrix}\text{third} \\ \text{integer}\end{pmatrix} = 126$$

$$x + x+1 + x+2 = 126$$

$$3x + 3 = 126$$

$$3x = 123$$

$$x = 41$$

first integer = 41

second integer $= x + 1 = 42$

third integer $= x + 2 = 43$

The integers are 41, 42, 43.

29) $x = \text{first even integer}$

$x+2 = \text{second even integer}$

$2(\text{first even integer}) = 16 + (\text{second even integer})$

$$2x = 16 + x + 2$$
$$2x = 18 + x$$
$$x = 18$$

first even integer $= 18$

second even integer $= x + 2 = 20$

The integers are 18, 20.

31) $x = \text{first odd integer}$

$x+2 = \text{second odd integer}$

$x+4 = \text{third odd integer}$

$$\begin{pmatrix}\text{first odd}\\ \text{integer}\end{pmatrix} + \begin{pmatrix}\text{second odd}\\ \text{integer}\end{pmatrix} + \begin{pmatrix}\text{third odd}\\ \text{integer}\end{pmatrix} = 5 + 4\begin{pmatrix}\text{third}\\ \text{odd integer}\end{pmatrix}$$
$$x + x+2 + x+4 = 5 + 4(x+4)$$
$$3x + 6 = 5 + 4x + 16$$
$$3x + 6 = 21 + 4x$$
$$6 = 21 + x$$
$$-15 = x$$

first odd integer $= -15$

second odd integer $= x + 2 = -13$

third odd integer $= x + 4 = -11$

The integers are -15,-13,-11.

33) $x = \text{first page number}$

$x+1 = \text{second page number}$

$$\begin{pmatrix}\text{first page}\\ \text{number}\end{pmatrix} + \begin{pmatrix}\text{second page}\\ \text{number}\end{pmatrix} = 215$$
$$x + x+1 = 215$$
$$2x = 214$$
$$x = 107$$

$x = \text{first page number} = 107$

$x+1 = \text{second page number} = 108$

The page numbers are 107 and 108.

35) Let x be the number of trout caught by Kelly
Then the number of trout caught by Jimmy is $x+6$

$$\begin{pmatrix} \text{Number of trout} \\ \text{caught by Kelly} \end{pmatrix} + \begin{pmatrix} \text{Number of trout} \\ \text{caught by Jimmy} \end{pmatrix} = 20$$

$$x \quad + \quad x+6 \quad = 20$$

$$2x+6=20$$

$$2x=14$$

$$x=7$$

$$\textit{number of trout caught by Kelly} = 7$$

$$\textit{number of trout caught by Jimmy} = x+6 = 7+6$$

$$=13$$

Jimmy: 13; Kelly:7

37) Let x be the length of the shorter piece
Then the length of the longer piece is $2x+1$

$$(\text{shorter piece}) + (\text{longer piece}) = 16$$

$$x \quad + \quad 2x+1 \quad = 16$$

$$3x+1=16$$

$$3x=15$$

$$x=5$$

$$\textit{shorter piece} = 5 \textit{ ft}$$

$$\textit{longer piece} = 2x+1$$

$$= 2\cdot 5+1$$

$$= 11 \textit{ ft}$$

The pieces are 5 ft and 11 ft.

39) The pieces are 1ft, 3ft and 5ft.Let x be the attendance at Bomaroo.
Then the attendance at the Lollapalooza is $3x+15,000$

$$(\text{attendance at Bomaroo})+(\text{attendance at Lollapalooza})=29,5000$$

$$x \quad + \quad 15000+3x \quad =29,5000$$

$$4x+15000=29,5000$$

$$4x=280,000$$

$$x=70,000$$

$$\text{attendance at Bomaroo}=70,000$$

$$\text{attendance at Lollapalooza}=3x+15,000$$

$$=3\cdot 70,000+15,000$$

$$=225,000$$

Bomaroo: 70,000; Lollapalooza: 225,000

41) Let x be the first page number.
$x+1=\text{second page number}$
$x+2=\text{third page number}$

$$\begin{pmatrix}\text{first page}\\ \text{number}\end{pmatrix}+\begin{pmatrix}\text{second page}\\ \text{number}\end{pmatrix}+\begin{pmatrix}\text{third page}\\ \text{number}\end{pmatrix}=174$$

$$x \quad + \quad x+1 \quad + \quad x+2 \quad =174$$

$$3x+3=174$$

$$3x=171$$

$$x=57$$

$$\text{first integer}=57$$

$$\text{second integer}=x+1=58$$

$$\text{third integer}=x+2=59$$

The page numbers are 57, 58, 59.

43) Let x be the amount weight lost by Helen.
Then the amount of weight lost by Tara is $x+15=$.
The amount of weight lost by Mike is $2x-73$.
$(Helen's\ weight\ loss)+(Tara's\ weight\ loss)+(Mike's\ weight\ loss)=502$

$$x \quad +x+15 \quad + \quad 2x-73 \quad =502$$
$$4x-58=502$$
$$4x=560$$
$$x=\frac{560}{4}$$
$$x=140$$
$$Helen's\ weight\ loss=140\,lbs$$
$$Tara's\ weight\ loss=x+15=140+15=155\,lbs$$
$$Mike's\ weight\ loss=2x-73=2\cdot 140-73$$
$$=280-73=207\,lbs$$

Helen: 140 lbs. Tara: 155 lbs. Mike: 207 lbs.

45) Let x be the length of the longest piece.
Then the length of the shortest piece is $\frac{1}{3}x$
Length of the medium-size piece is $x-12$

$$\begin{pmatrix}\text{length of}\\ \text{shortest piece}\end{pmatrix}+\begin{pmatrix}\text{length of}\\ \text{longest piece}\end{pmatrix}+\begin{pmatrix}\text{length of}\\ \text{medium-sized piece}\end{pmatrix}=72$$

$$\frac{1}{3}x \quad +x \quad + \quad x-12 \quad =72$$
$$\frac{7}{3}x-12=72$$
$$\frac{7}{3}x=84$$
$$x=\frac{3}{7}\cdot 84$$
$$x=36$$
$$length\ of\ longest\ piece=36\,in.$$
$$length\ of\ shortest\ piece=\frac{1}{3}x=\frac{1}{3}\cdot 36=12\,in.$$
$$length\ of\ medium-sized\ piece=x-12=36-12=24\,in.$$

Longest:36 in; medium sized: 24 in.; shortest: 12 in.

47) Let x be the number of copies in millions of Coldplay's album.
Then the number of copies of Lil Waynes's album is $x+0.75$
Number of copies of Taylor Swift's album is $x-0.04$

$$(\#of\ copies\ of\ Coldplay's\ album)+(\#of\ copies\ of\ Lil\,Wayne's\ album)+$$
$$(\#of\ copies\ of\ Taylor\ Swift's\ album)=7.14$$
$$x \quad +x+0.73 \quad + \quad x-0.04 \quad =7.14$$
$$3x+0.69=7.14$$
$$3x=7.14-0.69$$
$$3x=6.45$$
$$x=2.15$$
$$\#of\ copies\ of\ Coldplay's\ album=2.15\,million$$
$$\#of\ copies\ of\ Lil\ Wayne's\ album=x+0.73=2.15+0.73=2.88\,million$$
$$\#of\ copies\ of\ Taylor\ Swift's\ album=x-0.04=2\cdot 15-0.04=2.11\,million.$$

49) $x=$ first even integer
$x+2=$ second even integer
$x+4=$ third even integer

$$\frac{1}{6}\begin{pmatrix}\text{first even}\\ \text{integer}\end{pmatrix}=\frac{1}{10}\left[\begin{pmatrix}\text{second even}\\ \text{integer}\end{pmatrix}+\begin{pmatrix}\text{third even}\\ \text{integer}\end{pmatrix}\right]-3$$
$$\frac{1}{6}x \quad =\frac{1}{10}[\quad x+2 \quad + \quad x+4 \quad]-3$$
$$\frac{1}{6}x=\frac{1}{10}[2x+6]-3$$
$$30\left(\frac{1}{6}x\right)=30\left(\frac{1}{10}[2x+6]-3\right)$$
$$5x=3[2x+6]-90$$
$$5x=6x+18-90$$
$$5x=6x-72$$
$$-x=-72$$
$$x=72$$

$x=$ first even integer $=72$
$x+2=$ second even integer $=74$
$x+4=$ third even integer $=76$
The integers are 72, 74, 76.

Section 3.5: Exercises

1) $\text{Amount of Discount} = (\text{Rate of Discount})(\text{Original Price})$
$$= (0.15) \cdot (50.00)$$
$$= 7.50$$
$\text{Sale Price} = \text{Original Price} - \text{Amount of Discount}$
$$= 50.00 - 7.50$$
$$= 42.50$$
The sale price is \$42.50.

3) $\text{Amount of Discount} = (\text{Rate of Discount})(\text{Original Price})$
$$= (0.30) \cdot (29.50)$$
$$= 8.85$$
$\text{Sale Price} = \text{Original Price} - \text{Amount of Discount}$
$$= 29.50 - 8.85$$
$$= 20.65$$
The sale price is \$20.65.

5) $\text{Amount of Discount} = (\text{Rate of Discount})(\text{Original Price})$
$$= (0.60) \cdot (49.00)$$
$$= 29.40$$
$\text{Sale Price} = \text{Original Price} - \text{Amount of Discount}$
$$= 49.00 - 29.40$$
$$= 19.60$$
The sale price is \$19.60.

7) Let $x =$ the original price of the camera.
$\text{Sale Price} = \text{Original Price} - \text{Amount of Discount}$
$$119 = x - 0.15x$$
$$119 = 0.85x$$
$$140 = x$$
The original price was \$140.00.

9) Let $x =$ the original price of the calendar.
$\text{Sale Price} = \text{Original Price} - \text{Amount of Discount}$
$$4.40 = x - 0.75x$$
$$4.40 = 0.25x$$
$$17.60 = x$$
The original price was \$17.60.

11) Let $x =$ the original price of the coffe maker.
Sale Price = Original Price – Amount of Discount

$$40.08 = x - 0.40x$$
$$40.08 = 0.60x$$
$$66.80 = x$$

The original price was \$66.80.

13) Let $x =$ the number of acres of farmland in 2000.
farmland in 2009 = farmland in 2000-Amount of decrease

$$1224 = x - x(0.32)$$
$$1224 = 0.68x$$
$$1800 = x$$

There were 1800 acres of farmland in 2000.

15) Let $x =$ the number of Starbucks in 1996.
Starbucks in 2006 = Starbucks in 1996 + Amount of Increase

$$12,440 = x + x(11.26)$$
$$12,440 = 12.26x$$
$$1014.68 \approx x$$

There were about 1015 Starbucks stores in 1996.

17) Let $x =$ the number of employees at K-mart in 2001.
Number of employees in 2003 = employees in 2001-Amount of decrease

$$2900 = x - x(0.34)$$
$$2900 = 0.66x$$
$$4393.93 \approx x$$

There were about 4400 employees worked in 2001.

19) $I = PRT$
$P = 300,\ R = 0.03,\ T = 1$
$I = (300)(0.03)(1) = 9$
Kristie earned \$9 in interest.

21) $I = PRT$
$P = 6500,\ R = 0.07,\ T = 1$
$I = (6500)(0.07)(1) = 455$
Amount in account= P+I
=6500+455
=6955
There will be \$6955 in Jake's account.

Y?

23) $I = PRT$
$$\text{Total Interest Earned} = (3000)(0.065)(1) + (1500)(0.08)(1)$$
$$= 195 + 120 = 315$$
Rachel earned a total of \$315 in interest from the two accounts.

25) x = amount Amir invested in the 6% account.
$15,000 - x$ = amount Amir invested in the 7% account.
Total Interest Earned = Interest from 6% account + Interest from 7% account
$$960 = x(0.06)(1) + (15,000 - x)(0.07)(1)$$
$$100(960) = 100\left[x(0.06)(1) + (15,000 - x)(0.07)(1)\right]$$
$$96,000 = 6x + 7(15,000 - x)$$
$$96,000 = 6x + 105,000 - 7x$$
$$-9000 = -x$$
$$9000 = x$$
amount invested in 7% $= 15,000 - x = 15,000 - 9000 = 6000$
Amir invested \$9000 in the 6% account and \$6000 in the 7% account.

27) x = amount Barney invested in the 6% account.
$x + 450$ = amount Barney invested in the 5% account.
Total Interest Earned = Interest from 6% account + Interest from 5% account
$$204 = x(0.06)(1) + (x + 450)(0.05)(1)$$
$$100(204) = 100\left[x(0.06)(1) + (x + 450)(0.05)(1)\right]$$
$$20,400 = 6x + 5(x + 450)$$
$$20,400 = 6x + 5x + 2250$$
$$18,150 = 11x$$
$$1650 = x$$
amount invested in 5% account $= x + 450 = 1650 + 450 = 2100$
Barney invested \$1650 in the 6% account and \$2100 in the 5% account.

29) $x =$ amount Taz invested in the 9.5% account.
$7500 - x =$ amount Taz invested in the 6.5% account.
Total Interest Earned = Interest from 9.5% account + Interest from 6.5% account

$$577.50 = x(0.095)(1) + (7500-x)(0.065)(1)$$
$$1000(577.50) = 1000[x(0.095)(1)+(7500-x)(0.065)(1)]$$
$$577,500 = 95x + 65(7500-x)$$
$$577,500 = 95x + 487,500 - 65x$$
$$90,000 = 30x$$
$$3000 = x$$

amount invested in 6.5% $= 7500 - x = 7000 - 3000 = 4500$
Taz invested \$3000 in the 9.5% account and \$4500 in the 6.5% account.

31) oz of alcohol $= (0.06)(50) = 3$; There are 3 oz of alcohol in the 50 oz solution.

33) mL of acid $= (0.10)(75) + (0.025)(30) = 7.5 + 0.75 = 8.25$
There are 8.25 mL of acid in the mixture.

35) $x =$ number of oz of 4% acid solution
$24 - x =$ number of oz of 10% acid solution

Solution	Concentration	Number of oz of solution	Number of oz of acid in the solution
4%	0.04	x	$0.04x$
10%	0.10	$24-x$	$0.10(24-x)$
6%	0.06	24	$0.06(24)$

$$0.04x + 0.10(24-x) = 0.06(24)$$
$$100(0.04x + 0.10(24-x)) = 100(0.06(24))$$
$$4x + 10(24-x) = 6(24)$$
$$4x + 240 - 10x = 144$$
$$-6x = -96$$
$$x = 16$$

number of oz of 10% solution $= 24 - x = 24 - 16 = 8$
Mix 16 oz of the 4% solution and 8 oz of the 10% solution.

37) $x =$ number of liters of 25% antifreeze solution
$x + 4 =$ number of liters of 45% antifreeze solution

Solution	Concentration	Number of liters of solution	Number of liters of antifreeze in the solution
25%	0.25	x	$0.25x$
60%	0.6	4	$0.6(4)$
45%	0.45	$x+4$	$0.45(x+4)$

$$0.25x + 0.6(4) = 0.45(x+4)$$
$$100(0.25x + 0.6(4)) = 100(0.45(x+4))$$
$$25x + 60(4) = 45(x+4)$$
$$25x + 240 = 45x + 180$$
$$60 = 20x$$
$$3 = x$$

Add 3 liters of the 25% antifreeze solution.

39) $x =$ number of lbs of cashews
$x + 4 =$ number of lbs mixture

Nuts	Price per Pound	Number of lbs of Nuts	Value
cashews	\$7.00	x	$7x$
pistachios	\$4.00	4	$4(4)$
mixture	\$5.00	$x+4$	$5(x+4)$

$$7x + 4(4) = 5(x+4)$$
$$7x + 16 = 5x + 20$$
$$2x = 4$$
$$x = 2$$

Mix 2 lbs of the cashews with the pistachios.

41) $x =$ grams of a 50% silver alloy
$x + 500 =$ grams of a 20% silver alloy

Alloy	Concentration	Number of grams of alloy	Number of grams of silver in alloy
50%	0.50	x	$0.50x$
5%	0.05	500	$0.05(500)$
20%	0.20	$x+500$	$0.20(x+500)$

$$0.50x + 0.05(500) = 0.20(x+500)$$
$$100(0.50x + 0.05(500)) = 100(0.20(x+500))$$
$$50x + 5(500) = 20(x+500)$$
$$50x + 2500 = 20x + 10,000$$
$$30x = 7500$$
$$x = 250g$$

Mix 250g of the 50% silver alloy.

43) $x =$ number of gallons of pure acid solution.
$6 + x =$ number of gallons of 20% acid solution

Solution	Concentration	Number of gallons of solution	Number of gallons of acid in the solution
100%	1.00	x	$1.00x$
4%	0.04	6	$0.04(6)$
20%	0.20	$6+x$	$0.20(6+x)$

$$1.00x + 0.04(6) = 0.20(6+x)$$
$$100(1.00x + 0.04(6)) = 100(0.20(6+x))$$
$$100x + 4(6) = 20(6+x)$$
$$100x + 24 = 120 + 20x$$
$$80x = 96$$
$$x = 1\frac{1}{5}$$

$1\frac{1}{5}$ gallons of pure acid solution.

45) Let $x =$ Cheryl's Cost.
Price in store = Cheryl's Cost + Amount of Increase

$$14.00 = x + x(0.60)$$
$$14.00 = 1.60x$$
$$8.75 = x$$

Each stuffed animal cost Cheryl \$8.75.

47) Let $x =$ people collecting unemployment benefits in September 2009.
Benefits 2010 = Benefits 2009 − Amount of Decrease

$$8330 = x - x(0.02)$$
$$8330 = 0.98x$$
$$8500 = x$$

8500 people were getting unemployment benefits in September 2009.

49) $x =$ amount Erica invested in the 3% CD account.
$2x =$ amount Erica invested in the 4% IRA account.
$x + 1000 =$ amount Erica invested in the 5% mutual fund account.
Total Interest Earned = Interest from CD + Interest from IRA + Interest from mutual fund

$$370 = x(0.03)(1) + 2x(0.04)(1) + (x+1000)(0.05)(1)$$
$$100(370) = 100[x(0.03)(1) + 2x(0.04)(1) + (x+1000)(0.05)(1)]$$
$$37,000 = 3x + 8x + 5(x+1000)$$
$$37,000 = 3x + 8x + 5x + 5000$$
$$37,000 = 16x + 5000$$
$$32,000 = 16x$$
$$2000 = x$$

amount invested IRA $= 2x = x(2000) = 4000$

amount invested mutual fund $= x + 1000 = 2000 + 1000 = 3000$
Erica invested \$2000 in the CD, \$4000 in the IRA and \$3000 in the mutual fund.

51) Let $x =$ the original price of the desk lamp.
Sale Price = Original Price − Amount of Discount

$$25.60 = x - 0.20x$$
$$25.60 = 0.80x$$
$$32.00 = x$$

The original price was \$32.00.

53) Let x = Zoe's previous salary.

Current Salary = Previous Salaray + Amount of Increase

$$40{,}144 = x + x(0.04)$$
$$40{,}144 = 1.04x$$
$$38{,}600 = x$$

Zoe's salary was $38,600 last year.

55) x = number of oz of 9% alcohol solution

$12 - x$ = number of oz of 17% alcohol solution

Solution	Concentration	Number of oz of solution	Number of oz of alcohol in the solution
9%	0.09	x	$0.09x$
17%	0.17	$12-x$	$0.17(12-x)$
15%	0.15	12	$0.15(12)$

$$0.09x + 0.17(12-x) = 0.15(12)$$
$$100(0.09x + 0.17(12-x)) = 100(0.15(12))$$
$$9x + 17(12-x) = 15(12)$$
$$9x + 204 - 17x = 180$$
$$-8x = -24$$
$$x = 3$$

number of oz of 17% solution $= 12 - x = 12 - 3 = 9$

Mix 3 oz of the 9% solution and 9 oz of the 17% solution.

57) x = number of lbs of peanuts

$10 - x$ = number of lbs of cashews

Nuts	Price per Pound	Number of lbs of Nuts	Value
peanuts	$1.80	x	$1.80x$
cashews	$4.50	$10-x$	$4.50(10-x)$
mixture	$2.61	10	$2.61(10)$

$$
\begin{aligned}
1.80x + 4.50(10-x) &= 2.61(10) \\
100(1.80x) + 100(4.50(10-x)) &= 100(2.61(10)) \\
180x + 450(10-x) &= 261(10) \\
180x + 4500 - 450x &= 2610 \\
-270x &= -1890 \\
x &= 7 \\
\text{cashews} &= 10 - x = 10 - 7 = 3
\end{aligned}
$$

Mix 7 lbs of peanuts and 3 lbs of the cashews.

59) $x =$ amount Diego invested in the 4% account.
$20,000 - x =$ amount Diego invested in the 7% account.
Total Interest Earned = Interest from 4% account + Interest from 7% account

$$1130 = x(0.04)(1) + (20,000 - x)(0.07)(1)$$

$$
\begin{aligned}
100(1130) &= 100\left[x(0.04)(1) + (20,000-x)(0.07)(1)\right] \\
113,000 &= 4x + 7(20,000 - x) \\
113,000 &= 4x + 140,000 - 7x \\
-27,000 &= -3x \\
9000 &= x
\end{aligned}
$$

amount invested in 7% $= 20,000 - x = 20,000 - 9000 = 11,000$
Diego invested \$9000 in the 4% account and \$11,000 in the 7% account.

61) $x =$ number of ounces of pure orange juice.
$76 - x =$ number of ounces of 5% citrus fruit drink.

Solution	Concentration	Number of ounces of solution	Number of ounes of fruit juice in the solution
100%	1.00	x	$1.00x$
5%	0.05	$76 - x$	$0.05(76 - x)$
25%	0.25	76	$0.25(76)$

$$1.00x + 0.05(76 - x) = 0.25(76)$$
$$100(1.00x + 0.05(76 - x)) = 100(0.25(76))$$
$$100x + 5(76 - x) = 25(76)$$
$$100x + 380 - 5x = 1900$$
$$95x = 1520$$
$$x = 16$$

$$\textit{number of oz of 5\% citrus fruit drink} = 76 - x$$
$$= 76 - 16 = 60$$

16 ounces of pure orange juice and 60 ounces of fruit drink.

63) Let x = number of procedures performed in 1997.

Procedures in 2003 = Procedures in 1997 + 2.93(Procedures in 1997)

$$8,253,000 = x + 2.93x$$
$$8,253,000 = 3.93x$$
$$2,100,000 = x$$

2,100,000 procedures were performed in 1997.

Section 3.6: Exercises

1) No. The height of the triangle can not be a negative number.

3) Cubic centimeters

5) $A = lw$

$A = 44;\ l = 16$

$44 = 16w$

$\frac{11}{4} = w$

7) $I = PRT$

$240 = P(0.04)(2)$

$240 = 0.08P$

$3000 = P$

9) $d = rt$

$150 = (60)t$

$2.5 = t$

11) $C = 2\pi r$

$C = 2\pi(4.6)$

$C = 9.2\pi$

13) $P = 2l + 2w$

$11 = 2l + 2\left(\frac{3}{2}\right)$

$11 = 2l + 3$

$8 = 2l$

$4 = l$

15) $V = lwh$

$52 = (6.5)w(2)$

$52 = 13w$

$4 = w$

17) $V=\frac{1}{3}\pi r^2 h$

$48\pi=\frac{1}{3}\pi(4)^2 h$

$48\pi=\frac{1}{3}\pi(16)h$

$144\pi=16\pi h$

$9=h$

19) $S=2\pi r^2+2\pi rh$

$154\pi=2\pi(7)^2+2\pi(7)h$

$154\pi=2\pi(49)+2\pi(7)h$

$154\pi=98\pi+14\pi h$

$56\pi=14\pi h$

$4=h$

21) $A=\frac{1}{2}h(b_1+b_2)$

$136=\frac{1}{2}(16)(7+b_2)$

$136=8(7+b_2)$

$17=7+b_2$

$10=b_2$

23) $l=$ length of the tennis court

$A=2808,\ w=36$

$A=lw$

$2808=l\cdot 36$

$78=l$

The length of the tennis court is 78 ft.

25) $h=$ height of the box

$V=1232,\ l=22,\ w=7$

$V=lwh$

$1232=(22)(7)h$

$1232=154h$

$8=h$

The height of the flower box is 8 in.

27) $A=$ area of the center circle of a soccer field

$r=10$

$A=\pi r^2$

$A=\pi(10)^2$

$A=100\pi$

$A\approx 100(3.14)\approx 314\text{ yd}^2$

The area is about 314 yd^2.

29) r= average speed of Abbas

$d=rt$

$d=134;\ t=2$

$134=r(2)$

$67=r$

The average speed is 67 mph

31) $h=$ the height of the can

$V=864\pi,\ r=6$

$V=\pi r^2 h$

$864\pi=\pi(6)^2 h$

$864\pi=36\pi h$

$24=h$

The height of the can is 24 in.

33) $A=\frac{1}{2}bh$

$A=6$; $h=4$

$6=\frac{1}{2}b(4)$

$6=2b$

$3=b$

The length of the base is 3 ft.

35) $I=PRT$

$P=1500,\ I=75,\ T=2$

$75=(1500)R(2)$

$75=3000R$

$0.025=R$

$R=2.5\%$

Lelani received 2.5% interest rate.

37) Let x be the length of the frame.
Then the width of the frame is $x-10$.

$P=2l+2w$

$92=2x+2(x-10)$

$92=2x+2x-20$

$112=4x$

$28=x$

length of the frame is 28 in.

width of the frame is $x-10=18$ in.

39) Let x be the width of the lane court.
Then the length is $2x-5$

$P=2l+2w$

$62=2(2x-5)+2(x)$

$62=4x-10+2x$

$62=6x-10$

$72=6x$

$12=x$

width of the lane is 12 ft.

length of the lane is $2x-5$=19 ft.

41) Let $b_1=x$

then $b_2=3x+2$

The height is 5 in.

$A=\frac{1}{2}(b_1+b_2)h$

$25=\frac{1}{2}(x+3x+2)5$

$50=(4x+2)\cdot 5$

$50=20x+10$

$40=20x$

$2=x$

length of b_1 is 2 in.

length of the b_2 is $3x+2=8$ in.

43) Let $s_1=s_2=x$

then $s_3=x+1$

perimeter is 5.5 *ft*.

$P=s_1+s_2+s_3$

$5.5=x+x+x+1$

$5.5=3x+1$

$4.5=3x$

$1.5=x$

The sides are 1.5 ft, 1.5 ft, 2.5 ft.

45) $m\angle A=(x-27)^\circ$ $m\angle B=83^\circ$ $m\angle C=x^\circ$

$m\angle A+m\angle B+m\angle C=180$

$x-27\ +\ 83\ +\ x\ =180$

$2x+56=180$

$2x=124$

$x=62^\circ$

$m\angle A=x-27^\circ=35^\circ$

$m\angle C=x^\circ=62^\circ$

47) $m\angle A = x^\circ \quad m\angle B = (2x)^\circ \quad m\angle C = 102^\circ$

$$m\angle A + m\angle B + m\angle C = 180$$
$$x + 2x + 102 = 180$$
$$3x + 102 = 180$$
$$3x = 78$$
$$x = 26^\circ$$

$$m\angle A = 26^\circ$$
$$m\angle B = (2x)^\circ = 2(26) = 52^\circ$$

49) $m\angle A = x\left(\frac{1}{2}x + 10\right)^\circ \quad m\angle B = x^\circ \quad m\angle C = x^\circ$

$$m\angle A + m\angle B + m\angle C = 180$$
$$\left(\frac{1}{2}x + 10\right) + x + x = 180$$
$$\frac{5}{2}x + 10 = 180$$
$$\frac{5}{2}x = 170$$
$$x = \frac{2}{5} \cdot 170$$
$$x = 68^\circ$$

$$m\angle B = m\angle C = 68^\circ$$
$$m\angle A = \frac{1}{2}x + 10 = 44^\circ$$

51) $(x + 28)^\circ = (3x - 2)^\circ$

$$x + 28 = 3x - 2$$
$$30 = 2x$$
$$15 = x$$
$$(x + 28)^\circ = [15 + 28]^\circ = 43^\circ$$
$$(3x - 2)^\circ = [3(15) - 2]^\circ = 43^\circ$$

The labeled angles measure 43°.

53) $(4x - 12)^\circ = \left(\frac{5}{2}x + 57\right)^\circ$

$$4x - 12 = \frac{5}{2}x + 57$$
$$2(4x - 12) = 2\left(\frac{5}{2}x + 57\right)$$
$$8x - 24 = 5x + 114$$
$$3x = 138$$
$$x = 46$$
$$(4x - 12)^\circ = [184 - 12]^\circ = 172^\circ$$
$$\left(\frac{5}{2}x + 57\right)^\circ = \left[\frac{5}{2}(46) + 57\right]^\circ = 172^\circ$$

The labeled angles measure 172°.

55) $(3.5x + 3)^\circ = (3x + 8)^\circ$

$$3.5x + 3 = 3x + 8$$
$$0.5x = 5$$
$$x = 10$$
$$(3.5x + 3)^\circ = [3.5(10) + 3]^\circ = 38^\circ$$
$$(3x + 8)^\circ = [3(10) + 8]^\circ = 38^\circ$$

The labeled angles measure 38°.

57) $(4x)^\circ + (x)^\circ = 180^\circ$

$$(4x)^\circ + (x)^\circ = 180$$
$$5x = 180$$
$$x = 36$$
$$(4x)^\circ = [4(36)]^\circ = 144^\circ$$
$$(x)^\circ = 36^\circ$$

The labeled angles measure $144^\circ, 36^\circ$.

59) $(x)^{\circ}+\left(\frac{1}{2}x\right)^{\circ}=180^{\circ}$

$$x+\frac{1}{2}x=180$$
$$\frac{3}{2}x=180$$
$$x=120$$
$$(x)^{\circ}=120^{\circ}$$
$$\left(\frac{1}{2}x\right)^{\circ}=60^{\circ}$$

The labeled angles measure $120^{\circ}, 60^{\circ}$.

61) $(x+30)^{\circ}+(2x+21)^{\circ}=180^{\circ}$

$$(x+30)+(2x+21)=180$$
$$3x+51=129$$
$$3x=36$$
$$x=43$$
$$(x+30)^{\circ}=[43+30]^{\circ}=73^{\circ}$$
$$(2x+21)^{\circ}=[2(43)+21]=107^{\circ}$$

The labeled angles measure $73^{\circ}, 107^{\circ}$.

63) The supplement is $180-x$.

65) Let x = the measure of the angle.

$180-x$ = measure of the supplement

$90-x$ = measure of the complement

supplement$=2$(complement)$+63^{\circ}$

$$180-x=2(90-x)+63$$
$$180-x=180-2x+63$$
$$-x=-2x+63$$
$$x=63$$

The measure of the angle is 63°.

67) Let x = the measure of the angle.

$180-x$ = measure of the supplement

$6x$ = supplement-12°

$$6x=(180-x)-12$$
$$6x=168-x$$
$$7x=168$$
$$24=x$$

The measure of the angle is 24°.

69) Let x = the measure of the angle.

$180-x$ = measure of the supplement

$90-x$ = measure of the complement

4(complement) = 2(supplement) − 40

$$4(90-x)=2(180-x)-40$$
$$360-4x=360-2x-40$$
$$360-4x=320-2x$$
$$360=320+2x$$
$$40=2x$$
$$20=x$$

The measure of the angle, complement, and supplement are 20°, 70°, and 160°.

71) Let x = the measure of the angle.

$180-x$ = measure of the supplement

$90-x$ = measure of the complement

$x+\frac{1}{2}$(supplement) = 7(complement)

$$x+\frac{1}{2}(180-x)=7(90-x)$$
$$90+\frac{1}{2}x=630-7x$$
$$180+x=1260-14x$$
$$15x=1080$$
$$x=72$$

The measure of the angle is 72°.

73) Let x = the measure of the angle.
$90-x$ = measure of the complement

$$4x+2(\text{complement})=270^\circ$$
$$4x+2(90-x)=270$$
$$4x+180-2x=270$$
$$2x=90$$
$$x=45$$

The measure of the angle is 45°.

75)

a) $$x+16=37$$
$$x+16-16=37-16$$
$$x=21$$

b) $$x+h=y$$
$$x+h-h=y-h$$
$$x=y-h$$

c) $$x+r=c$$
$$x+r-r=c-r$$
$$x=c-r$$

77)

a) $$8c=56$$
$$\frac{8c}{8}=\frac{56}{8}$$
$$c=7$$

b) $$ac=d$$
$$\frac{ac}{a}=\frac{d}{a}$$
$$c=\frac{d}{a}$$

c) $$mc=v$$
$$\frac{mc}{m}=\frac{v}{m}$$
$$c=\frac{v}{m}$$

79) a) $$\frac{a}{4}=11$$
$$4\cdot\frac{a}{4}=11\cdot 4$$
$$a=44$$

b) $$\frac{a}{y}=r$$
$$y\cdot\frac{a}{y}=r\cdot y$$
$$a=ry$$

c) $$\frac{a}{w}=d$$
$$w\left(\frac{a}{w}\right)=(d)w$$
$$a=dw$$

81) a) $$8d-7=17$$
$$8d-7+7=17+7$$
$$8d=24$$
$$\frac{8d}{8}=\frac{24}{8}$$
$$d=3$$

b) $$kd-a=z$$
$$kd-a+a=z+a$$
$$kd=z+a$$
$$\frac{kd}{k}=\frac{z+a}{k}$$
$$d=\frac{z+a}{k}$$

83) a) $$9h+23=17$$
$$9h+23-23=17-23$$
$$9h=-6$$
$$\frac{9h}{9}=\frac{-6}{9}$$
$$h=-\frac{2}{3}$$

b)
$$\begin{aligned} qh+v&=n \\ qh+v-v&=n-v \\ qh&=n-v \\ \frac{qh}{q}&=\frac{n-v}{q} \\ h&=\frac{n-v}{q} \end{aligned}$$

85)
$$\begin{aligned} F&=ma \\ \frac{F}{a}&=\frac{ma}{a} \\ \frac{F}{a}&=m \end{aligned}$$

87)
$$\begin{aligned} n&=\frac{c}{v} \\ n\cdot v&=\frac{c}{v}\cdot v \\ nv&=c \end{aligned}$$

89)
$$\begin{aligned} E&=\sigma T^4 \\ \frac{E}{T^4}&=\frac{\sigma T^4}{T^4} \\ \frac{E}{T^4}&=\sigma \end{aligned}$$

91)
$$\begin{aligned} V&=\frac{1}{3}\pi r^2 h \\ 3\cdot V&=3\cdot\frac{1}{3}\pi r^2 h \\ 3V&=\pi r^2 h \\ \frac{3V}{\pi r^2}&=\frac{\pi r^2 h}{\pi r^2} \\ \frac{3V}{\pi r^2}&=h \end{aligned}$$

93)
$$\begin{aligned} R&=\frac{E}{I} \\ I\cdot R&=I\cdot\frac{E}{I} \\ IR&=E \end{aligned}$$

95)
$$\begin{aligned} I&=PRT \\ I&=PRT \\ \frac{I}{PT}&=\frac{PRT}{PT} \\ \frac{I}{PT}&=R \end{aligned}$$

97)
$$\begin{aligned} P&=2l+2w \\ P-2w&=2l+2w-2w \\ P-2w&=2l \\ \frac{P-2w}{2}&=\frac{2l}{2} \\ \frac{P-2w}{2}&=l \end{aligned}$$

99)
$$\begin{aligned} H&=\frac{D^2N}{2.5} \\ 2.5\cdot H&=2.5\cdot\frac{D^2N}{2.5} \\ 2.5H&=D^2N \\ \frac{2.5H}{D^2}&=\frac{D^2N}{D^2} \\ \frac{2.5H}{D^2}&=N \end{aligned}$$

101) $A=\frac{1}{2}h(b_1+b_2)$

$2\cdot A=2\cdot\frac{1}{2}h(b_1+b_2)$

$2A=h(b_1+b_2)$

$\frac{2A}{h}=\frac{h(b_1+b_2)}{h}$

$\frac{2A}{h}=b_1+b_2$

$\frac{2A}{h}-b_1=b_1-b_1+b_2$

$\frac{2A}{h}-b_1=b_2$ or $b_2=\frac{2A-hb_1}{h}$

103) $S=\frac{\pi}{4}(4h^2+c^2)$

$\frac{4}{\pi}\cdot S=\frac{4}{\pi}\cdot\frac{\pi}{4}(4h^2+c^2)$

$\frac{4S}{\pi}=4h^2+c^2$

$\frac{4S}{\pi}-c^2=4h^2+c^2-c^2$

$\frac{4S}{\pi}-c^2=4h^2$

$\frac{4S}{4\pi}-\frac{c^2}{4}=\frac{4h^2}{4}$

$\frac{S}{\pi}-\frac{c^2}{4}=h^2$

105) a) $P=2l+2w$

$P-2l=2l-2l+2w$

$P-2l=2w$

$\frac{P-2l}{2}=\frac{2w}{2}$

$\frac{P-2l}{2}=w$

b) $w=\frac{P-2l}{2}$

$w=\frac{28-2(11)}{2}$

$w=\frac{28-22}{2}=\frac{6}{2}=3$ cm

107) a) $C=\frac{5}{9}(F-32)$

$\frac{9}{5}\cdot C=\frac{9}{5}\cdot\frac{5}{9}(F-32)$

$\frac{9}{5}C=F-32$

$\frac{9}{5}C+32=F-32+32$

$\frac{9}{5}C+32=F$

b) $F=\frac{9}{5}C+32$

$F=\frac{9}{5}(20)+32=36+32=68^\circ$

Section 3.7: Exercises

1) Answers may vary, but some possible answers are

$\frac{6}{8},\frac{9}{12},\frac{12}{16}$

3) Yes. A percent can be written as a fraction with a denominator of 100. For example 25% can be written as $\frac{13}{20}$ or $\frac{25}{100}$ or $\frac{1}{4}$.

5) $\frac{16}{12}=\frac{4}{3}$

7) $\frac{4}{50}=\frac{2}{25}$

9) $\frac{20}{80} = \frac{1}{4}$

11) $\frac{2\,ft}{36\,in} = \frac{2(12)\,in}{36\,in}$

$\frac{24}{36} = \frac{2}{3}$

13) $\frac{18\,hours}{2\,days} = \frac{18\,hours}{2 \cdot 24\,hours}$

$\frac{18}{48} = \frac{3}{8}$

15) Unit price for size 8 package $= \frac{\$6.29}{8} \approx 0.78625$

Unit price for size 16 package $= \frac{\$12.99}{16} \approx 0.811875$

Package of 8 : \$0.786 per battery $\approx$ the best buy.

17) Unit price for 8 oz jar $= \frac{\$2.69}{8} \approx 0.336$

Unit price for 15 oz jar $= \frac{\$3.59}{15} \approx 0.239$

Unit price for 48 oz jar $= \frac{\$8.49}{48} \approx 0.177$

48 oz jar : \$0.177 per oz is the best buy.

19) Unit price for 11 oz box $= \frac{\$4.49}{11} \approx 0.408$

Unit price for 16 oz box $= \frac{\$5.15}{16} \approx 0.322$

Unit price for 24 oz box $= \frac{\$6.29}{24} \approx 0.262$

24 oz box : \$0.262 per oz is the best buy.

21) A ratio is a quotient of two quantities.

A proportion is a statement that two ratios are equal.

23) True
$$4 \cdot 35 = 20 \cdot 7$$
$$140 = 140$$

25) False
$$72 \cdot 7 = 8 \cdot 54$$
$$504 \neq 432$$

27) True
$$8 \cdot \frac{5}{2} = 2 \cdot 10$$
$$20 = 20$$

29) $\frac{8}{36} = \frac{c}{9}$
$$72 = 36c$$
$$2 = c \qquad \{2\}$$

31) $\frac{w}{15} = \frac{32}{12}$
$$12w = 480$$
$$w = 40 \qquad \{40\}$$

33) $\frac{40}{24} = \frac{30}{a}$
$$40a = 720$$
$$a = 18 \qquad \{18\}$$

35) $\frac{2}{k} = \frac{9}{12}$
$$24 = 9k$$
$$\frac{24}{9} = k$$
$$\frac{8}{3} = k \qquad \left\{\frac{8}{3}\right\}$$

37) $\frac{3z+10}{14} = \frac{2}{7}$
$$7(3z+10) = 2 \cdot 14$$
$$21z + 70 = 28$$
$$21z = -42$$
$$z = -2 \qquad \{-2\}$$

39) $\frac{r+7}{9} = \frac{r-5}{3}$
$$3(r+7) = 9(r-5)$$
$$3r + 21 = 9r - 45$$
$$66 = 6r$$
$$11 = r \qquad \{11\}$$

41) $\frac{3h+15}{16} = \frac{2h+5}{4}$
$$4(3h+15) = 16(2h+5)$$
$$12h + 60 = 32h + 80$$
$$-20 = 20h$$
$$-1 = h \qquad \{-1\}$$

43) $\frac{4m-1}{6} = \frac{6m}{10}$
$$10(4m-1) = 6(6m)$$
$$40m - 10 = 36m$$
$$-10 = -4m$$
$$\frac{-10}{-4} = m$$
$$\frac{5}{2} = m \qquad \left\{\frac{5}{2}\right\}$$

45) Let $x =$ cost of 6 containers of yogurt.

$$\frac{2.36}{4}=\frac{x}{6}$$
$$6(2.36)=4x$$
$$14.16=4x$$
$$3.54=x$$

The cost of 6 contaiers of yogurt is \$3.54

47) Let $x =$ amount of orange juice.

$$\frac{2}{3}=\frac{\frac{1}{3}}{x}$$
$$2x=\left(\frac{1}{3}\right)3$$
$$2x=1$$
$$x=\frac{1}{2}$$

The amount of orange juice is $\frac{1}{2}$ cup.

49) Let $x =$ caffeine in an 18 oz serving

$$\frac{12}{55}=\frac{18}{x}$$
$$12x=990$$
$$x=82.5$$

There are 82.5 mg of caffeine in an 18 oz serving of Mountain Dew.

51) Let $x =$ number of smokers who started smoking before they were 21.

$$\frac{9}{10}=\frac{x}{400}$$
$$9(400)=10x$$
$$3600=10x$$
$$360=x$$

The number of smokers is 360.

53) Let x be the number of pounds of kitchen scraps.

$$\frac{5}{2}=\frac{20}{x}$$
$$5x=20(2)$$
$$5x=40$$
$$x=8$$

The number of spounds of kitchen scrap is 8 lb.

55) Let x bet the number of Euros.

$$\frac{20}{14.30}=\frac{50.00}{x}$$
$$20x=50(14.3)$$
$$20x=715$$
$$x=\frac{715}{20}$$
$$x=35.75$$

The number of Euros is 35.75

57) $\frac{7}{5}=\frac{14}{x}$

$$7x=70$$
$$x=10$$

59) $\frac{12}{20}=\frac{x}{\frac{65}{3}}$

$$260=20x$$
$$13=x$$

61) $\dfrac{x}{28}=\dfrac{45}{20}$

$20x=1260$

$x=63$

63) a) $(\$0.10)\cdot 7=\$0.70$

b) $10¢\cdot 7=70¢$

65) a) $(\$0.01)\cdot 422=\$4.22$

b) $(1¢)\cdot 422=422¢$

67) a) $(\$0.05)\cdot 9+(\$0.25)\cdot 7$

$=\$0.45+\$1.75=\$2.20$

b) $(5¢)\cdot 9+(25¢)\cdot 7$

$=45¢+175¢=220¢$

69) a) $0.25\cdot q=0.25q$

b) $25\cdot q=25q$

71) a) $0.10\cdot d=0.10d$ b) $10\cdot d=10d$

73) a) $0.01\cdot p+0.05\cdot n=0.01p+0.05n$

b) $1\cdot p+5\cdot n=p+5n$

75) $x=$ number of dimes; $8+x=$ number of quarters

Value of Dimes + Value of Quarters = Total Value

$$0.10x \quad + \quad 0.25(x+8) \quad = \quad 5.15$$

$$100(0.10x+0.25(x+8))=100(5.15)$$

$$10x+25(x+8)=515$$

$$10x+25x+200=515$$

$$35x=315$$

$$x=9$$

quarters $=x+8=9+8=17$

There are 9 dimes and 17 quarters.

77) $x=$ number of \$1 bills; $29-x=$ number of \$5 bills

Value of \$1 bills + Value of \$5 bills = Total Value

$$1x \quad + \quad 5(29-x) \quad = \quad 73.00$$

$$x+145-5x=73$$

$$-4x=-72$$

$$x=18$$

\$5 bills $=29-x=29-18=11$

There are 11-\$5 bills and 18-\$1 bills.

79) $x =$ number of adult tickets; $2x =$ number of children's ticket

Rev. from adult tickets + Rev. from children's tickets = Total Revenue

$$9x \quad + \quad 7(2x) \quad = \quad 437.00$$

$$9x + 14x = 437$$

$$23x = 437$$

$$x = 19$$

children's tickets $= 2x = 2(19) = 38$

There were 19 adult tickets and 38 children's tickets sold.

81) $x =$ cost of ticket for Marc Anthony concert

$x - 19.50 =$ cost of ticket for Santana concert

Cost of Marc Anthony concert + Cost of Santana concert = Total Amount Paid

$$5x \quad + \quad 2(x - 19.5) \quad = \quad 563$$

$$5x + 2x - 39 = 563$$

$$7x = 602$$

$$x = 86$$

cost of Santana concert ticket $= x - 19.5 = \$66.50$

Marc Anthony: \$86; Santana: \$66.50

83) Miles

85) $r =$ the rate of the North bound plane.

Then the rate of the South bound plane is $r + 50$.

	d =	r	t
Northbound	$2r$	r	2
Southbound	$2(r+50)$	$r+50$	2

Northbound distance+Southbound distance=900

$$2r + 2(r+50) = 900$$
$$2r + 2r + 100 = 900$$
$$4r + 100 = 900$$
$$4r = 800$$
$$r = 200$$

Speed of Southbound plane= $r + 50 = 250$

Northbound: 200 mph; Southbound: 250 mph

87) t = the amount of time traveling until they are 200 miles apart

	d	= r	· t
Lance	$22t$	22	t
Danica	$18t$	18	t

$$\text{Lance's distance} + \text{Danica's distance} = 200$$
$$22t + 18t = 200$$
$$40t = 200$$
$$t = 5 \text{ hours}$$

89) t = the time Ahmad traveling

$t - \frac{20}{60} = t - \frac{1}{3}$ is the time Davood has been traveling

	d	= r	· t
Ahmad	$30t$	30	t
Davood	$36\left(t - \frac{1}{3}\right)$	36	$\left(t - \frac{1}{3}\right)$

$$\text{Ahmad's distance} = \text{Davood's distance}$$

$$30t = 36\left(t - \frac{1}{3}\right)$$
$$30t = 36t - \frac{36}{3}$$
$$30t = 36t - 12$$
$$-6t = -12$$
$$t = 2 \text{ hours}$$

Ahmad has been traveling for 2 hours.

Davood has been traveling for $2 - \frac{1}{3} = 1\frac{2}{3}$ hours.

So Davood catches up with Ahmad in $1\frac{2}{3}$ hours.

91) $t =$ the amount of time traveling until they are 6 miles apart

Truck's distance $+6 =$ Car's distance

	d	= r	· t
Truck	$35t$	35	t
Car	$45t$	45	t

$$35t + 6 = 45t$$

$$6 = 10t$$

$$\frac{3}{5} = t \qquad \frac{3}{5} \cdot 60 = 36 \text{ min}$$

93) $r =$ Nick's speed

Then Scott's speed is $r - 2$

	d	= r	· t
Nick	$\frac{1}{2}r$	r	$\frac{1}{2}$
Scott	$\frac{1}{2}(r-2)$	$r-2$	$\frac{1}{2}$

Nick's distance + Scott's distance = 13

$$\frac{1}{2}r + \frac{1}{2}(r-2) = 13$$

$$2 \cdot \frac{1}{2}r + 2 \cdot \frac{1}{2}(r-2) = 2 \cdot 13$$

$$r + r - 2 = 26$$

$$2r = 28$$

$$r = 14$$

Nick's speed= 14 mph

Scott's speed= 12 mph

95) $r =$ Speed of Freight train

Then speed of Passenger train is $r + 20$

	d	= r ·	t
Freight train	$5r$	r	5
Passenger train	$5(r+20)$	$r+20$	5

Freight train's distance + Passenger train's distance = 400

$$5r + 5(r+20) = 400$$
$$5r + 5r + 100 = 400$$
$$10r + 100 = 400$$
$$10r = 300$$
$$r = 30$$

Freight train's speed = 30 mph
Passenger train's speed = 50 mph

97) Let x be the number of Yen.

$$\frac{4.00}{442} = \frac{70.00}{x}$$
$$4x = 442(70)$$
$$4x = 30{,}940$$
$$x = \frac{30{,}940}{4}$$
$$x = 7735 \text{ Yen}$$

The number of Yen is 7735.

99) t = the amount of time traveling until they are 6 miles apart
Bill's distance + Sherri's distance=6

	d	= r	· t
Bill	$14t$	14	t
Sherri	$10t$	10	t

$$14t + 10t = 6$$
$$24t = 6$$
$$t = \frac{6}{24}$$
$$t = \frac{1}{4} \text{ hour} = 15 \text{ minutes}$$

So they will be 6 miles apart in 15 minutes.

101) x = number of quarters; $11 + x$ = number of dimes

Value of Quarters + Value of Dimes = Total Value

$$0.25x + 0.10(11+x) = 6.70$$
$$100(0.25x + 0.10(11+x)) = 100(6.70)$$
$$25x + 10(11+x) = 670$$
$$25x + 110 + 10x = 670$$
$$35x = 560$$
$$x = 16$$

dimes $= x + 11 = 16 + 11 = 27$

There are 27 dimes and 16 quarters.

103) $r =$ the speed of the small plane
$2r =$ the speed of the jet

	d	= r	· t
Small Plane	$\frac{3}{4}r$	r	$\frac{45}{60}=\frac{3}{4}$
Jet	$\frac{3}{4}(2r)$	2r	$\frac{3}{4}$

Small Plane's distance + 150 = Jet's distance

$$\frac{3}{4}r + 150 = \frac{3}{4}(2r)$$
$$3r + 600 = 6r$$
$$600 = 3r$$
$$200 = r$$

the speed of the jet $= 2r = 2(200) = 400$

The jet is traveling at 400 mph, and the small plane is traveling at 200 mph.

105) Let $x =$ the number of girls that are babysitters.

$$\frac{3}{5} = \frac{x}{400}$$
$$400 \cdot 3 = 5x$$
$$1200 = 5x$$
$$240 = x$$

There are 240 girls that are babysitters.

Section 3.8: Exercises

1) You use brackets when there is a $\leq$ or $\geq$ symbol.

3) $(-\infty, 4)$

5) $[-3, \infty)$

7) −3 −2 −1 0 1 2 3
a) $\{k \mid k \leq 2\}$ b) $(-\infty, 2]$

9) −3 −2 −1 0 1 2 3
a) $\left\{c \mid c < \frac{5}{2}\right\}$ b) $\left(-\infty, \frac{5}{2}\right)$

11) −6 −5 −4 −3 −2 −1 0 1 2 3
a) $\{a \mid a \geq -4\}$ b) $[-4, \infty)$

13) When you multiply or divide an inequality by a negative number

15) $k+9\geq 7$

$k+9-9\geq 7-9$

$k\geq -2$

−3 −2 −1 0 1 2 3 4

a) $\{k \mid k\geq -2\}$ b) $[-2, \infty)$

17) $c-10\leq -6$

$c-10+10\leq -6+10$

$c\leq 4$

−3 −2 −1 0 1 2 3 4 5

a) $\{c \mid c\leq 4\}$ b) $(-\infty, 4]$

19) $-3+d<-4$

$-3+d+3<-4+3$

$d<-1$

−4 −3 −2 −1 0 1 2 3 4

a) $\{d \mid d<-1\}$ b) $(-\infty, -1)$

21) $16<z+11$

$16-11<z+11-11$

$5<z$

$z>5$

0 1 2 3 4 5 6 7

a) $\{z \mid z>5\}$ b) $(5, \infty)$

23) $5m>15$

$\frac{5m}{5}>\frac{15}{5}$

$m>3$

−4 −3 −2 −1 0 1 2 3 4

a) $\{m \mid m>3\}$ b) $(3, \infty)$

25) $12x<-21$

$\frac{12x}{12}<\frac{-21}{12}$

$x<-\frac{7}{4}$

−5 −4 −3 −2 −1 0 1 2 3

a) $\left\{x \mid x<-\frac{7}{4}\right\}$ b) $\left(-\infty, -\frac{7}{4}\right)$

27) a) $\left\{y \mid y<-\frac{11}{3}\right\}$ b) $\left(-\infty, -\frac{11}{3}\right)$

29) $-7b\geq 21$

$\frac{-7b}{-7}\leq\frac{21}{-7}$

$b\leq -3$

−5 −4 −3 −2 −1 0 1 2

a) $\{b \mid b\leq -3\}$ b) $(-\infty, -3]$

31) $-12n>-36$

$\frac{-12n}{-12}<\frac{-36}{-12}$

$n<3$

−3 −2 −1 0 1 2 3 4

a) $\{n \mid n<3\}$ b) $(-\infty, 3)$

33) $\frac{1}{2}w<-3$

$2\cdot\frac{1}{2}w<2\cdot(-3)$

$w<-6$

−7 −6 −5 −4 −3 −2 −1 0

a) $\{w \mid w<-6\}$ b) $(-\infty,-6)$

35) $-\frac{7}{2}d \geq 35$

$-\frac{2}{7}\cdot\left(-\frac{7}{2}d\right) \geq -\frac{2}{7}\cdot 35$

$d \leq -10$

−12 −11 −10 −9 −8 −7 −6

a) $\{d \mid d \leq -10\}$ b) $(-\infty, -10]$

37) $6y+5 > -13$

$6y+5-5 > -13-5$

$6y > -18$

$y > -3$

−4 −3 −2 −1 0 1 2 3 4

$(-3, \infty)$

39) $17-7x \geq 20$

$17-17-7x \geq 20-17$

$-7x \geq 3$

$\frac{-7x}{-7} \leq \frac{3}{-7}$

$x \leq -\frac{3}{7}$

−3 −2 −1 0 1 2 3

$\left(-\infty, -\frac{3}{7}\right]$

41) $\frac{1}{2}k-3 < -2$

$2\cdot\left(\frac{1}{2}k\right)-2\cdot 3 < 2\cdot(-2)$

$k-6 < -4$

$k < 2$

−2 −1 0 1 2 3 4

$(-\infty, 2)$

43) $a+2 < 2a+3$

$2-3 < 2a-a$

$-1 < a$

−3 −2 −1 0 1 2 3

$(-1, \infty)$

45) $-6-(t+8) \leq 2(11-3t)+4$

$-6-t-8 \leq 22-6t+4$

$-14-t \leq 26-6t$

$-14-t+14 \leq 26+14-6t$

$-t \leq 40-6t$

$-t+6t \leq 40-6t+6t$

$5t \leq 40$

$t \leq 8$

0 1 2 3 4 5 6 7 8 9

$(-\infty, 8]$

47) $\frac{11}{6}+\frac{3}{2}(d-2) > \frac{2}{3}(d+5)+\frac{1}{2}d$

$6\cdot\frac{11}{6}+6\cdot\left(\frac{3}{2}(d-2)\right) > 6\cdot\left(\frac{2}{3}(d+5)\right)+6\cdot\left(\frac{1}{2}d\right)$

$11+9(d-2) > 4(d+5)+3d$

$11+9d-18 > 4d+20+3d$

$9d-7 > 7d+20$

$9d > 7d+27$

$2d > 27$

$d \geq \frac{27}{2}$

10 11 12 13 14 15

$\left[\frac{27}{2}, \infty\right)$

49) $0.02c+0.1(30)<0.08(30+c)$

$$100(0.02c)+100(0.1(30))<100(0.08(30+c))$$
$$2c+10(30)<8(30+c)$$
$$2c+300<240+8c$$
$$-6c+300<240$$
$$-6c<-60$$
$$c>10$$

0 2 4 6 8 10 12 14

$(10,\infty)$

51) $(1,4)$

53) $[-2,5)$

55)

−2 −1 0 1 2 3 4 5

a) $\{t \mid 1\le t\le 4\}$ b) $[1, 4]$

57)

−3 −2 −1 0 1 2 3

a) $\{p \mid -2<p<1\}$ b) $(-2, 1)$

59)

−4 −3 −2 −1 0 1 2 3 4

a) $\{a \mid -2\le a<3\}$ b) $[-2, 3)$

61) $4<k+9<10$

$$4-9<k+9-9<10-9$$
$$-5<k<1$$

−6 −5 −4 −3 −2 −1 0 1 2

$(-5, 1)$

63) $-5\le 5m\le -2$

$$\frac{-5}{5}\le\frac{5m}{5}\le\frac{-2}{5}$$
$$-1\le d\le -\frac{2}{5}$$

−5 −4 −3 −2 −1 0 1 2 3 4 5

$\left[-1, -\frac{2}{5}\right]$

65) $-4<2y-7<-1$

$$-4+7<2y-7+7<-1+7$$
$$3<2y<6$$
$$\frac{3}{2}<y<3$$

−3 −2 −1 0 1 2 3 4

$\left(\frac{3}{2}, 3\right)$

67) $2\le\frac{1}{2}n+3\le 5$

$$2\cdot 2\le 2\cdot\frac{1}{2}n+2\cdot 3\le 2\cdot 5$$
$$4\le n+6\le 10$$
$$4-6\le n+6-6\le 10-6$$
$$-2\le n\le 4$$

−5 −4 −3 −2 −1 0 1 2 3 4 5

$[-2, 4]$

69) $-4\le 3w-1\le 3$

$$-4+1\le 3w-1+1\le 3+1$$
$$-3\le 3w\le 4$$
$$-1\le w\le\frac{4}{3}$$

−5 −4 −3 −2 −1 0 1 2 3 4 5

$\left[-1, \frac{4}{3}\right]$

71) $0 < \dfrac{5t+2}{3} < \dfrac{7}{3}$

$$3 \cdot 0 < 3\left(\frac{5t+2}{3}\right) < 3 \cdot \frac{7}{3}$$

$$0 < 5t + 2 \leq 7$$

$$0 - 2 \leq 5t + 2 - 2 \leq 7 - 2$$

$$-2 < 5t < 5$$

$$-\frac{2}{5} < t < 1$$

−4 −3 −2 −1 0 1 2 3 4

$$\left(-\frac{2}{5}, 1\right)$$

73) $-9 < 7 - 4m < 9$

$$-9 - 7 < 7 - 7 - 4m < 9 - 7$$

$$-16 < -4m < 2$$

$$\frac{-16}{-4} > m > \frac{2}{-4}$$

$$4 > m > -\frac{1}{2}$$

$$-\frac{1}{2} < m < 4$$

−4 −3 −2 −1 0 1 2 3 4 5

$$\left(-\frac{1}{2}, 4\right)$$

75) $6 \leq 4 - 3b < 10$

$$6 - 4 \leq 4 - 4 - 3b < 10 - 4$$

$$2 \leq -3b < 6$$

$$\frac{2}{-3} \geq \frac{-3b}{-3} > \frac{6}{-3}$$

$$-\frac{2}{3} \geq b > -2$$

$$-2 < b \leq -\frac{2}{3}$$

−5 −4 −3 −2 −1 0 1 2 3 4 5

$$\left(-2, -\frac{2}{3}\right]$$

77) Let x = the number of $\frac{1}{2}$ hour intervals after the first 4 hours.

$$\frac{1}{2}x + 4 = \text{Total Time}$$

$$\begin{pmatrix}\text{Total}\\ \text{Cost}\end{pmatrix} = \begin{pmatrix}\text{Cost of First}\\ \text{4 hours}\end{pmatrix} + \begin{pmatrix}\text{Cost of}\\ \text{Add. } \frac{1}{2} \text{ hour}\end{pmatrix}$$

$$= \quad 36 \quad + \quad 3x$$

$$36 + 3x \leq 50$$

$$3x \leq 14$$

$$x \leq \frac{14}{3}$$

$$x \leq 4\frac{2}{3}$$

Edwardo can afford at most 4 intervals.

$$\text{Total Time} = \frac{1}{2}x + 4 = \frac{1}{2}(4) + 4$$

$$= 6 \text{ hours}$$

79) Let x = the number of people he can invite for the meeting.

$$\begin{pmatrix}\text{Total}\\ \text{Cost}\end{pmatrix} = \begin{pmatrix}\text{Cost to rent}\\ \text{the room}\end{pmatrix} + \begin{pmatrix}\text{Cost of}\\ \text{snacks and}\\ \text{beverages}\\ \text{per person}\end{pmatrix}$$

$$= \quad 500 \quad + \quad 8x$$

$$500 + 8x \leq 1000$$

$$8x \leq 500$$

$$x \leq \frac{500}{8}$$

$$x \leq 62.5$$

The greatest number of people who can attend the meeting is 62.

81) Let $x =$ the number of $\frac{1}{4}$ mi.

$$\frac{1}{4}x = \text{Total Mileage}$$

$$\begin{pmatrix}\text{Total}\\\text{Cost}\end{pmatrix} = \begin{pmatrix}\text{Initial}\\\text{Cost}\end{pmatrix} + \begin{pmatrix}\text{Cost per}\\\text{Mile}\end{pmatrix}$$

$$= \quad 2.00 \quad + \quad 0.30x$$

$$2.00 + 0.30x \le 14.00$$
$$0.30x \le 12$$
$$x \le 40$$

$$\text{Total Mileage} = \frac{1}{4}x = \frac{1}{4}(40) = 10 \text{ miles}$$

83) Let $x =$ the grade she needs to make
$92 + 85 + 96 + x =$ Total of four scores

$$\frac{92 + 85 + 96 + x}{4} \ge 90$$
$$4 \cdot \frac{273 + x}{4} \ge 4 \cdot 90$$
$$273 + x \ge 360$$
$$x \ge 87$$

Eliana must make an 87 or higher.

Chapter 3 Review

1) No.

$$\frac{3}{2}k - 5 = 1$$
$$\frac{3}{2}(-4) - 5 = 1$$
$$-6 - 5 = 1$$
$$-11 \ne 1$$

3) The variables are eliminated and you get a false statement like $5 = 13$.

5)
$$h + 14 = -5$$
$$h + 14 - 14 = -5 - 14$$
$$h = -19$$

The solution set is $\{-19\}$.

7)
$$-7g = 56$$
$$\frac{-7g}{-7} = \frac{56}{-7}$$
$$g = -8$$

The solution set is $\{-8\}$.

9)
$$4 = \frac{c}{9}$$
$$9 \cdot 4 = 9 \cdot \frac{c}{9}$$
$$36 = c$$

The solution set is $\{36\}$.

11)
$$23 = 4m - 7$$
$$23 + 7 = 4m$$
$$30 = 4m$$
$$\frac{30}{4} = \frac{4m}{4}$$
$$\frac{15}{2} = m$$

The solution set is $\left\{\frac{15}{2}\right\}$.

13)
$$4c + 9 + 2(c - 12) = 15$$
$$4c + 9 + 2c - 24 = 15$$
$$6c - 15 = 15$$
$$6c - 15 + 15 = 15 + 15$$
$$6c = 30$$
$$c = 5$$

The solution set is $\{5\}$.

15)
$$\begin{aligned} 2z+11&=8z+15\\ 2z-8z+11&=8z-8z+15\\ -6z+11&=15\\ -6z+11-11&=15-11\\ -6z&=4\\ z&=\frac{4}{-6}\\ z&=-\frac{2}{3} \end{aligned}$$
The solution set is $\left\{-\frac{2}{3}\right\}$.

17)
$$\begin{aligned} k+3(2k-5)&=4(k-2)-7\\ k+6k-15&=4k-8-7\\ 7k-15&=4k-15\\ 7k-15+15&=4k-15+15\\ 7k&=4k\\ 7k-4k&=4k-4k\\ 3k&=0\\ k&=0 \end{aligned}$$
The solution set is $\{0\}$.

19)
$$\begin{aligned} 0.18a+0.1(20-a)&=0.14(20)\\ 100(0.18a+0.1(20-a))&=100(0.14(20))\\ 18a+10(20-a)&=14(20)\\ 18a+200-10a&=280\\ 8a+200&=280\\ 8a&=80\\ a&=10 \end{aligned}$$
The solution set is $\{10\}$.

21)
$$\begin{aligned} 16&=-\frac{12}{5}d\\ -\frac{5}{12}\cdot 16&=\left(-\frac{5}{12}\right)\left(-\frac{12}{5}\right)d\\ -\frac{20}{3}&=m \end{aligned}$$
The solution set is $\left\{-\frac{20}{3}\right\}$.

23)
$$\begin{aligned} 3(r+4)-r&=2(r+6)\\ 3r+12-r&=2r+12\\ 2r+12&=2r+12\\ 12&=12 \end{aligned}$$
The solution set is $\{\text{all real numbers}\}$.

25) Let x = a number
$$\begin{aligned} 2x-9&=25\\ 2x-9+9&=25+9\\ 2x&=34\\ x&=17 \end{aligned}$$
The number is 17.

27) Let x be the number of emails received on Thursday.
Then the number of
emails received on Friday $= x - 24$

$$\begin{pmatrix}\text{number of emails}\\ \text{received}\\ \text{on Thursday}\end{pmatrix} + \begin{pmatrix}\text{number of emails}\\ \text{received}\\ \text{on Friday}\end{pmatrix} = 126$$

$$x \quad + x - 24 = 126$$
$$2x - 24 = 126$$
$$2x = 150$$
$$x = 75$$

Emails received on Friday $= x - 24$
$= 75 - 24$
$= 51$

Kendrick received 75 emails on Thursday and 51 on Friday.

29) Let x be the length of the shorter pipe.
Then the length of the longer pipe is $x + 8$

$$(\text{length of shorter pipe}) + (\text{length of longer pipe}) = 36$$

$$x \quad + x + 8 = 36$$
$$2x + 8 = 36$$
$$2x = 28$$
$$x = 14$$

length of the longer pipe $= x + 8$
$= 14 + 8$
$= 22$

The pipes are 14 in. and 22 in.

31) Let x be the weight of implant 20 years ago.
Then the weight of today's implant is $0.5x$

$$\begin{pmatrix}\text{weight of toda'ys}\\ \text{implant}\end{pmatrix} = 3$$

$$0.5x = 3$$
$$x = 6$$

weight of implant 20 years ago is $x = 6$ lb.

33) $x =$ amount Jerome invested in the 2% account.
$6000 - x =$ amount Jerome invested in the 4% account.
Total Interest Earned = Interest from 2% account + Interest from 4% account

$$210 = x(0.02)(1) + (6000 - x)(0.04)(1)$$
$$100(210) = 100[x(0.02)(1) + (3000 - x)(0.04)(1)]$$
$$21{,}000 = 2x + 4(24{,}000 - x)$$
$$-3000 = -2x$$
$$1500 = x$$

amount invested in 4% $= 6000 - x = 6000 - 1500 = 4500$
Jose invested \$1500 in the 2% account and \$4500 in the 4% account.

33) $P = 2l + 2w$
$$32 = 2(9) + 2w$$
$$32 = 18 + 2w$$
$$14 = 2w$$
$$7 = w$$

35) $A = \frac{1}{2}bh$
$$42 = \frac{1}{2}(12)h$$
$$42 = 6h$$
$$7 = h$$
The height of the triangle is 7 inches.

37) $m\angle A = (x)^\circ \quad m\angle B = (x)^\circ$
$$m\angle C = (x+15)^\circ$$
$$m\angle A + m\angle B + m\angle C = 180$$
$$x + x + (x+15) = 180$$
$$3x + 15 = 180$$
$$3x = 165$$
$$x = 55$$

$$m\angle A = (x)^\circ = 55^\circ$$
$$m\angle B = (x)^\circ = 55^\circ$$
$$m\angle C = (x+15)^\circ = (55+15)^\circ = 70^\circ$$

39) $(6x+7)^\circ = (9x-20)^\circ$
$$6x + 7 = 9x - 20$$
$$27 = 3x$$
$$9 = x$$

$$(6x+7)^\circ = (6(9)+7)^\circ = 61^\circ$$
$$(9x-20)^\circ = (9(9)-20)^\circ = 61^\circ$$

41) $p - n = z$
$$p - n + n = z + n$$
$$p = z + n$$

43) $A = \frac{1}{2}bh$
$$2A = bh$$
$$\frac{2A}{h} = b$$

45) Yes, it can be written as $\frac{15}{100}$ or $\frac{3}{20}$

47) $\frac{12}{15} = \frac{4}{5}$

49) $\frac{x}{15}=\frac{8}{10}$

$10x=120$

$x=12$

$\{12\}$

51) Let x be the number of students that have used alcohol.

$\frac{x}{2500}=\frac{9}{20}$

$20x=2500\cdot 9$

$20x=22,500$

$x=1125$

$\{1125\}$

53) $x=$ number of \$10 bills;

$2+x=$ number of \$20 bills

Value of \$10 bills + Value of \$20 bills = Total Value

$10x+20(2+x)=340.00$

$10x+40+20x=340$

$30x=300$

$x=10$

\$20 bills $=2+x=2+10=12$

There are 12-\$20 bills and 10-\$10 bills.

55) $x=$ Meg's speed. Then Jared's speed is $x-1$

	d =	r ·	t
Meg	$1(x)$	x	1
Jared	$1(x-1)$	$x-1$	1

Meg's distance + Jared's distance $=11$

$x + x-1=11$

$2x-1=11$

$2x=12$

$x=6$ mph

Meg's speed: 6 mph

Jared's speed: x - 1 = 5 mph

57) $w+8>5$

$w>-3$

−4 −3 −2 −1 0 1 2 3 4

$(-3, \infty)$

59) $5x-2\le 18$

$5x\le 20$

$x\le 4$

−3 −2 −1 0 1 2 3 4 5

$(-\infty, 4]$

61) $-19\le 7p+9\le 2$

$-19-9\le 7p+9-9\le 2-9$

$-28\le 7p\le -7$

$-4\le p\le -1$

−6 −5 −4 −3 −2 −1 0 1 2

$[-4, -1]$

63) $\frac{1}{2} < \frac{1-4t}{6} < \frac{3}{2}$

$$6\left(\frac{1}{2}\right) < 6 \cdot \frac{1-4t}{6} < 6 \cdot \frac{3}{2}$$
$$3 < 1-4t < 9$$
$$3-1 < 1-4t-1 < 9-1$$
$$2 < -4t < 8$$
$$\frac{2}{-4} > t > \frac{8}{-4}$$
$$-\frac{1}{2} > t > -2$$
$$-2 < t < -\frac{1}{2}$$

−4 −3 −2 −1 0 1 2 3 4

$\left(-2, -\frac{1}{2}\right)$

65) $-8k+13=-7$

$$-8k+13-13=-7-13$$
$$-8k=-20$$
$$\frac{-8k}{-8}=\frac{-20}{-8}$$
$$k=\frac{5}{2} \qquad \left\{\frac{5}{2}\right\}$$

67) $29=-\frac{4}{7}m+5$

$$7(29)=7\left(-\frac{4}{7}\right)m+7(5)$$
$$203=-4m+35$$
$$203-35=-4m$$
$$\frac{168}{-4}=m$$
$$-42=m \qquad \{-42\}$$

69) $10p+11=5(2p+3)-1$

$$10p+11=10p+15-1$$
$$10p+11=10p+14$$
$$10p+11-10p=10p+14-10p$$
$$11 \neq 14$$

The solution set is $\varnothing$.

71) $\frac{2x+9}{5}=\frac{x+1}{2}$

$$2(2x+9)=5(x+1)$$
$$4x+18=5x+5$$
$$4x+18-5=5x+5-5$$
$$4x+13=5x+5-5$$
$$4x+13=5x$$
$$4x-4x+13=5x-4x$$
$$13=x \qquad \{13\}$$

73) $\frac{5}{6}-\frac{3}{4}(r+2)=\frac{1}{2}r+\frac{7}{12}$

$$12\left(\frac{5}{6}\right)-12\left(\frac{3}{4}(r+2)\right)=12\left(\frac{1}{2}r\right)+12\left(\frac{7}{12}\right)$$
$$2(5)-9(r+2)=6r+7$$
$$10-9r-18=6r+7$$
$$-9r-8=6r+7$$
$$-9r-8+9r=6r+9r+7$$
$$-8=15r+7$$
$$-8-7=15r$$
$$-15=15r$$
$$-1=r$$

The solution set is $\{-1\}$.

75) $x=$ number of oz of 5% alcohol solution
$60+x=$ number of oz of 9% alcohol solution

Solution	Concentration	Number of oz of solution	Number of oz of alcohol in the solution
5%	0.05	x	$0.05x$
17%	0.17	60	$0.17(60)$
9%	0.09	$60+x$	$0.09(60+x)$

$$0.05x+0.17(60)=0.09(60+x)$$
$$100(0.05x+0.17(60))=100(0.09(60+x))$$
$$5x+17(60)=9(60+x)$$
$$5x+1020=540+9x$$
$$-4x=-480$$
$$x=120$$

120 oz of the 5% solution is needed.

77) Let x be the first odd integer.
$x+2=$ second odd integer

$$\begin{pmatrix}\text{first odd}\\\text{integer}\end{pmatrix}+\begin{pmatrix}\text{second odd}\\\text{integer}\end{pmatrix}=3\begin{pmatrix}\text{larger}\\\text{integer}\end{pmatrix}-21$$
$$x \quad + \quad x+2 \quad =3(x+2)-21$$
$$2x+2=3x+6-21$$
$$2x+2=3x-15$$
$$2x=3x-15-2$$
$$2x=3x-17$$
$$2x-3x=3x-17-3x$$
$$-x=-17$$
$$x=17$$

$x=$ first odd integer $=17$
$x+2=$ second odd integer $=17+2=19$
The integers are 17 and 19.

79) Let x be the length of the shortest side.

Then the length of the other side is $x+3$

Then length of the longest side is $2x$.

shortest side + other side + longest side = Perimeter

$$x+(x+3)+2x=35$$
$$4x+3=35$$
$$4x=32$$
$$x=8$$

shortest side is 8 cm

other side is $x+3=8+3=11$ cm

longest side is $2x=2(8)=16$ cm

81) Let x = number of polled residents who want to secede.

$$\frac{9}{25}=\frac{x}{1000}$$
$$9(1000)=25x$$
$$9000=25x$$
$$\frac{9000}{25}=x$$
$$360=x$$

360 residents out of 1000 want to secede.

Chapter 3 Test

1)
$$-18y=14$$
$$\frac{-18y}{-18}=\frac{14}{-18}$$
$$y=-\frac{7}{9}$$

The solution set is $\left\{-\frac{7}{9}\right\}$

3)
$$\frac{8}{3}n-11=5$$
$$3\left(\frac{8}{3}n\right)-3\cdot 11=3\cdot 5$$
$$8n-33=15$$
$$8n=48$$
$$n=6$$

The solution set is $\{6\}$

5) $$\frac{1}{2}-\frac{1}{6}(x-5)=\frac{1}{3}(x+1)+\frac{2}{3}$$

$$6\cdot\frac{1}{2}-6\left(\frac{1}{6}(x-5)\right)=6\left(\frac{1}{3}(x+1)\right)+6\cdot\frac{2}{3}$$

$$3-1(x-5)=2(x+1)+4$$

$$3-x+5=2x+2+4$$

$$8-x=2x+6$$

$$2=3x$$

$$\frac{2}{3}=x$$

The solution set is $\left\{\frac{2}{3}\right\}$

7) $$\frac{9-w}{4}=\frac{3w+1}{2}$$

$$2(9-w)=4(3w+1)$$

$$18-2w=12w+4$$

$$14=14w$$

$$1=w$$

The solution set is $\{1\}$

9) Let $x=$ first even integer.
Then the second even integer is $x+2$
Third even integer is $x+4$

$$x+x+2+x+4=114$$

$$3x+6=114$$

$$3x=108$$

$$x=36$$

Scond even intger=38
Third even integer=40
The integers are 36, 38, 40.

11) Let $x=$ amount spent by Debra.

$$\frac{14}{40.60}=\frac{11}{x}$$

$$14x=11(40.60)$$

$$14x=446.60$$

$$x=31.90$$

Debra spent \$31.90.

13) $x=$ speed of eastbound car. Then speed of westbound car is $x+6$.

	d =	r ·	t
westbound	$2.5(x)$	x	2.5
eastbound	$2.5(x+6)$	$x+6$	2.5

westbound car's distance + eastbound car's distance = 345

$$2.5\,x \quad + \quad 2.5(x+6) \;=345$$

$$2.5x+2.5x+15=345$$

$$5x+15=345$$

$$5x=330$$

$$x=66$$

westbound car's speed: 66 mph
eastbound car's speed: $x+6=$ 7 mph

15) $S = 2\pi r^2 + 2\pi rh$

$S - 2\pi r^2 = 2\pi r^2 - 2\pi r^2 + 2\pi rh$

$S - 2\pi r^2 = 2\pi rh$

$\dfrac{S - 2\pi r^2}{2\pi r} = \dfrac{2\pi rh}{2\pi r}$

$\dfrac{S - 2\pi r^2}{2\pi r} = h$

or $h = \dfrac{S}{2\pi r} - r$

17) $6m + 19 \le 7$

$6m \le -12$

$m \le -2$

−4 −3 −2 −1 0 1 2 3 4 $(-\infty, -2]$

19) $-\dfrac{5}{6} < \dfrac{4c-1}{6} \le \dfrac{3}{2}$

$-5 < 4c - 1 \le 9$

$-4 < 4c \le 10$

$-1 < c \le \dfrac{5}{2}$

−4 −3 −2 −1 0 1 2 3 4 $\left(-1, \dfrac{5}{2}\right]$

Cumulative Review: Chapter 1-3.

1) $\dfrac{3}{8} - \dfrac{5}{6} = \dfrac{9}{24} - \dfrac{20}{24} = -\dfrac{11}{24}$

3) $26 - 14 \div 2 + 5 \cdot 7 = 26 - 7 + 35 = 54$

5) $-39 - |7 - 15| = -39 - |-8| = -39 - 8 = -47$

7) $\{-15, 0, 9\}$

9) $\{0, 9\}$

11) No. For example, $10 - 3 \ne 3 - 10$

13) $\dfrac{35r^{16}}{28r^4} = \dfrac{5}{4}r^{16-4} = \dfrac{5}{4}r^{12}$

15) $\left(-12z^{10}\right)\left(\dfrac{3}{8}z^{-16}\right) = -\dfrac{9}{2}z^{10-16} = -\dfrac{9}{2}z^{-6} = -\dfrac{9}{2z^6}$

17) 0.00000895: 0.00000895

$= 8.95 \times 10^{-6}$

19) $\dfrac{3}{2}n + 14 = 20$

$2 \cdot \dfrac{3}{2}n + 2 \cdot 14 = 2 \cdot 20$

$3n + 28 = 40$

$3n = 12$

$n = 4$

The solution set is $\{4\}$

21) $\dfrac{x+3}{10} = \dfrac{2x-1}{4}$

$4(x+3) = 10(2x-1)$

$4x + 12 = 20x - 10$

$22 = 16x$

$\dfrac{22}{16} = x$

$\dfrac{11}{8} = x$

The soltution set is $\left\{\dfrac{11}{8}\right\}$

23) $x =$ speed of the train
speed of the car $= x - 10$

	d	= r	· t
train	140	x	$\frac{140}{x}$
car	120	$x-10$	$\frac{120}{x-10}$

time by train = time by car

$$\frac{140}{x} = \frac{120}{x-10}$$
$$140(x-10) = 120x$$
$$140x - 1400 = 120x$$
$$-1400 = -20x$$
$$70 = x$$

speed of train = 70 mph
speed of car $= 70 - 10 =$ 60 mph

25) $-17 < 6b - 11 < 4$

$$-6 < 6b < 15$$
$$-1 < b < \frac{5}{2}$$
$$\left(-1, \frac{5}{2}\right)$$

Chapter 4: Linear Equations in Two Variables

Section 4.1 Exercises

1) 16.1 gallons

3) 2004 and 2006: 15.9 gallons

5) Consumption was increasing

7) New Jersey: 86.3%

9) Florida's graduation rate is about 32.4% less than New Jersey's.

11) Answers may vary

13) Yes

$$2x+5y=1$$
$$2(-2)+5(1)=1$$
$$-4+5=1$$
$$1=1$$

15) Yes

$$-3x-2y=-15$$
$$-3(7)-2(-3)=-15$$
$$-21+6=-15$$
$$-15=-15$$

17) No

$$y=-\frac{3}{2}x-7$$
$$(5)=-\frac{3}{2}(8)-7$$
$$5=-12-7$$
$$5\neq-19$$

19) Yes

$$y=-7$$
$$-7=-7$$

21)

$$y=3x-7$$
$$y=3(4)-7$$
$$y=12-7$$
$$y=5$$

23)

$$2x-15y=13$$
$$2x-15\left(-\frac{4}{3}\right)=13$$
$$2x+20=13$$
$$2x=-7$$
$$x=-\frac{7}{2}$$

25) $x=5$

27) $y=2x-4$

$$y=2(0)-4$$
$$y=0-4$$
$$y=-4$$

$$y=2(-1)-4$$
$$y=-2-4$$
$$y=-6$$

$$y=2(1)-4$$
$$y=2-4$$
$$y=-2$$

$$y=2(-2)-4$$
$$y=-4-4$$
$$y=-8$$

x	y
0	−4
1	−2
−1	−6
−2	−8

29) $y=4x$

$y=4(0)$

$y=0$

$y=4\left(\frac{1}{2}\right)$

$y=2$

$12=4x \quad -20=4x$

$\frac{12}{4}=x \qquad \frac{-20}{4}=x$

$3=x \qquad -5=x$

x	y
0	0
$\frac{1}{2}$	2
3	12
−5	−20

31) $5x+4y=-8$

$5(0)+4y=-8$

$4y=-8$

$y=\frac{-8}{4}=-2$

$5x+4(0)=-8$

$5x+0=-8$

$5x=-8$

$x=-\frac{8}{5}$

$5(1)+4y=-8$

$5+4y=-8$

$4y=-13$

$y=-\frac{13}{4}$

$5x+4y=-8$

$5\left(-\frac{12}{5}\right)+4y=-8$

$-12+4y=-8$

$4y=4$

$y=\frac{4}{4}=1$

x	y
0	−2
$-\frac{8}{5}$	0
1	$-\frac{13}{4}$
$-\frac{12}{5}$	1

33) $y=-2$

x	y
0	−2
−3	−2
8	−2
17	−2

35) Answers may vary.

37) A: $(-2, 1)$ quadrant II

B: $(5, 0)$ no quadrant

C: $(-2, -1)$ quadrant III

D: $(0, -1)$ no quadrant

E: $(2, -2)$ quadrant IV

F: $(3, 4)$ quadrant I

39–42)

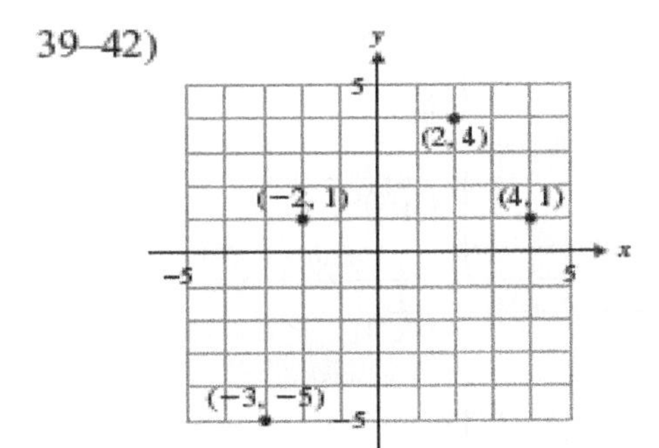

55–56)

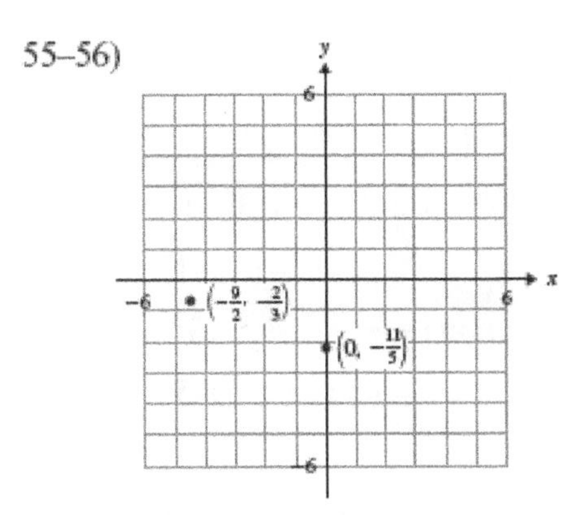

43–46)

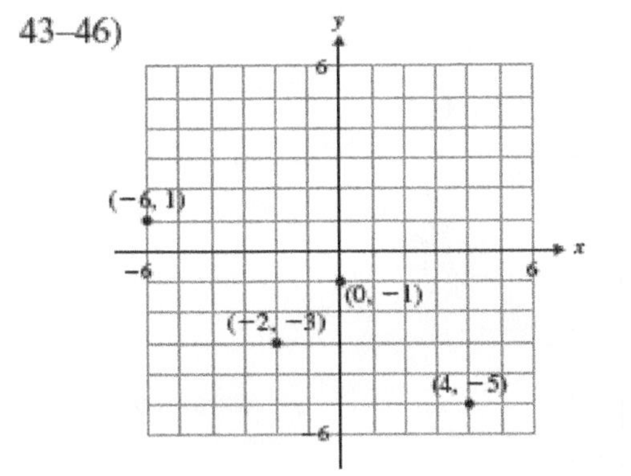

47–50)

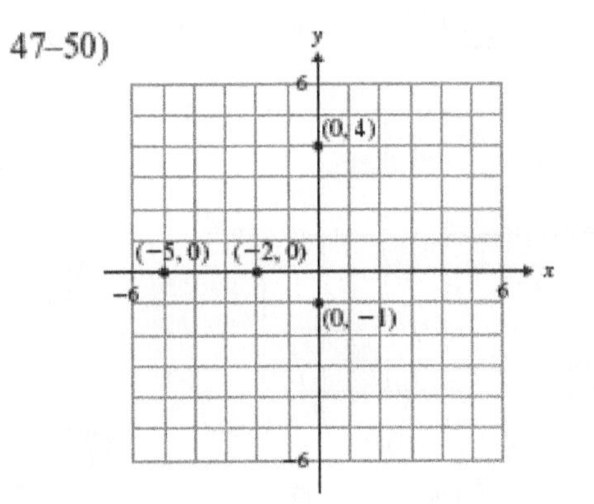

51–54)

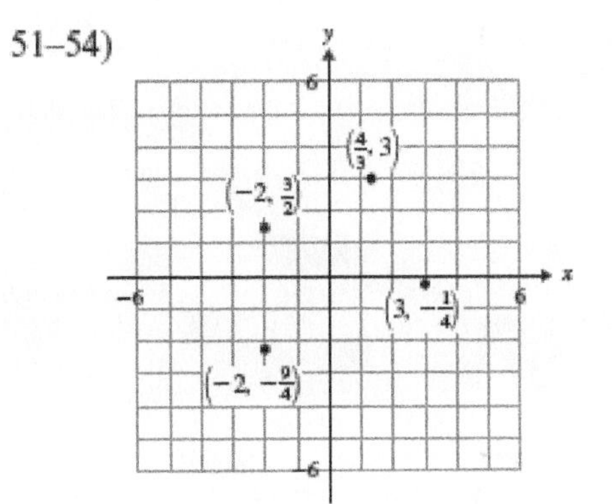

57) $y = -4x + 3$

$y = -4(0) + 3$ $\quad$ $(0) = -4x + 3$

$y = 0 + 3$ $\quad$ $-3 = -4x$

$y = 3$ $\quad$ $\dfrac{3}{4} = x$

$y = -4(2) + 3$ $\quad$ $(7) = -4x + 3$

$y = -8 + 3$ $\quad$ $4 = -4x$

$y = -5$ $\quad$ $-1 = x$

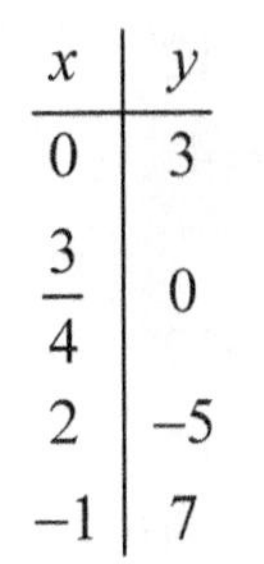

x	y
0	3
$\frac{3}{4}$	0
2	−5
−1	7

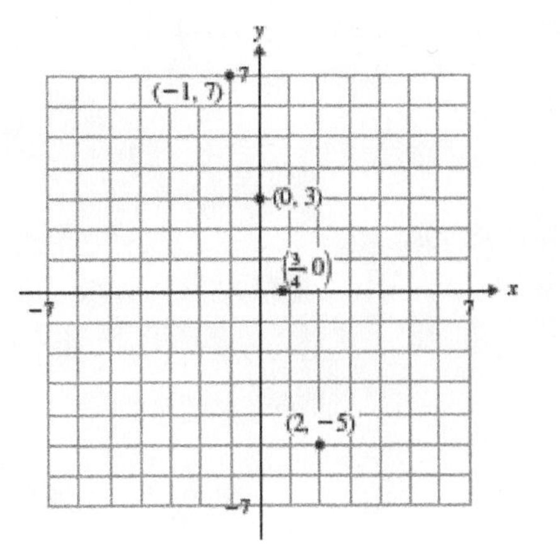

59) $y = x$

$y = (0)$ $\quad$ $(3) = x$

$y = 0$ $\quad$ $3 = x$

$y = (-1)$ $\quad$ $(-5) = x$

$y = -1$ $\quad$ $-5 = x$

x	y
0	0
−1	−1
3	3
−5	−5

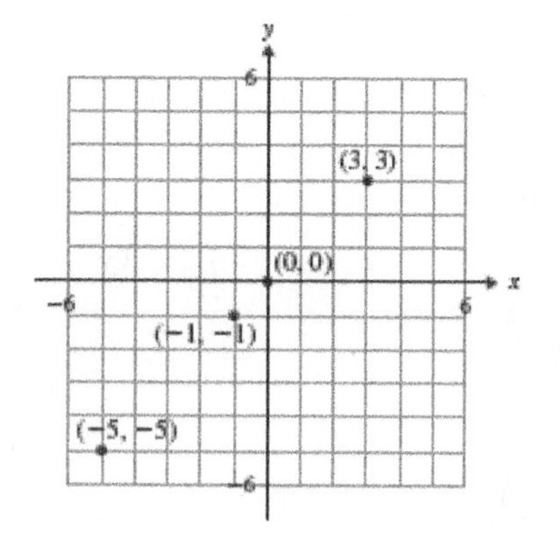

61) $3x + 4y = 12$

$3(0) + 4y = 12$

$4y = 12$

$y = 3$

$3x + 4(0) = 12$

$3x = 12$

$x = 4$

$3(1) + 4y = 12$

$3 + 4y = 12$

$4y = 9$

$y = \dfrac{9}{4}$

$3x + 4(6) = 12$

$3x + 24 = 12$

$3x = -12$

$x = -4$

x	y
0	3
4	0
1	$\frac{9}{4}$
−4	6

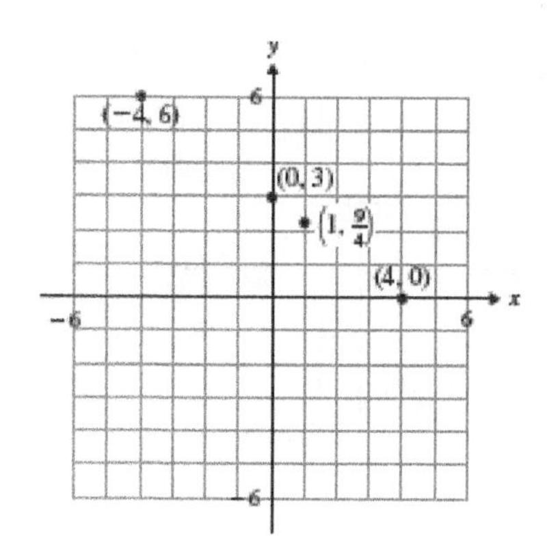

63) $y + 1 = 0$

$y = -1$

x	y
0	−1
1	−1
−3	−1
−1	−1

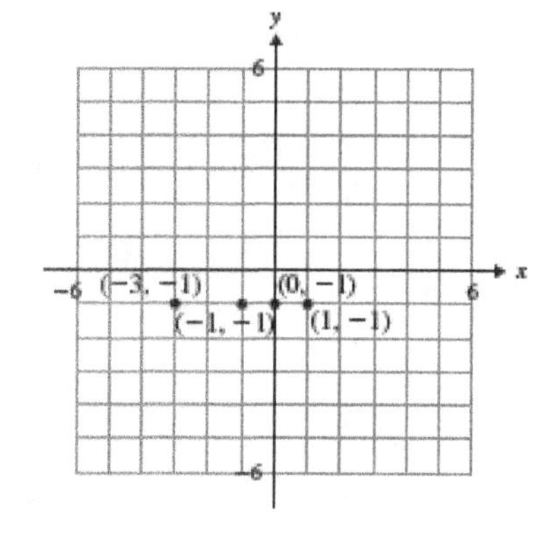

65) $y = \frac{1}{4}x + 2$

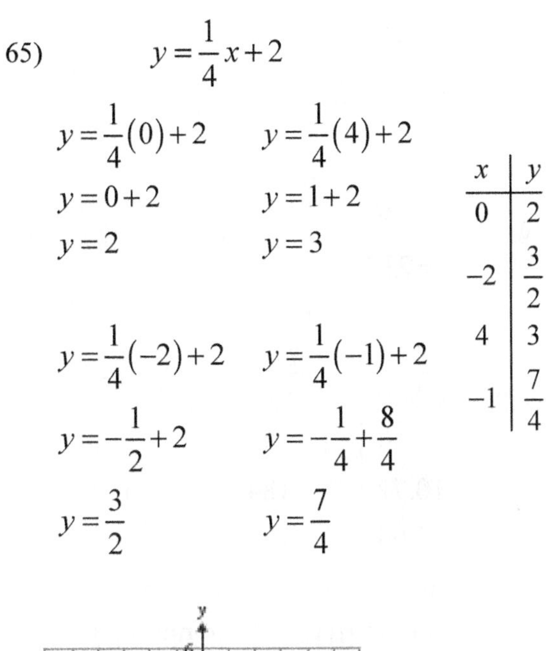

$y = \frac{1}{4}(0) + 2$ $\quad$ $y = \frac{1}{4}(4) + 2$

$y = 0 + 2$ $\quad$ $y = 1 + 2$

$y = 2$ $\quad$ $y = 3$

$y = \frac{1}{4}(-2) + 2$ $\quad$ $y = \frac{1}{4}(-1) + 2$

$y = -\frac{1}{2} + 2$ $\quad$ $y = -\frac{1}{4} + \frac{8}{4}$

$y = \frac{3}{2}$ $\quad$ $y = \frac{7}{4}$

x	y
0	2
−2	$\frac{3}{2}$
4	3
−1	$\frac{7}{4}$

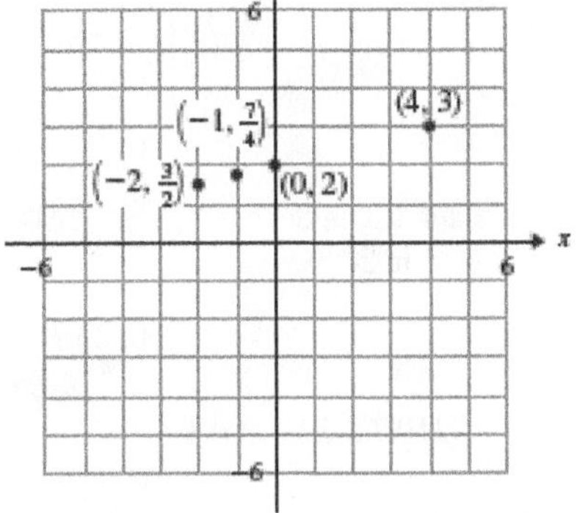

67) $y = \frac{2}{3}x - 7$

a) $x = 3$ $\quad$ $y = \frac{2}{3}(3) - 7$

$y = 2 - 7$

$y = -5$

$x = 6$ $\quad$ $y = \frac{2}{3}(6) - 7$

$y = 4 - 7$

$y = -3$

$x = -3$ $\quad$ $y = \frac{2}{3}(-3) - 7$

$y = -2 - 7$

$y = -9$

$(3, -5), (6, -3), (-3, -9)$

b) $x = 1$ $\quad$ $y = \frac{2}{3}(1) - 7$

$y = \frac{2}{3} - \frac{21}{3}$

$y = -\frac{19}{3}$

$x = 5$ $\quad$ $y = \frac{2}{3}(5) - 7$

$y = \frac{10}{3} - \frac{21}{3}$

$y = -\frac{11}{3}$

$x = -2$ $\quad$ $y = \frac{2}{3}(-2) - 7$

$y = -\frac{4}{3} - \frac{21}{3}$

$y = -\frac{25}{3}$

$\left(1, -\frac{19}{3}\right), \left(5, -\frac{11}{3}\right), \left(-2, -\frac{25}{3}\right)$

c) The x-values in part a) are multiples of the denominator of $\frac{2}{3}$. When you multiply $\frac{2}{3}$ by a multiple of 3, the fraction is eliminated.

69) negative

71) negative

73) positive

75) zero

77) a) x represents the year
y represents the number of visitors in millions

b) In 2004 there were 37.4 million visitors to Las Vegas.

c) 38.9 million

d) 2005

e) 2 million

f) (2007, 39.2)

79) a) $(1985, 52.9), (1990, 50.6), (1995, 42.4)$
$(2000, 41.4), (2005, 40.0)$

b)

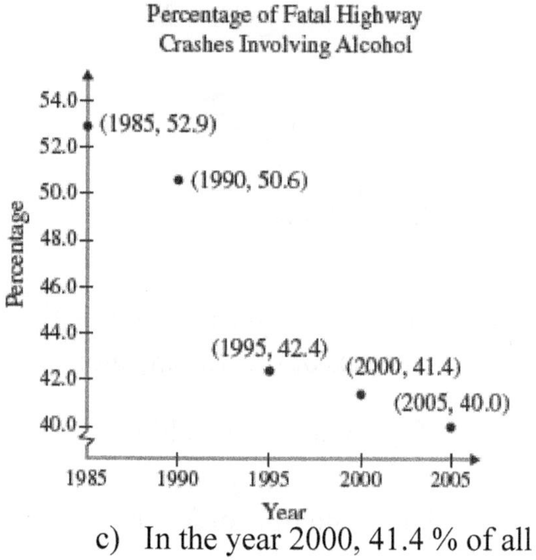

c) In the year 2000, 41.4 % of all fatal accidents involved alcohol.

81) a) $y = 0.095x$

$y = 0.095(100.00)$

$y = 9.50$

$y = 0.095(140.00)$

$y = 13.30$

$y = 0.095(210.72)$

$y = 20.0184$

$y = 0.095(250.00)$

$y = 23.75$

x	y
100.00	9.50
140.00	13.30
210.72	20.0184
250.00	23.75

$(100, 9.50), (140.00, 13.30),$
$(210.72, 20.0184), (250.00, 23.75)$

b)

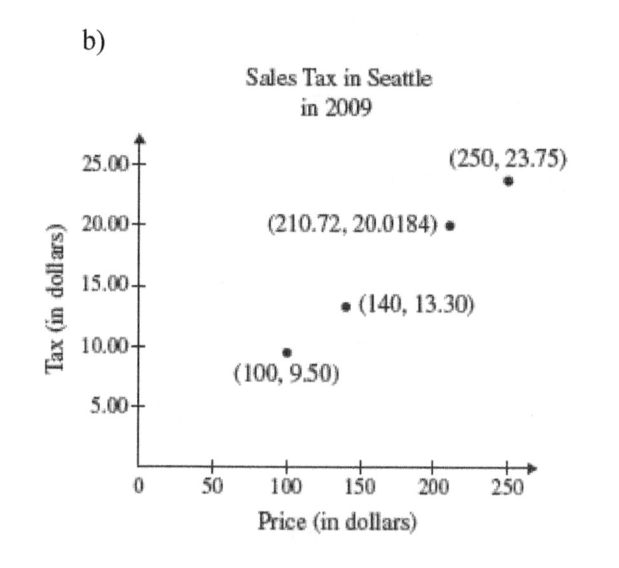

c) If the bill totals \$140.00, then sales tax will be \$13.30.

d) \$20.02

e) Yes, they lie on a straight line.

f) $y = 0.095x$

$$19.00 = 0.095x$$
$$1000(19.00) = 1000(0.95x)$$
$$19000 = 95x$$
$$\frac{19000}{95} = x$$
$$\$200.00 = x$$

The cost of the item purchased is \$20.00.

It will take Kyle about 13 hours to drive to Oklahoma City.

Section 4.2 Exercises

1) line

3) $y = -2x + 4$

$$y = -2(0) + 4 \qquad y = -2(2) + 4$$
$$y = 0 + 4 \qquad y = -4 + 4$$
$$y = 4 \qquad y = 0$$

$$y = -2(-1) + 4 \qquad y = -2(3) + 4$$
$$y = 2 + 4 \qquad y = -6 + 4$$
$$y = 6 \qquad y = -2$$

x	y
0	4
–1	6
2	0
3	–2

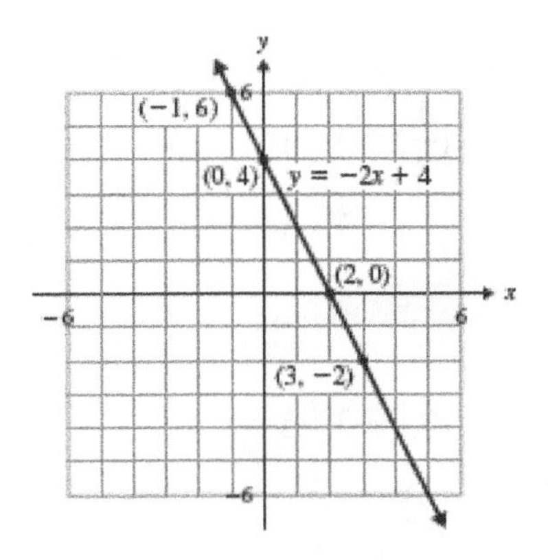

5) $y = \frac{3}{2}x + 7$

$$y = \frac{3}{2}(0) + 7 \qquad y = \frac{3}{2}(-2) + 7$$
$$y = 0 + 7 \qquad y = -3 + 7$$
$$y = 7 \qquad y = 4$$

$$y = \frac{3}{2}(2) + 7 \qquad y = \frac{3}{2}(-4) + 7$$
$$y = 3 + 7 \qquad y = -6 + 7$$
$$y = 10 \qquad y = 1$$

x	y
0	7
2	10
–2	4
–4	1

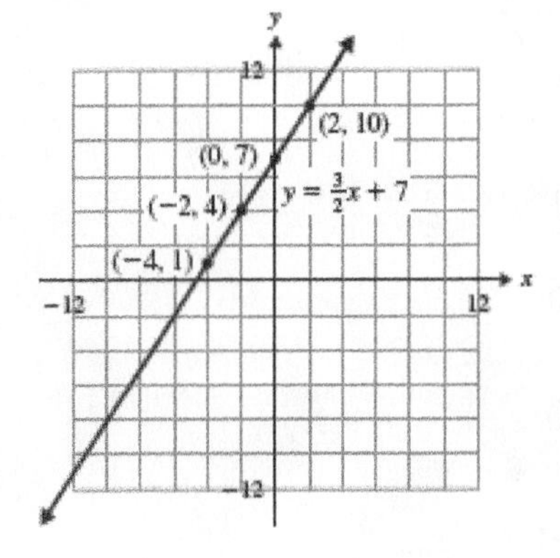

7) $2x = 3 - y$

$$2x = 3 - 0 \qquad 2\left(\frac{1}{2}\right) = 3 - y$$
$$x = \frac{3}{2} \qquad 1 = 3 - y$$
$$2 = y$$

$$2(0) = 3 - y \qquad 2x = 3 - 5$$
$$0 = 3 - y \qquad 2x = -2$$
$$3 = y \qquad x = -1$$

x	y
$\frac{3}{2}$	0
0	3
$\frac{1}{2}$	2
-1	5

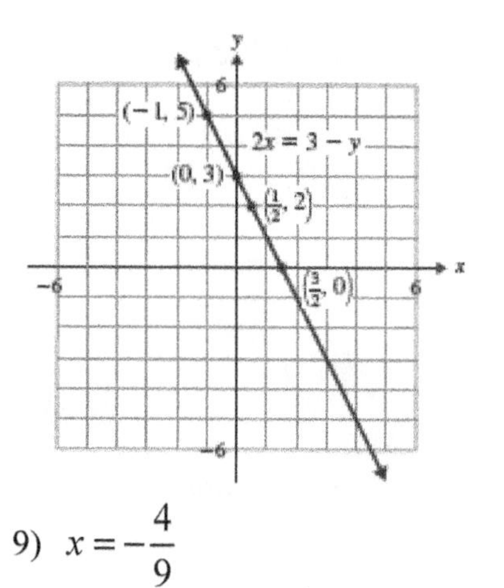

9) $x = -\frac{4}{9}$

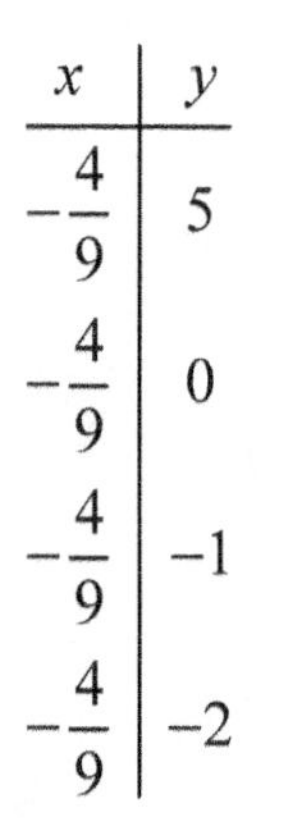

x	y
$-\frac{4}{9}$	5
$-\frac{4}{9}$	0
$-\frac{4}{9}$	-1
$-\frac{4}{9}$	-2

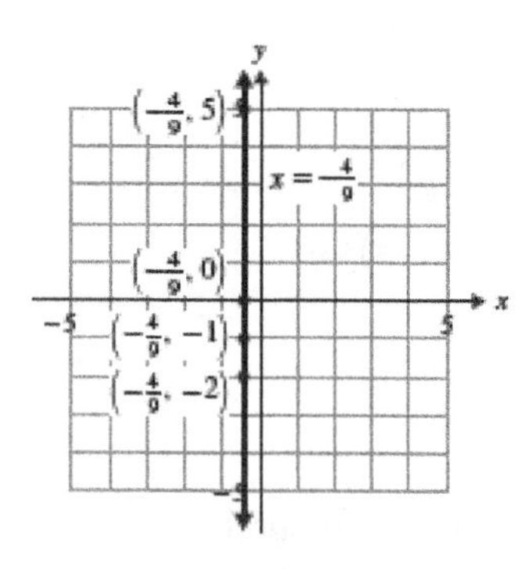

11) It is the point where the graph intersects the y-axis.

Let $x = 0$, and solve for y.

13) $y = x - 1$

x-int: Let $y = 0$, and solve for x.

$(0) = x - 1$

$0 = x - 1$

$1 = x \qquad (1, 0)$

y-int: Let $x = 0$, and solve for y.

$y = (0) - 1$

$y = -1 \qquad (0, -1)$

Let $y = 1$.

$(1) = x - 1$

$2 = x \qquad (2, 1)$

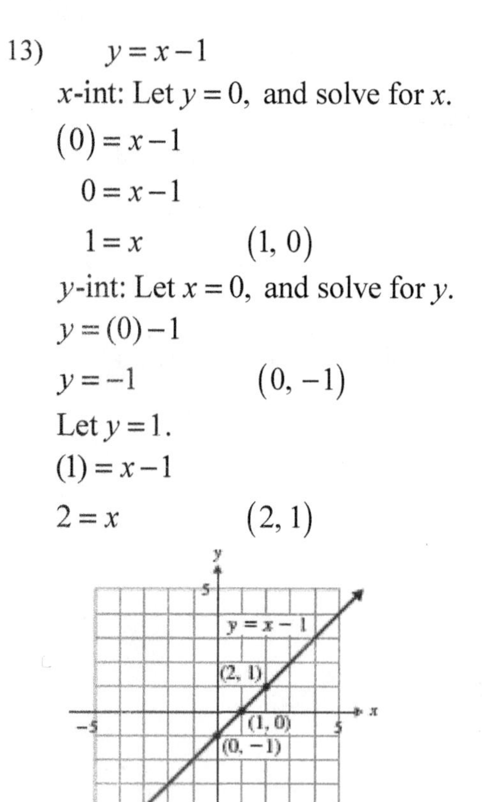

15) $3x-4y=12$

x-int: Let $y=0,$ and solve for x.

$$\begin{aligned} 3x-4(0)&=12 \\ 3x-0&=12 \\ 3x&=12 \\ x&=4 \qquad (4,\ 0) \end{aligned}$$

y-int: Let $x=0,$ and solve for y.

$$\begin{aligned} 3(0)-4y&=12 \\ 0-4y&=12 \\ -4y&=12 \\ y&=-3 \qquad (0,\ -3) \end{aligned}$$

Let $x=2$.

$$\begin{aligned} 3(2)-4y&=12 \\ 6-4y&=12 \\ -4y&=6 \\ y&=-\frac{3}{2} \qquad \left(2,\ -\frac{3}{2}\right) \end{aligned}$$

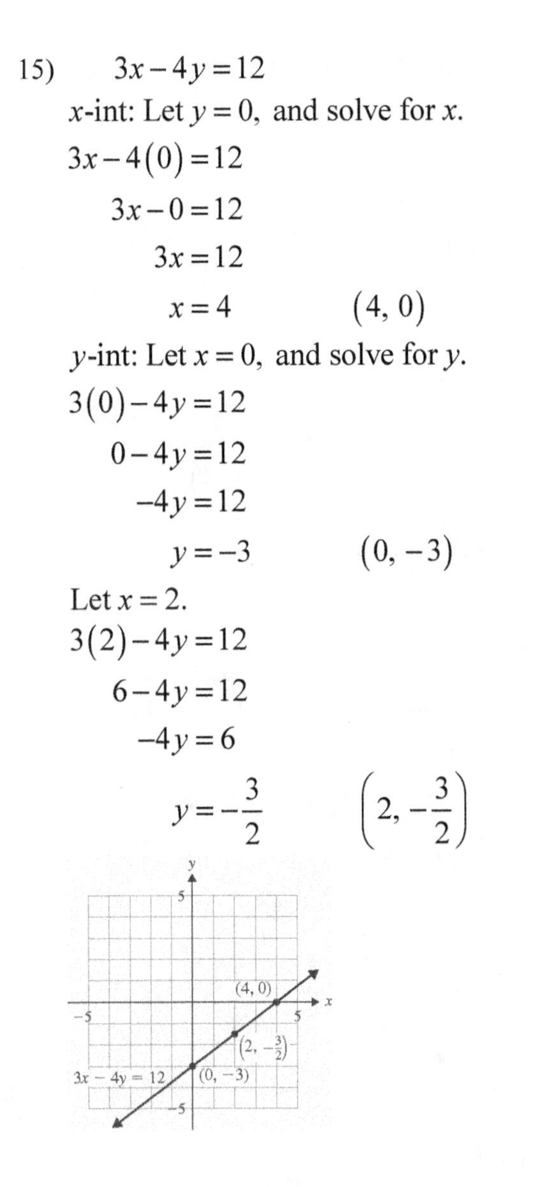

17) $x=-\frac{4}{3}y-2$

x-int: Let $y=0,$ and solve for x.

$$\begin{aligned} x&=-\frac{4}{3}(0)-2 \\ x&=0-2 \\ x&=-2 \qquad (-2,\ 0) \end{aligned}$$

y-int: Let $x=0,$ and solve for y.

$$\begin{aligned} (0)&=-\frac{4}{3}y-2 \\ 2&=-\frac{4}{3}y \\ -\frac{3}{2}&=y \qquad \left(0,\ -\frac{3}{2}\right) \end{aligned}$$

Let $y=-3$.

$$\begin{aligned} x&=-\frac{4}{3}(-3)-2 \\ x&=4-2 \\ x&=2 \qquad (2,\ -3) \end{aligned}$$

19) $2x-y=8$

x-int: Let $y=0,$ and solve for x.

$$\begin{aligned} 2x-(0)&=8 \\ 2x&=8 \\ x&=4 \qquad (4,\ 0) \end{aligned}$$

y-int: Let $x=0,$ and solve for y.

$$\begin{aligned} 2(0)-y&=8 \\ 0-y&=8 \\ -y&=8 \\ y&=-8 \qquad (0,\ -8) \end{aligned}$$

Let $x=2$.

$$\begin{aligned} 2(2)-y&=8 \\ 4-y&=8 \\ -y&=4 \\ y&=-4 \qquad (2,\ -4) \end{aligned}$$

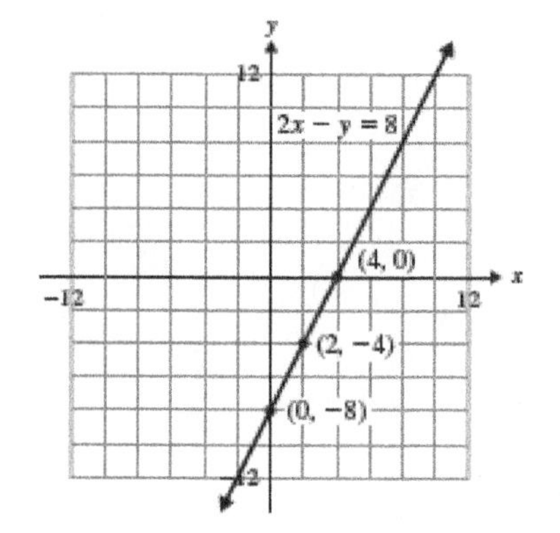

21) $y = -x$

x-int: Let $y = 0$, and solve for x.

$(0) = -x$

$0 = x \qquad (0, 0)$

y-int: Let $x = 0$, and solve for y.

$y = -(0)$

$y = 0 \qquad (0, 0)$

Let $x = 1$.

$y = -(1)$

$y = -1 \qquad (1, -1)$

Let $x = -1$.

$y = -(-1)$

$y = 1 \qquad (-1, 1)$

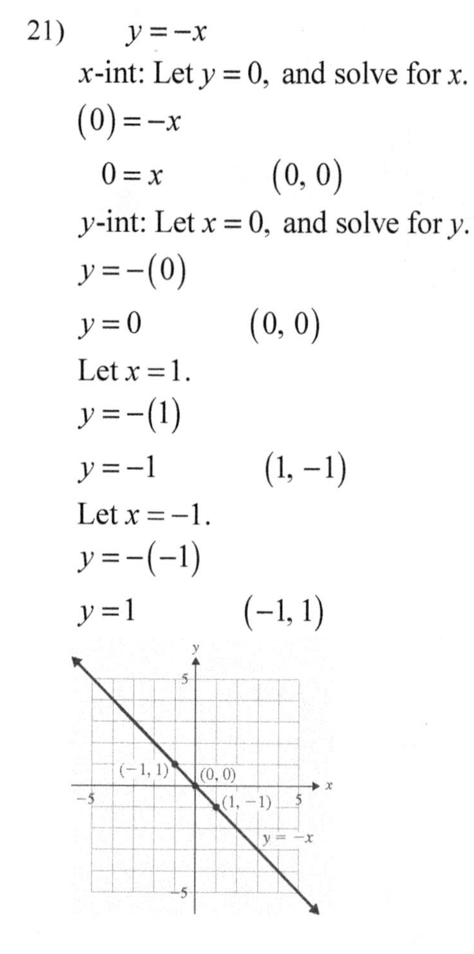

23) $4x - 3y = 0$

x-int: Let $y = 0$, and solve for x.

$4x - 3(0) = 0$

$4x - 0 = 0$

$4x = 0$

$x = 0 \qquad (0, 0)$

y-int: Let $x = 0$, and solve for y.

$4(0) - 3y = 0$

$0 - 3y = 0$

$3y = 0$

$y = 0 \qquad (0, 0)$

Let $x = 3$.

$4(3) - 3y = 0$

$12 - 3y = 0$

$12 = 3y$

$4 = y \qquad (3, 4)$

Let $x = -3$.

$4(-3) - 3y = 0$

$-12 - 3y = 0$

$-12 = 3y$

$-4 = y \qquad (-3, -4)$

25) $x = 5$

$(5, 0), (5, 2), (5, -1)$

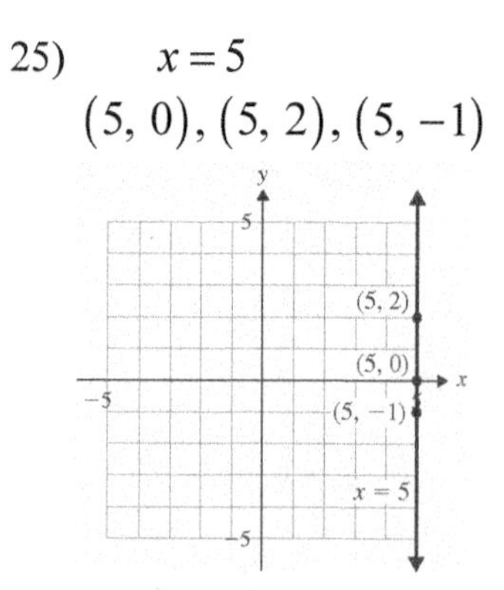

27) $y=0$

$(0,\ 0),\ (1,\ 0),\ (-2,\ 0)$

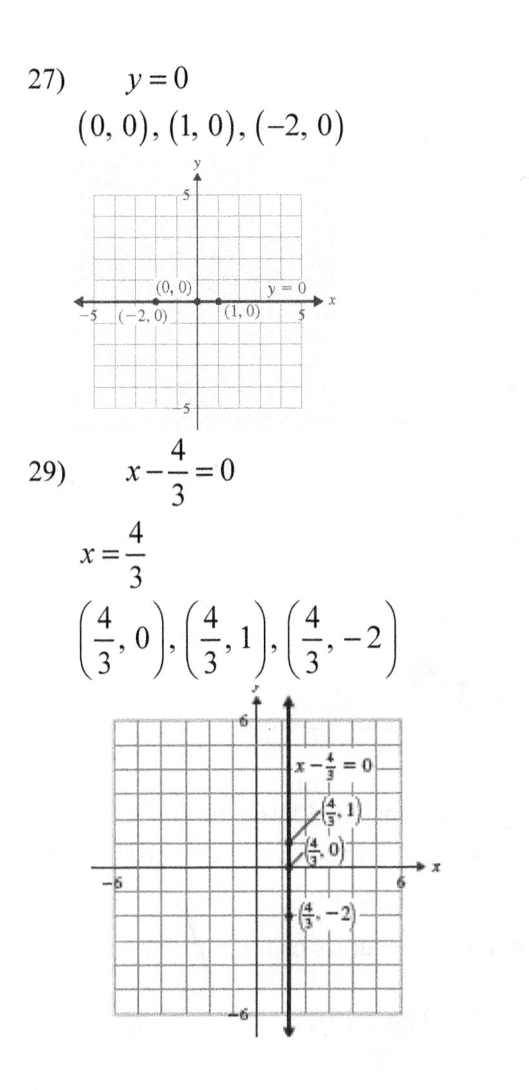

29) $x-\dfrac{4}{3}=0$

$x=\dfrac{4}{3}$

$\left(\dfrac{4}{3},\ 0\right),\ \left(\dfrac{4}{3},\ 1\right),\ \left(\dfrac{4}{3},\ -2\right)$

31) $4x-y=9$

x-int: Let $y=0,$ and solve for $x.$

$4x-(0)=9$

$4x=9$

$x=\dfrac{9}{4}$ $\quad\left(\dfrac{9}{4},\ 0\right)$

y-int: Let $x=0,$ and solve for $y.$

$4(0)-y=9$

$0-y=9$

$y=-9$ $\quad(0,\ -9)$

Let $y=3.$

$4x-(3)=9$

$4x=12$

$x=3$ $\quad(3,\ 3)$

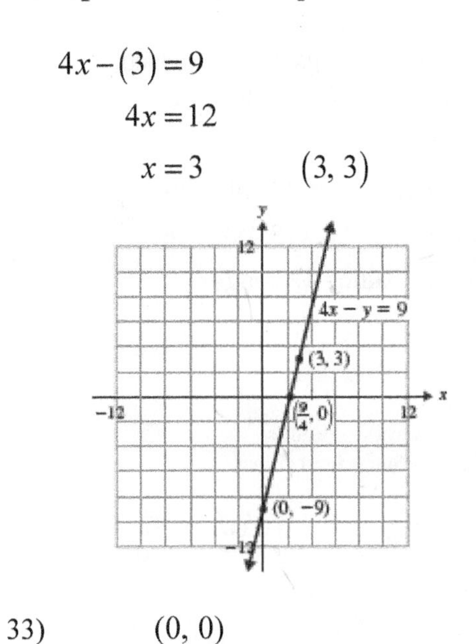

33) $(0,\ 0)$

35) $y=1.29x$

a)

$x=0$ $\quad y=1.29(0)$

$y=0$

$x=4$ $\quad y=1.29(4)$

$y=5.16$

$x=7$ $\quad y=1.29(7)$

$y=9.03$

$x=12$ $\quad y=1.29(12)$

$y=15.48$

x	y
0	0
4	5.16
7	9.03
12	15.48

$(0,\ 0),\ (4,\ 5.16),$
$(7,\ 9.03),\ (12,\ 15.48)$

b) (0, 0): If no songs are purchased, the cost is \$0. (4, 5.16): The cost of downloading 4 songs is \$5.16. (7, 9.03): The cost of downloading 7 songs is \$9.03. (12, 15.48): The cost of downloading 12 songs is \$15.48.

c)

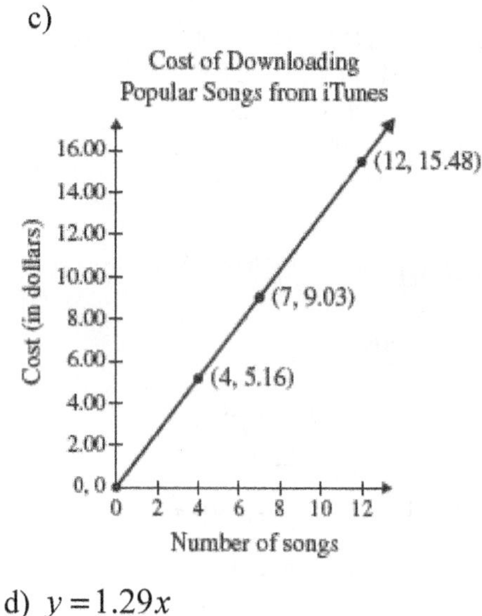

d) $y = 1.29x$

$11.61 = 1.29x$

$\dfrac{11.61}{1.29} = x$

$9 = x$

9 songs can be downloaded for \$11.61.

37) a) $2004: 26,275;\ 2007: 31,801$

b) $y = 1662x + 24,916$

$x = 1$ for 2004

$y = 1662(1) + 24,916$

$y = 1662 + 24,916$

$y = 26,578$

$x = 4$ for 2007

$y = 1662(4) + 24,916$

$y = 6648 + 24,916$

$y = 31,564$

Yes, they are close.

c)

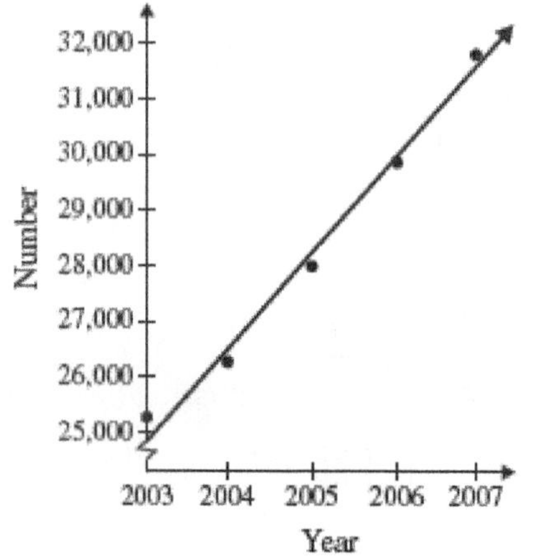

d) The y-intercept is 24,916. In 2003, approximately 24,916 science and engineering doctorates were awarded. It looks like it is within about 300 units of the plotted point.

e) $y = 1662x + 24,916$

$x = 9$ for 2012

$y = 1662(9) + 24,916$

$y = 14,958 + 24,916$

$y = 39,874$

Section 4.3 Exercises

1) The slope of a line is the ratio of vertical change to horizontal change.

It is $\dfrac{\text{change in } y}{\text{change in } x}$ or $\dfrac{\text{rise}}{\text{run}}$ or $\dfrac{y_2 - y_1}{x_2 - x_1}$ where (x_1, y_1) and (x_2, y_2) are points on the line.

3) It slants upward from left to right.

5) undefined

7) a) Vertical change: 3 units
Horizontal change: 4 units

$\text{Slope} = \frac{3}{4}$

b) $(x_1, y_1) = (1, -1)$

$(x_2, y_2) = (5, 2)$

$m = \frac{y_2 - y_1}{x_2 - x_1} = \frac{2-(-1)}{5-1} = \frac{3}{4}$

9) a) Vertical change: -2 units
Horizontal change: 3 units

$\text{Slope} = \frac{-2}{3} = -\frac{2}{3}$

b) $(x_1, y_1) = (1, 5)$

$(x_2, y_2) = (4, 3)$

$m = \frac{y_2 - y_1}{x_2 - x_1} = \frac{3-5}{4-1} = \frac{-2}{3}$

$= -\frac{2}{3}$

11) a) Vertical change: -3 units
Horizontal change: 1 unit

$\text{Slope} = \frac{-3}{1} = -3$

b) $(x_1, y_1) = (2, 3)$

$(x_2, y_2) = (3, 0)$

$m = \frac{y_2 - y_1}{x_2 - x_1} = \frac{0-3}{3-2} = \frac{-3}{1} = -3$

13) a) Vertical change: 6 units
Horizontal change: 0 units

$\text{Slope} = \frac{6}{0}$ is undefined.

b) $(x_1, y_1) = (-3, -4)$

$(x_2, y_2) = (-3, 2)$

$m = \frac{y_2 - y_1}{x_2 - x_1} = \frac{2-(-4)}{-3-(-3)} = \frac{6}{0}$

Slope is undefined.

15)

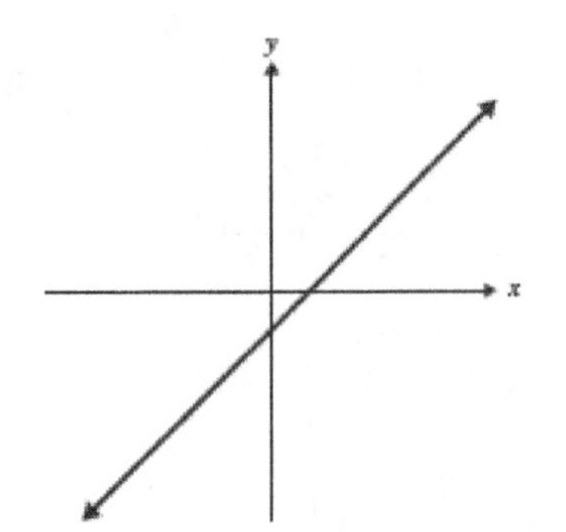

17) $(x_1, y_1) = (2,1)$

$(x_2, y_2) = (0, -3)$

$m = \frac{y_2 - y_1}{x_2 - x_1} = \frac{-3-1}{0-2} = \frac{-4}{-2} = 2$

19) $(x_1, y_1) = (2, -6)$

$(x_2, y_2) = (-1, 6)$

$m = \frac{y_2 - y_1}{x_2 - x_1} = \frac{6-(-6)}{-1-2} = \frac{12}{-3} = -4$

21) $(x_1, y_1) = (-4, 3)$

$(x_2, y_2) = (1, -8)$

$$m = \frac{y_2 - y_1}{x_2 - x_1} = \frac{-8-3}{1-(-4)} = \frac{-11}{5} = -\frac{11}{5}$$

23) $(x_1, y_1) = (-2, -2)$

$(x_2, y_2) = (-2, 7)$

$$m = \frac{y_2 - y_1}{x_2 - x_1} = \frac{7-(-2)}{(-2)-(-2)} = \frac{9}{0}$$

Slope is undefined.

25) $(x_1, y_1) = (3, 5)$

$(x_2, y_2) = (-1, 5)$

$$m = \frac{y_2 - y_1}{x_2 - x_1} = \frac{5-5}{-1-3} = \frac{0}{-4} = 0$$

27) $(x_1, y_1) = \left(\frac{2}{3}, \frac{5}{2}\right)$

$(x_2, y_2) = \left(-\frac{1}{2}, 2\right)$

$$m = \frac{y_2 - y_1}{x_2 - x_1} = \frac{2-\left(\frac{5}{2}\right)}{-\frac{1}{2}-\frac{2}{3}}$$

$$= \frac{\frac{4}{2}-\frac{5}{2}}{-\frac{3}{6}-\frac{4}{6}} = \frac{-\frac{1}{2}}{-\frac{7}{6}}$$

$$= -\frac{1}{2} \div \left(-\frac{7}{6}\right) = -\frac{1}{2} \cdot \left(-\frac{6}{7}\right) = \frac{3}{7}$$

29) $(x_1, y_1) = (3.5, -1.4)$

$(x_2, y_2) = (7.5, 1.6)$

$$m = \frac{y_2 - y_1}{x_2 - x_1} = \frac{1.6-(-1.4)}{7.5-3.5}$$

$$= \frac{3.0}{4.0} = 0.75$$

31) $m = \frac{\text{rise}}{\text{run}} = \frac{-60}{395} = \frac{-12}{79}$

33) $m = \frac{\text{rise}}{\text{run}} = \frac{6}{9} = \frac{2}{3} = 0.\overline{6}$

No. The slope of the slide is $0.\overline{6}$. This is more than the recommended slope of 0.577.

35) $m = \frac{\text{rise}}{\text{run}} = \frac{0.75}{20} = 0.0375$

Yes. The slope of the driveway is 0.0375.

This is less than the maximum allowed slope of 0.05.

37) $m = \frac{\text{rise}}{\text{run}} = \frac{12}{26} = \frac{6}{13}$. Slope is $\frac{6}{13}$.

39) a) 2003: 2.89 million; 2005: 2.70 million

b) The line slants downward from left to right; therefore it has a negative slope.

c) The number of injuries is decreasing over time.

d)

$$m = \frac{\text{rise}}{\text{run}} = \frac{2.49-2.89}{2007-2003} = \frac{-0.4}{4} = -0.1$$

The number of injuries is decreasing by 0.1 million or 100,000 per year.

41)

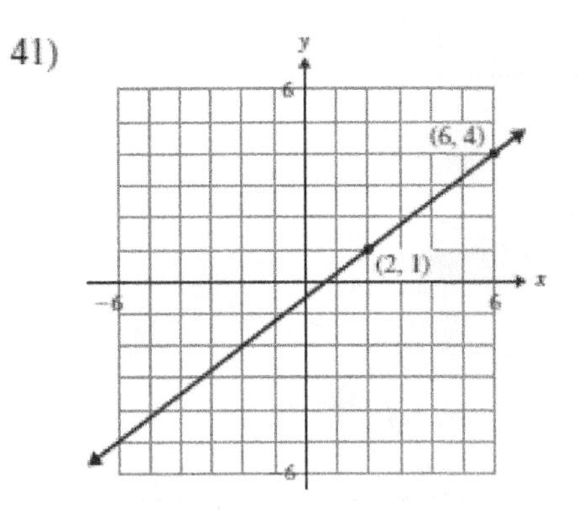

43)

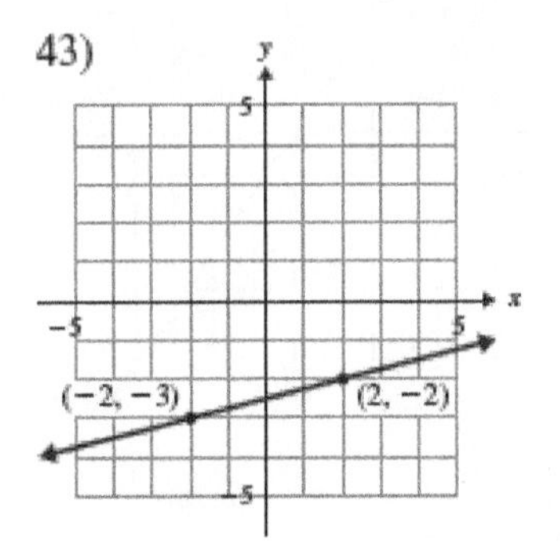

45)

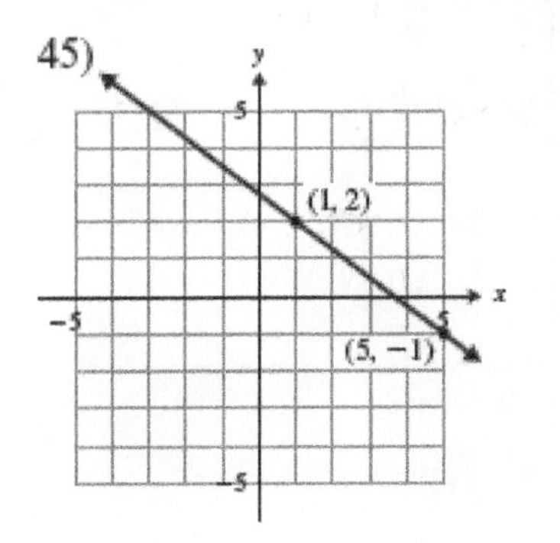

47)

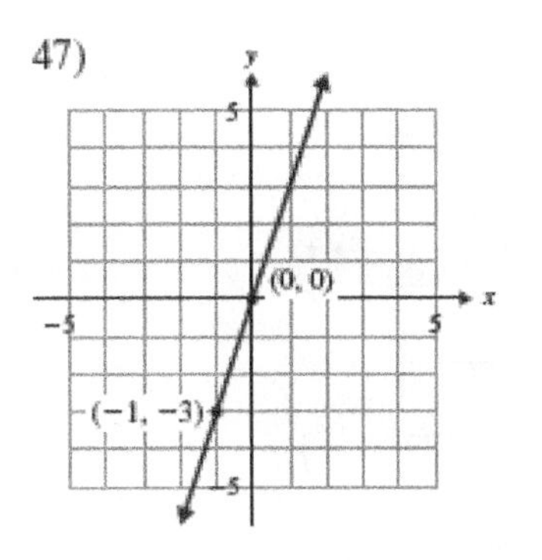

49)

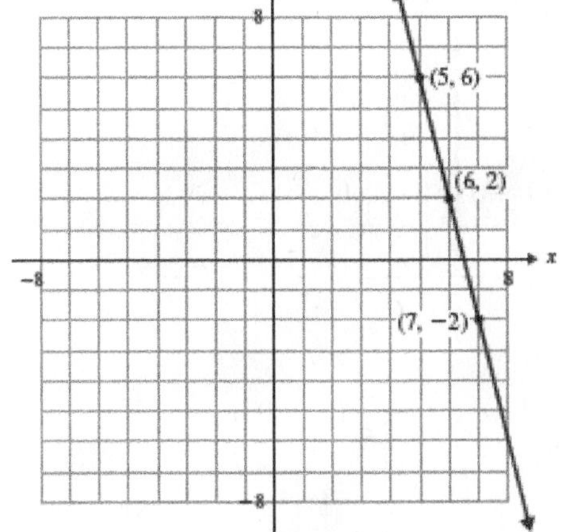

51)

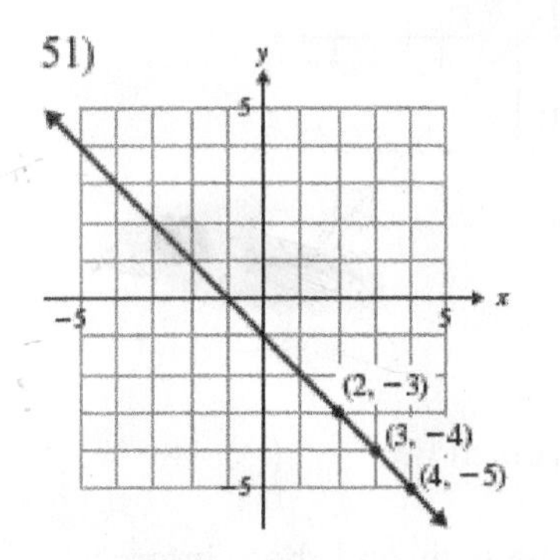

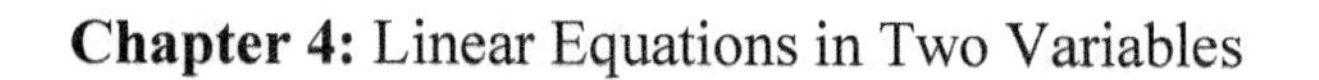

53)

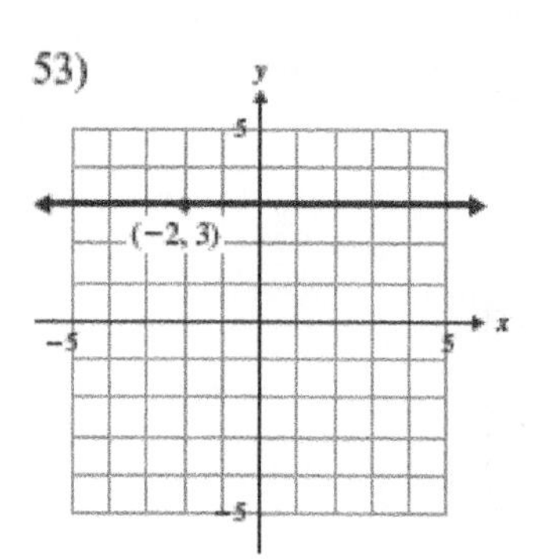

55)

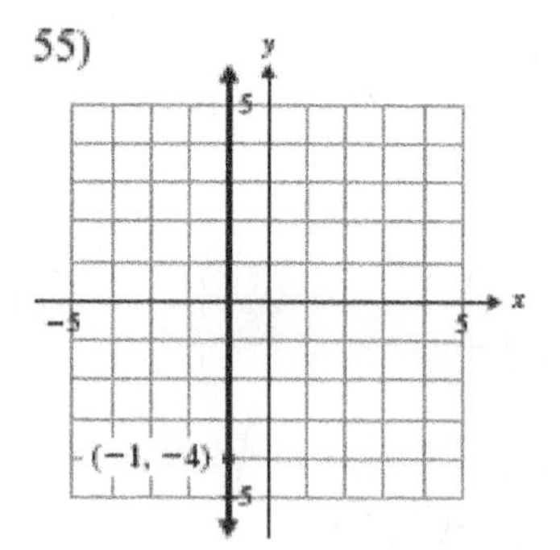

57)

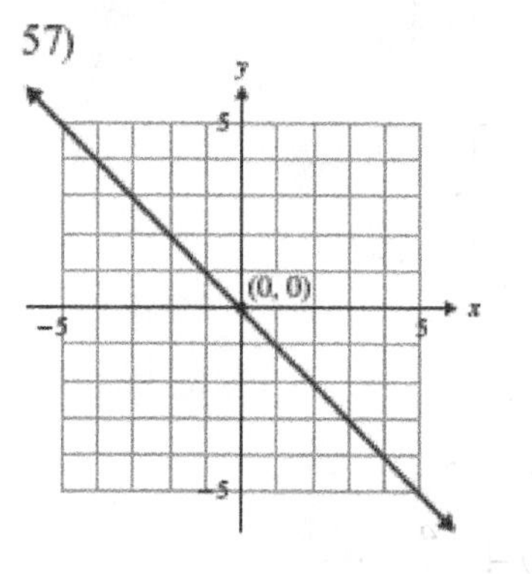

Section 4.4 Exercises

1) The slope is m, and the y-intercept is $(0, b)$.

3) $m = \frac{2}{5}$, $y\text{-int} = (0, -6)$

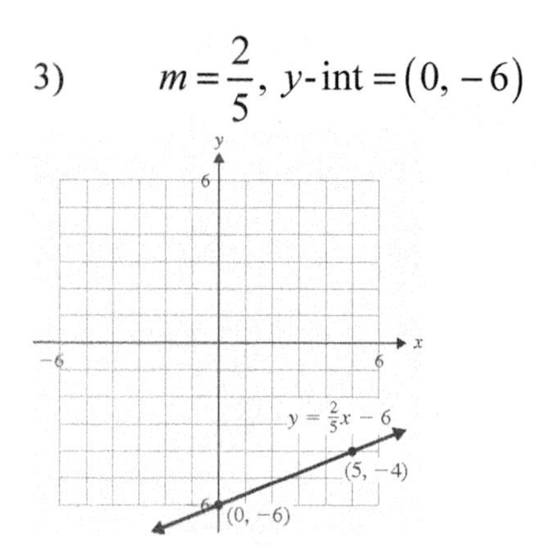

5) $m = -\frac{3}{2}$, y-int: $(0, 3)$

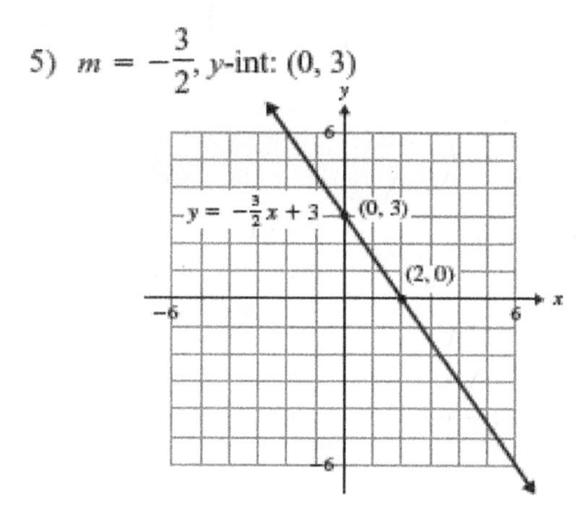

7) $m = \frac{3}{4}$, y-int: $(0, 2)$

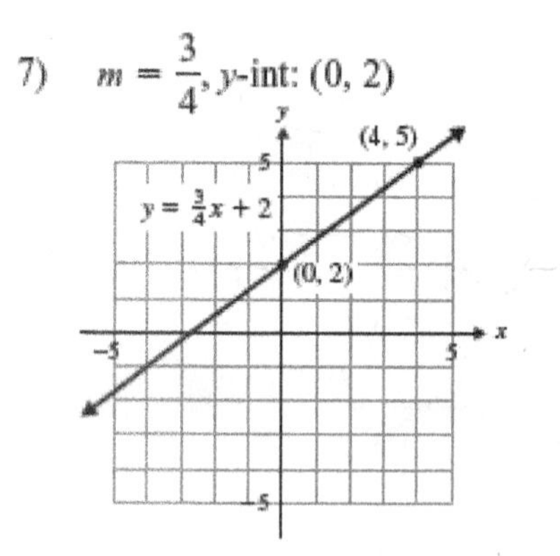

9) $m = -2$, y-int: $(0, -3)$

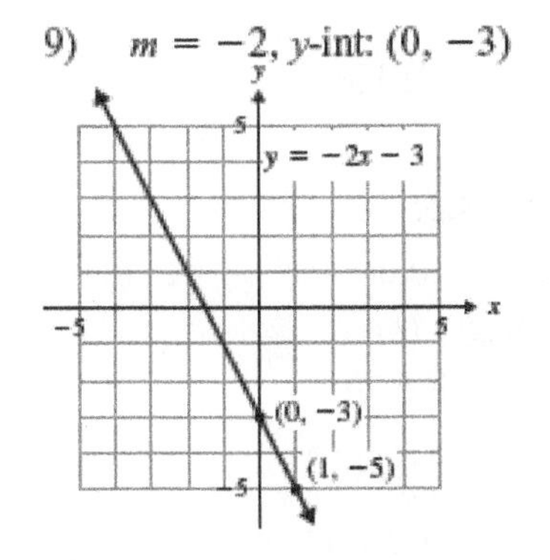

11) $m = 5$, y-int: $(0, 0)$

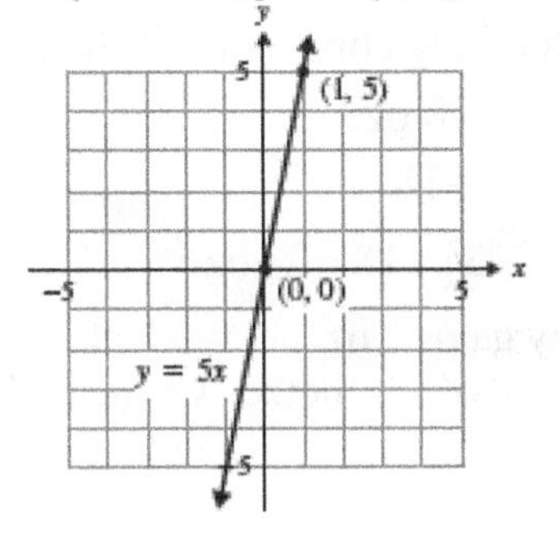

13) $m = -\frac{3}{2}$, y-int: $\left(0, -\frac{7}{2}\right)$

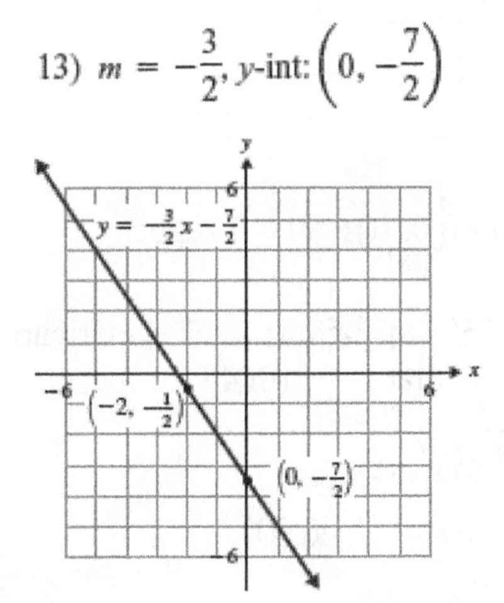

15) $m = 0$, y-int: $(0, 6)$

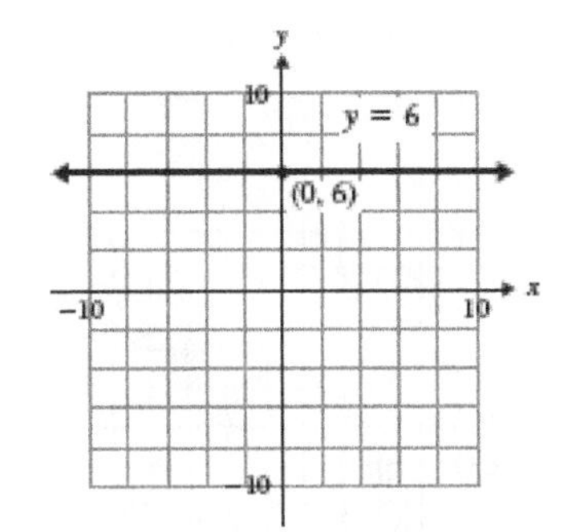

17)
$$x + 3y = -6$$
$$3y = -x - 6$$
$$y = -\frac{1}{3}x - 2$$
$$m = -\frac{1}{3},\ y\text{-int}: (0, -2)$$

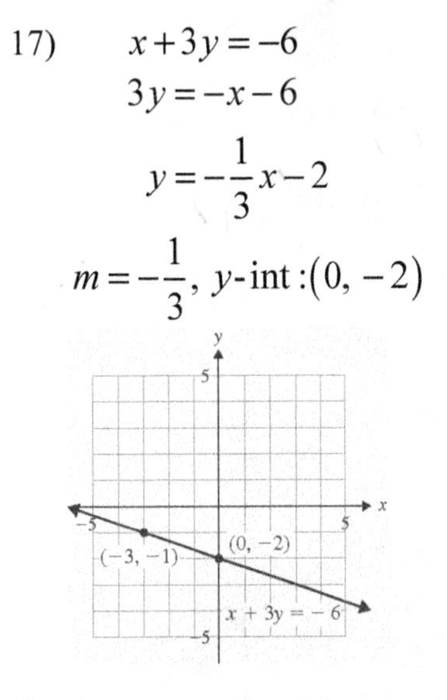

19)
$$4x + 3y = 21$$
$$3y = -4x + 21$$
$$y = -\frac{4}{3}x + 7$$
$$m = -\frac{4}{3},\ y\text{-int}: (0, 7)$$

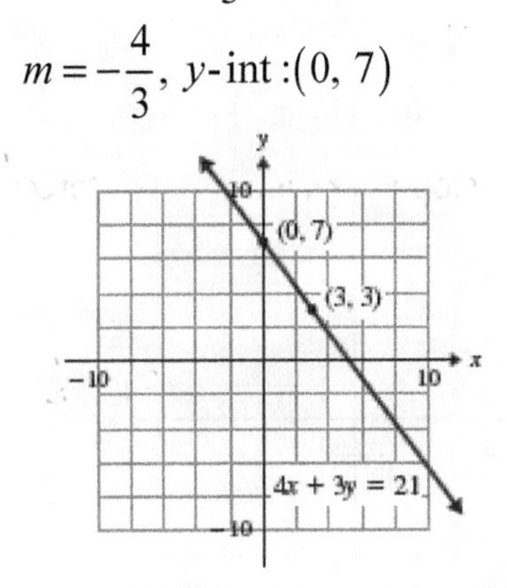

21) $2 = x + 3$
$-1 = x$
The slope is undefined,
and no y-intercept

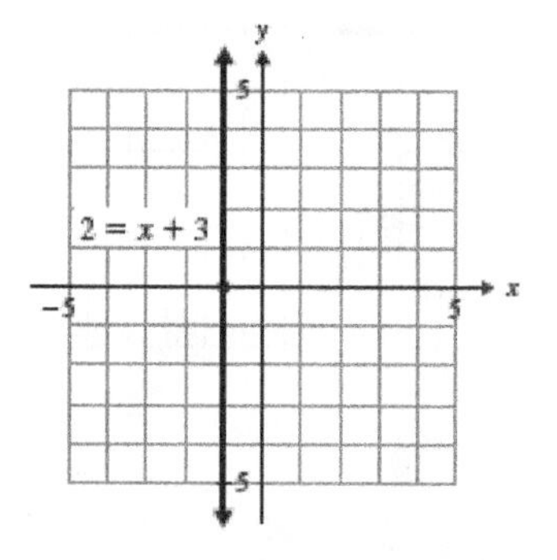

23) $2x = 18 - 3y$
$3y = -2x + 18$
$$y = -\frac{2}{3}x + 6$$

$m = -\frac{2}{3}$, y-int :$(0, 6)$

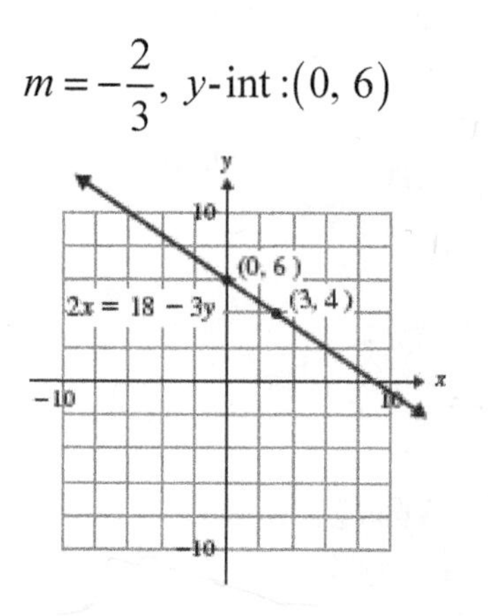

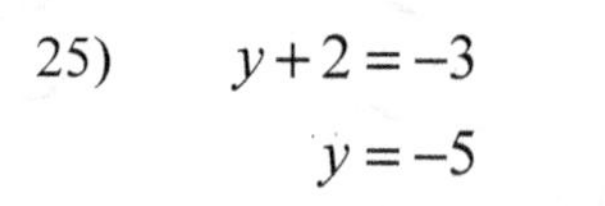

25) $y + 2 = -3$
$y = -5$
$m = 0$, y-int :$(0, -5)$

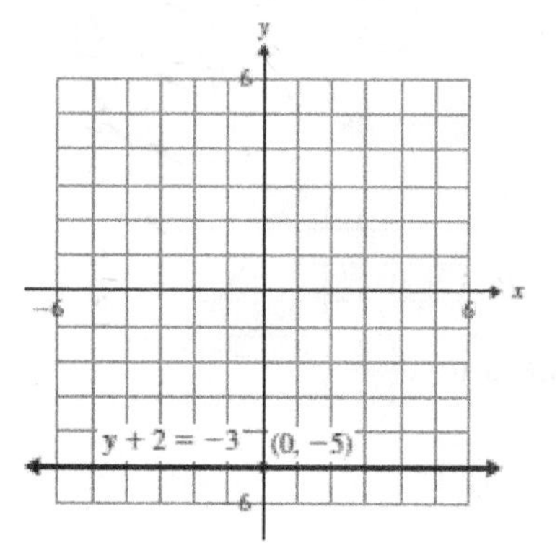

27)
a) $(0,0)$; If Kolya works 0 hours, he earns \$0.
b) $m = 8.5$;
Kolya earns \$8.50 per hour.
c) \$102.00
$P = 8.50h$
$P = 8.50(12)$
$P = 102.00$

29) a) $(0, 18)$;
when the joey comes out of the pouch he weighs 18 oz.
b) $y = 2x + 18$
$y = 2(3) + 18 = 24\text{oz}$
c) A joey gains 2 oz. per week after coming out of it's mother's pouch.

d) $y = 2x + 18$
$32 = 2x + 18$
$14 = 2x$
$7 = x$

31) a) $(0, 0)$; $\$0 = 0$ rupees

b) $m = 48.2$: each American dollar is worth 48.2 rupees.

c) $r = 48.2d$
$r = 48.2(80.00)$
$r = 3856$
she gets 3856 rupees
d) $r = 46.2d$

$r = 46.2d$

$2410 = 46.2d$

$\frac{2410}{46.2} = d$

$d = 50$

She gets $50 for 2410 rupees.

33) $y = mx + b \quad m = -4;\ b = 7$

$y = -4x + 7$

35) $y = mx + b \quad m = \frac{9}{5};\ b = -3$

$y = \frac{9}{5}x - 3$

37) $y = mx + b \quad m = -\frac{5}{2};\ b = -1$

$y = -\frac{5}{2}x - 1$

39) $(x_1, y_1) = (0, 2);\ m = 1;\ b = 2$

$y = mx + b$

$y = x + 2$

41) $y = mx + b \quad m = 0;\ b = 0$

$y = 0$

43) Their slopes are negative reciprocals, or one line is vertical and one line is horizontal.

45) $y = -x - 5 \qquad y = x + 8$

$m = -1 \qquad m = 1$

perpendicular

47) $y = \frac{2}{9}x + 4 \qquad 4x - 18y = 9$

$-18y = -4x + 9$

$y = \frac{2}{9}x - \frac{1}{2}$

$m = \frac{2}{9} \qquad m = \frac{2}{9}$

parallel

49) $3x - y = 4 \qquad 2x - 5y = -9$

$-y = 3x + 4 \qquad -5y = -2x - 9$

$y = -3x - 4 \qquad y = \frac{2}{5}x + \frac{9}{5}$

$m = -3 \qquad m = \frac{2}{5}$

neither

51) $-x + y = -21 \qquad y = 2x + 5$

$y = x - 21$

$m = 1 \qquad m = 2$

neither

53) $x + 7y = 4 \qquad y - 7x = 4$

$7y = -x + 4 \qquad y = 7x + 4$

$y = -\frac{1}{7}x + \frac{4}{7}$

$m = -\frac{1}{7} \qquad m = 7$

perpendicular

55) $y = -\frac{1}{2}x \qquad x + 2y = 4$

$2y = -x + 4$

$y = -\frac{1}{2}x + 2$

$m = -\frac{1}{2} \qquad m = -\frac{1}{2}$

parallel

57) $x = -1$ $y = 6$

$m = \text{undefined}$ $m = 0$

perpendicular

59) $x = -4.3$ $x = 0$

$m = \text{undefined}$

$m = \text{undefined}$

parallel

61) $L_1: m = \dfrac{8-(-7)}{2-(-1)} = \dfrac{15}{3} = 5$

$L_2: m = \dfrac{4-2}{0-10} = \dfrac{2}{-10} = -\dfrac{1}{5}$

perpendicular

63) $L_1: m = \dfrac{8-10}{3-1} = \dfrac{-2}{2} = -1$

$L_2: m = \dfrac{-17-4}{-5-2} = \dfrac{-21}{-7} = 3$

neither

65) $L_1: m = \dfrac{-1-6}{4-(-3)} = \dfrac{-7}{7} = -1$

$L_2: m = \dfrac{-1-(-5)}{-10-(-6)} = \dfrac{4}{-4} = -1$

parallel

67) $L_1: m = \dfrac{1-2}{-6-(-6)} = \dfrac{-1}{0} = \text{undefined}$

$L_2: m = \dfrac{-5-0}{4-4} = \dfrac{-5}{0} = \text{undefined}$

parallel

69) $L_1: m = \dfrac{5-2}{7-7} = \dfrac{3}{0} = \text{undefined}$

$L_2: m = \dfrac{0-0}{1-(-2)} = \dfrac{0}{3} = 0$

perpendicular

Section 4.5 Exercises

1) $y = -2x - 4$

$2x + y = -4$

3) $x = y + 1$

$x - y = 1$

5) $y = \dfrac{4}{5}x - 1$

$5 \cdot y = 5\left(\dfrac{4}{5}x - 1\right)$

$5y = 4x - 5$

$-4x + 5y = -5$

$4x - 5y = 5$

7) $y = -\dfrac{1}{3}x - \dfrac{5}{4}$

$12 \cdot y = 12\left(-\dfrac{1}{3}x - \dfrac{5}{4}\right)$

$12y = -4x - 15$

$4x + 12y = -15$

9) Use $y = mx + b$ and substitute the slope and y-intercept values into the equation.

11) $y = mx + b$ $m = -7;\ b = 2$

$y = -7x + 2$

13) $y = mx + b$ $m = -4;\ b = 6$

$y = -4x + 6$

$4x + y = 6$

15) $y = mx + b \quad m = \frac{2}{7};\ b = -3$

$$y = \frac{2}{7}x - 3$$
$$7 \cdot y = 7\left(\frac{2}{7}x - 3\right)$$
$$7y = 2x - 21$$
$$-2x + 7y = -21$$
$$2x - 7y = 21$$

17) $y = mx + b \quad m = -1;\ b = 0$

$$y = -x$$

19) a) $y - y_1 = m(x - x_1)$

b)

Use the point-slope formula and substitute the point on the line and the slope into the equation.

21) $(x_1, y_1) = (5, 7);\ m = 1$

$$y - y_1 = m(x - x_1)$$
$$y - 7 = 1(x - 5)$$
$$y - 7 = x - 5$$
$$y = x + 2$$

23) $(x_1, y_1) = (4, -1);\ m = -5$

$$y - y_1 = m(x - x_1)$$
$$y - (-1) = -5(x - 4)$$
$$y + 1 = -5(x - 4)$$
$$y + 1 = -5x + 20$$
$$y = -5x + 19$$

25) $(x_1, y_1) = (-2, -1);\ m = 4$

$$y - y_1 = m(x - x_1)$$
$$y - (-1) = 4(x - (-2))$$
$$y + 1 = 4x + 8$$
$$-4x + y = 7$$
$$4x - y = -7$$

27) $(x_1, y_1) = (-5, 8);\ m = \frac{2}{5}$

$$y - y_1 = m(x - x_1)$$
$$y - 8 = \frac{2}{5}(x - (-5))$$
$$y - 8 = \frac{2}{5}x + 2$$
$$5(y - 8) = 5\left(\frac{2}{5}x + 2\right)$$
$$5y - 40 = 2x + 10$$
$$-2x + 5y = 50$$
$$2x - 5y = -50$$

29) $(x_1, y_1) = (5, 1);\ m = -\frac{5}{4}$

$$y - y_1 = m(x - x_1)$$
$$y - 1 = -\frac{5}{4}(x - 5)$$
$$y - 1 = -\frac{5}{4}x + \frac{25}{4}$$
$$y = -\frac{5}{4}x + \frac{25}{4} + 1$$
$$y = -\frac{5}{4}x + \frac{29}{4}$$

31) $(x_1, y_1) = (-3, 0);\ m = \frac{5}{6}$

$$y - y_1 = m(x - x_1)$$
$$y - 0 = \frac{5}{6}(x - (-3))$$
$$y = \frac{5}{6}x + \frac{5}{2}$$
$$6 \cdot y = 6\left(\frac{5}{6}x + \frac{5}{2}\right)$$
$$6y = 5x + 15$$
$$-5x + 6y = 15$$
$$5x - 6y = -15$$

33) Use the points to find the slope of the line, and then use the slope and either one of the points in the point-slope formula.

35) $m = \frac{-5-7}{3-(-1)} = \frac{-12}{4} = -3$

$$(x_1, y_1) = (-1, 7)$$
$$y - y_1 = m(x - x_1)$$
$$y - 7 = -3(x - (-1))$$
$$y - 7 = -3(x + 1)$$
$$y = -3x - 3 + 7$$
$$y = -3x + 4$$

37) $m = \frac{11-5}{7-4} = \frac{6}{3} = 2$

$$(x_1, y_1) = (4, 5)$$
$$y - y_1 = m(x - x_1)$$
$$y - 5 = 2(x - 4)$$
$$y - 5 = 2x - 8$$
$$y = 2x - 3$$

39) $m = \frac{3-4}{1-(-2)} = \frac{-1}{3} = -\frac{1}{3}$

$$(x_1, y_1) = (-2, 4)$$
$$y - y_1 = m(x - x_1)$$
$$y - 4 = -\frac{1}{3}(x - (-2))$$
$$y - 4 = -\frac{1}{3}(x + 2)$$
$$y - 4 = -\frac{1}{3}x - \frac{2}{3}$$
$$y = -\frac{1}{3}x + \frac{10}{3}$$

41) $m = \frac{-2-1}{4-(-5)} = \frac{-3}{9} = -\frac{1}{3}$

$$(x_1, y_1) = (-5, 1)$$
$$y - y_1 = m(x - x_1)$$
$$y - 1 = -\frac{1}{3}(x - (-5))$$
$$y - 1 = -\frac{1}{3}x - \frac{5}{3}$$
$$3(y - 1) = 3\left(-\frac{1}{3}x - \frac{5}{3}\right)$$
$$3y - 3 = -x - 5$$
$$x + 3y = -2$$

43) $m=\frac{-1-(-11)}{3-(-3)}=\frac{10}{6}=\frac{5}{3}$

$(x_1, y_1)=(-3, -11)$

$$y-y_1=m(x-x_1)$$
$$y-(-11)=\frac{5}{3}(x-(-3))$$
$$y+11=\frac{5}{3}x+5$$
$$3(y+11)=3\left(\frac{5}{3}x+5\right)$$
$$3y+33=5x+15$$
$$3y-5x=-18$$
$$5x-3y=18$$

45) $m=\frac{-13.9-8.3}{5.1-(-2.3)}=\frac{-22.2}{7.4}=-3.0$

$(x_1, y_1)=(-2.3, 8.3)$

$$y-y_1=m(x-x_1)$$
$$y-8.3=-3.0(x-(-2.3))$$
$$y-8.3=-3.0x-6.9$$
$$y=-3.0x+1.4$$

47) $m=\frac{-4-(-1)}{(-4)-0}=\frac{-3}{-4}=\frac{3}{4}$

$y=mx+b \quad m=\frac{3}{4};\ b=-1$

$y=\frac{3}{4}x-1$

49) $m=\frac{-7-2}{1-(-2)}=\frac{-9}{3}=-3$

$(x_1, y_1)=(-2, 2)$

$$y-y_1=m(x-x_1)$$
$$y-2=-3(x-(-2))$$
$$y-2=-3x-6$$
$$y=-3x-4$$

51) $y=3$

53) $m=\frac{-1-7}{2-(-4)}=\frac{-8}{6}=-\frac{4}{3}$

$(x_1, y_1)=(-4, 7)$

$$y-y_1=m(x-x_1)$$
$$y-7=-\frac{4}{3}(x-(-4))$$
$$y-7=-\frac{4}{3}(x+4)$$
$$y-7=-\frac{4}{3}x-\frac{16}{3}$$
$$y=-\frac{4}{3}x-\frac{16}{3}+7$$
$$y=-\frac{4}{3}x-\frac{16}{3}+\frac{21}{3}$$
$$y=-\frac{4}{3}x+\frac{5}{3}$$

55) $m=1$

$(x_1, y_1)=(3, 5)$

$$y-y_1=m(x-x_1)$$
$$y-5=1(x-3)$$
$$y-5=x-3$$
$$y=x+2$$

57) $m=7;\ b=6$
$y=mx+b$
$y=7x+6$

59) Vertical Line
$(c,\ d)=(3,\ 5)$
$x=c$
$x=3$

61) Horizontal Line
$(c,\ d)=(2,\ 3)$
$y=d$
$y=3$

62)
$y=d$
$y=-4$

63) $m=-4;\ b=-4$
$y=mx+b$
$y=-4x-4$

65) $m=-3;$
$(x_1,\ y_1)=(10,\ -10)$
$y-y_1=m(x-x_1)$
$y-(-10)=-3(x-10)$
$y+10=-3x+30$
$y=-3x+20$

67) $m=\dfrac{-1-(-4)}{2-(-4)}=\dfrac{3}{6}=\dfrac{1}{2}$
$(x_1,\ y_1)=(-4,\ -4)$
$y-y_1=m(x-x_1)$
$y-(-4)=\dfrac{1}{2}(x-(-4))$
$y+4=\dfrac{1}{2}x+2$
$y=\dfrac{1}{2}x-2$

69)
They have the same slopes and different y-intercepts.

71) $y=mx+b \qquad m=4;\ b=2$
$y=4x+2$

73) $(x_1,\ y_1)=(-1,\ -4);\ m=4$
$y-y_1=m(x-x_1)$
$y-(-4)=4(x-(-1))$
$y+4=4x+4$
$y=4x$
$4x-y=0$

75) Determine the slope.
$x+2y=22$
$2y=-x+22$
$y=-\dfrac{1}{2}x+11$

$(x_1, y_1) = (-4, 7);\ m = -\frac{1}{2}$

$$y - y_1 = m(x - x_1)$$
$$y - 7 = -\frac{1}{2}(x - (-4))$$
$$y - 7 = -\frac{1}{2}(x + 4)$$
$$2(y - 7) = 2\left(-\frac{1}{2}x - 2\right)$$
$$2y - 14 = -x - 4$$
$$x + 2y = 10$$

77) Determine the slope.

$$15x - 3y = 1$$
$$-3y = -15x + 1$$
$$y = 5x - \frac{1}{3}$$

$(x_1, y_1) = (-2, -12);\ m = 5$

$$y - (-12) = 5(x - (-2))$$
$$y + 12 = 5x + 10$$
$$y = 5x - 2$$

79) $(x_1, y_1) = (4, 2);\ m_{\text{perp}} = \frac{3}{2}$

$$y - y_1 = m(x - x_1)$$
$$y - 2 = \frac{3}{2}(x - 4)$$
$$y - 2 = \frac{3}{2}x - 6$$
$$y = \frac{3}{2}x - 4$$

81) $(x_1, y_1) = (10, 0);\ m_{\text{perp}} = \frac{1}{5}$

$$y - y_1 = m(x - x_1)$$
$$y - 0 = \frac{1}{5}(x - 10)$$
$$5 \cdot y = 5\left(\frac{1}{5}x - 2\right)$$
$$5y = x - 10$$
$$-x + 5y = -10$$
$$x - 5y = 10$$

83) $(x_1, y_1) = (4, -9);\ m_{\text{perp}} = -1$

$$y - y_1 = m(x - x_1)$$
$$y - (-9) = -1(x - 4)$$
$$y + 9 = -x + 4$$
$$y = -x - 5$$

85) Determine the slope.

$$x + 3y = 18$$
$$3y = -x + 18;$$
$$y = -\frac{1}{3}x + 6$$
$$m = -\frac{1}{3}$$

$(x_1, y_1) = (4, 2);\ m_{perp} = 3$
$y - y_1 = m(x - x_1)$
$y - 2 = 3(x - 4)$
$y - 2 = 3x - 12$
$y = 3x - 10$
$y - 3x = -10$
$3x - y = 10$

87) Determine the slope.
$3x + y = 8$
$y = -3x + 8$
$m = -3$
$(x_1, y_1) = (-4, 0);\ m = -3$
$y - y_1 = m(x - x_1)$
$y - 0 = -3(x - (-4))$
$y = -3x - 12$

89) Determine the slope.
$y = x - 2$
$m = 1$
$(x_1, y_1) = (2,9);\ m_{perp} = -1$
$y - y_1 = m(x - x_1)$
$y - 9 = -1(x - 2)$
$y - 9 = -x + 2$
$y = -x + 11$

91) Determine the slope.
$y = 1$
$m = 0$
$(x_1, y_1) = (-3, 4);\ m = 0$
$y - y_1 = m(x - x_1)$
$y - 4 = 0(x - 4)$
$y - 4 = 0$
$y = 4$

93) Determine the slope.
$x = 0$
$m = \text{undefined}$

$(x_1, y_1) = (9, 2);\ m_{perp} = 0$

$y - y_1 = m(x - x_1)$
$y - 2 = 0(x - 9)$
$y - 2 = 0$
$y = 2$

95) Determine the slope.
$21x - 6y = 2$
$-6y = -21x + 2$
$\frac{-6y}{-6} = \frac{-21}{-6}x + \frac{2}{-6}$
$y = \frac{7}{2}x - \frac{1}{3} \qquad m = \frac{7}{2}$
$(x_1, y_1) = (4, -1);\ m_{perp} = -\frac{2}{7}$
$y - y_1 = m(x - x_1)$
$y - (-1) = -\frac{2}{7}(x - 4)$
$y + 1 = -\frac{2}{7}x + \frac{8}{7}$
$y = -\frac{2}{7}x + \frac{1}{7}$

97) Determine the slope.

$y=0$

$m=0$

$(x_1, y_1)=\left(4, -\frac{3}{2}\right); \quad m=0$

$$y-y_1=m(x-x_1)$$
$$y-\left(-\frac{3}{2}\right)=0(x-4)$$
$$y+\frac{3}{2}=0$$
$$y=-\frac{3}{2}$$

99) a)

$2005:(0, 81{,}150); \quad 2008:(3, 94{,}960)$

$$m=\frac{94{,}960-81{,}150}{3-0}=\frac{13{,}810}{3}=4603.3$$

$b=81{,}150$

$$y=mx+b$$
$$y=4603.3x+81{,}150$$

b)

The average salary of a mathematician is increasing by \$4603.3 per year.

c) $y=4603.3x+81{,}150$

In 2014, $x=9$

$$y=4603.3(9)+81{,}150$$
$$y=41{,}429.70+81{,}150$$
$$y=\$122{,}579.70$$

101) a) $m=-15{,}000$

The year 2007 corresponds to $x=0$.
Then 500,000 corresponds to $y=500{,}000$.
A point on the line is $(0, 500{,}000)=(x_1, y_1)$.

$$y-y_1=m(x-x_1)$$
$$y-500{,}000=-15{,}000(x-0)$$
$$y=-15{,}000x+500{,}000$$

b)

The budget is cut by \$15,000 per year.

c) $y=-15000x+500{,}000$

In 2010, $x=3$

$$y=-15{,}000(3)+500{,}000$$
$$y=-45{,}000+500{,}000$$
$$=\$455{,}000$$

The budget in 2010 \$455,000.

d) $y=-15000x+500{,}000$

$$365{,}000=-15{,}000(x)+500{,}000$$
$$365{,}000-500{,}000=-15{,}000x$$
$$-135{,}000=-15{,}000x$$
$$\frac{-135{,}000}{-15{,}000}=x$$
$$9=x$$

The year is $2007+9=2016$.

103) a) $(0, 100)$; $m=8g$; $b=100$

$y=mx+b$

$y=8x+100$

b)

A kitten gains about 8 g per day.

c) $y=8x+100$

$x=5$

$y=8(5)+100$

$y=40+100$

$y=140$ g

$x=2$ weeks $=14$ days

$y=8(14)+100$

$y=112+100$

$y=212$ g

d) $y=mx+b$

$284=8x+100$

$184=8x$

$23=x$

It would take 23 days to reach a weight of 28 g

105) a) $(6,38)$; $(8.5,42)$

$m=\dfrac{42-38}{8.5-6}=\dfrac{4}{2.5}=1.6;$

$y-y=m(x-x_1)$

$y=E;\ x=A$

$E-38=1.6(A-6)$

$E-38=1.6A-9.6$

$E=1.6A+28.4$

b) $A=7.5$

$E=1.6(7.5)+28.4$

$E=12+28.4$

$E=40.4$

Size 40

Section 4.6 Exercises

1) a) any set of ordered pairs

b) A relation in which each element of the domain corresponds to exactly one element of the range.

c) Answers may vary.

3) Domain: $\{-4, 1, 4, 8\}$

Range: $\{-4, -3, 2, 10\}$

Function

5) Domain: $\{1, 9, 25\}$

Range: $\{-3, -1, 1, 5, 7\}$

Not a function

7) Domain: $\{-4, -3, -1, 0\}$

Range: $\left\{-2, -\dfrac{1}{2}\right\}$

Function

9) Domain: $\{-4, 1, -2.3, 3.0\}$

Range: $\{-5.7, 3.1, 6.2, 7.8\}$

Not a function

11) Domain: {Hawaii, New York, Miami}

Range: {State, City}

not a function

13) Domain: $\{-5,-2,0,6\}$
Range: $\{-11,-5,4\}$
function

15) Domain: $(-\infty, \infty)$
Range: $(-\infty, \infty)$
Function

17) Domain: $[-4,\infty)$
Range: $(-\infty,\infty)$
Not a function

19) Domain: $(-\infty, \infty)$
Range: $(-\infty, 6]$
Function

21) Domain: $(-\infty, \infty)$

23) Domain: $(-\infty, \infty)$

25) Domain: $(-\infty, \infty)$

27) Domain: $(-\infty,\infty)$

29) $x \neq 0$
Domain: $(-\infty, 0)\cup(0, \infty)$

31) $x-5=0$
$x=5$
Domain: $(-\infty, 5)\cup(5, \infty)$

33) $x+1=0$
$x=-1$
Domain: $(-\infty, -1)\cup(-1, \infty)$

35) $x-20=0$
$x=20$
Domain: $(-\infty, 20)\cup(20, \infty)$

37) y is a function, and y is a function of x.

39) a) $y=2(6)-1$
$y=12-1$
$y=11$
b) $f(6)=2(6)-1$
$f(6)=12-1$
$f(6)=11$

41) a) $y=-4(3)+3$
$y=-12+3$
$y=-9$
b) $f(3)=-4(3)+3$
$f(3)=-12+3$
$f(3)=-9$

43) $f(4)=2(4)-11$
$f(4)=8-11$
$f(4)=-3$

45) $f(0)=2(0)-11$
$f(0)=0-11$
$f(0)=-11$

47) $g(2)=(2)^2-4(2)+2$
$g(2)=4-8+2$
$g(2)=-2$

49) $g(1)=(1)^2-4(1)+2$
$g(1)=1-4+2$
$g(1)=-1$

51) $g\left(\frac{1}{2}\right)=\left(\frac{1}{2}\right)^2-4\left(\frac{1}{2}\right)+2$

$g\left(\frac{1}{2}\right)=\frac{1}{4}-2+2$

$g\left(\frac{1}{2}\right)=\frac{1}{4}$

53) $f(-2)=12$

$f(5)=4$

55) $f(-2)=4$

$f(5)=2$

57) $f(-2)=-2$

$f(5)=2$

59) $f(-2)=-2$

$f(5)=-2$

61) $10=-3x-2$

$12=-3x$

$-4=x$

63) $1=-\frac{3}{2}x-5$

$6=-\frac{3}{2}x$

$\left(-\frac{2}{3}\right)\cdot 6=\left(-\frac{2}{3}\right)\left(-\frac{3}{2}\right)x$

$-4=x$

65) $m=-4$, y-int:$(0,-1)$

67) $m=-\frac{1}{3}$, y-int:$(0,-4)$

69) $m=5$, y-int:$(0,-1)$

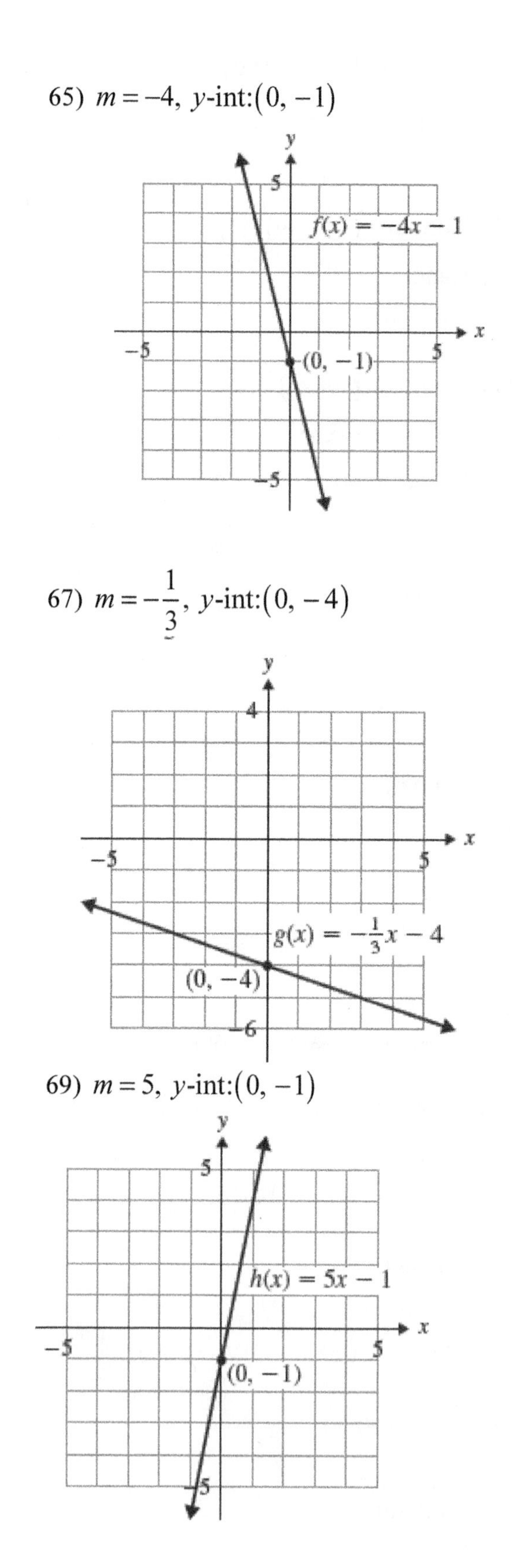

71) $m = \frac{3}{2}$, y-int: $\left(0, -\frac{5}{2}\right)$

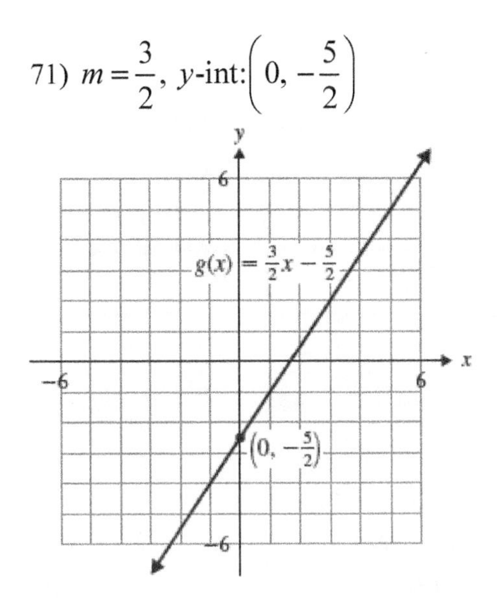

73) a) $A(t) = 9t$

$A(4) = 9(4) = 36$

She owes \$36.00

b) $A(6.5) = 9(6.5) = 58.50$

She owes \$58.50

c) $A(t) = 9t$

$76.50 = 9t$

$\frac{76.50}{9} = t$

$8.5 = t$

She could hire her for 8.5 hours

d) $m = 9$

Fiona pays the babysitter \$9 per hour.

75) a) $C(t) = 59.50t + 12.00$

$C(2) = 59.50(2) + 12.00$

$= 119.00 + 12.00$

$= 131.00$

The cost of two tickets is \$131.00

b) $C(4) = 59.50(4) + 12.00$

$= 238.00 + 12.00$

$= 250.00$

The cost of four tickets is \$250.00

c) $C(t) = 59.50t + 12.00$

$428.5 = 59.50t + 12.00$

$428.50 - 12.00 = 59.50t$

$416.50 = 59.50t$

$\frac{416.50}{59.50} = t$

$7 = t$

Yoshiko bought 7 tickets.

77) $h(t) = 2.39t + 81.45$

$h(41.6) = 2.39(41.6) + 81.45$

$= 99.424 + 81.45$

$= 180.874$

He is 180.9 cm or about 5' 11" tall.

79) $h(L) = 3.08L + 64.67$

$h(33.5) = 3.08(33.5) + 64.67$

$= 103.18 + 64.67$

$= 167.85$

She is 167.9 cm or about 5' 6" tall.

Chapter 4 Review

1) Yes

$$5x-y=13$$
$$5(2)-(-3)=13$$
$$10+3=13$$
$$13=13$$

3) Yes

$$y=-\frac{4}{3}x+\frac{7}{3}$$
$$(-3)=-\frac{4}{3}(4)+\frac{7}{3}$$
$$-3=-\frac{16}{3}+\frac{7}{3}$$
$$-3=\frac{-9}{3}$$
$$-3=-3$$

5) $y=-2x+4$

$y=-2(-5)+4$

$y=10+4$

$y=14$

$(-5,\ 14)$

7) $y=-9$

$(7,\ -9)$

9) $y=x-14$

$y=(0)-14$ $\quad y=(-3)-14$

$y=-14$ $\quad y=-17$

$y=(6)-14$ $\quad y=(-8)-14$

$y=-8$ $\quad y=-22$

x	y
0	−14
6	−8
−3	−17
−8	−22

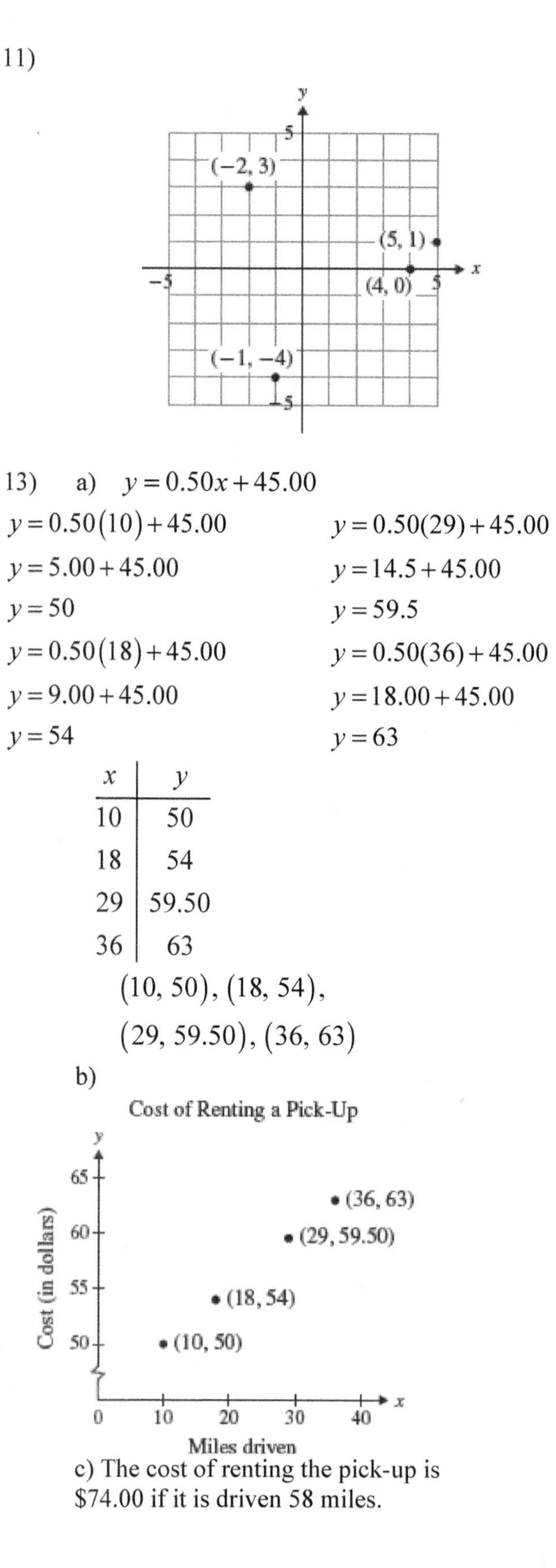

13) a) $y=0.50x+45.00$

$y=0.50(10)+45.00$

$y=5.00+45.00$

$y=50$

$y=0.50(18)+45.00$

$y=9.00+45.00$

$y=54$

$y=0.50(29)+45.00$

$y=14.5+45.00$

$y=59.5$

$y=0.50(36)+45.00$

$y=18.00+45.00$

$y=63$

x	y
10	50
18	54
29	59.50
36	63

$(10,\ 50),\ (18,\ 54),$
$(29,\ 59.50),\ (36,\ 63)$

b)

c) The cost of renting the pick-up is \$74.00 if it is driven 58 miles.

15) $y=-2x+4$

$y=-2(0)+4$
$y=0+4$
$y=4$

$y=-2(1)+4$
$y=-2+4$
$y=2$

$y=-2(2)+4$
$y=-4+4$
$y=0$

$y=-2(3)+4$
$y=-6+4$
$y=-2$

x	y
0	4
1	2
2	0
3	−2

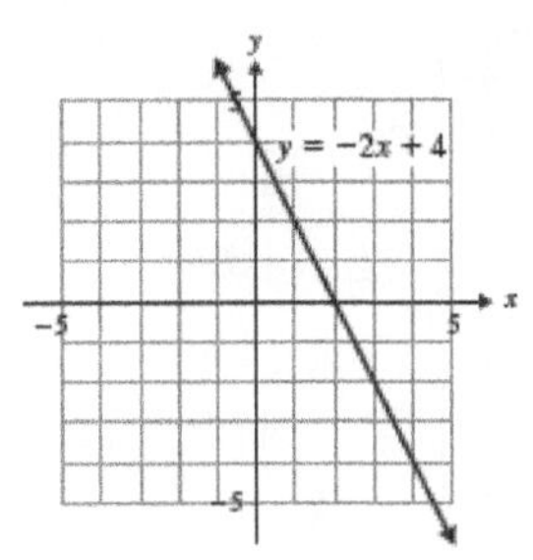

17) $x-2y=2$

x-int: Let $y=0$, and solve for x.

$x-2(0)=2$
$x-0=2$
$x=2$ $\quad$ (2, 0)

y-int: Let $x=0$, and solve for y.

$(0)-2y=2$
$-2y=2$
$y=-1$ $\quad$ (0, −1)

Let $x=4$.

$(4)-2y=2$
$-2y=-2$
$y=1$ $\quad$ (4, 1)

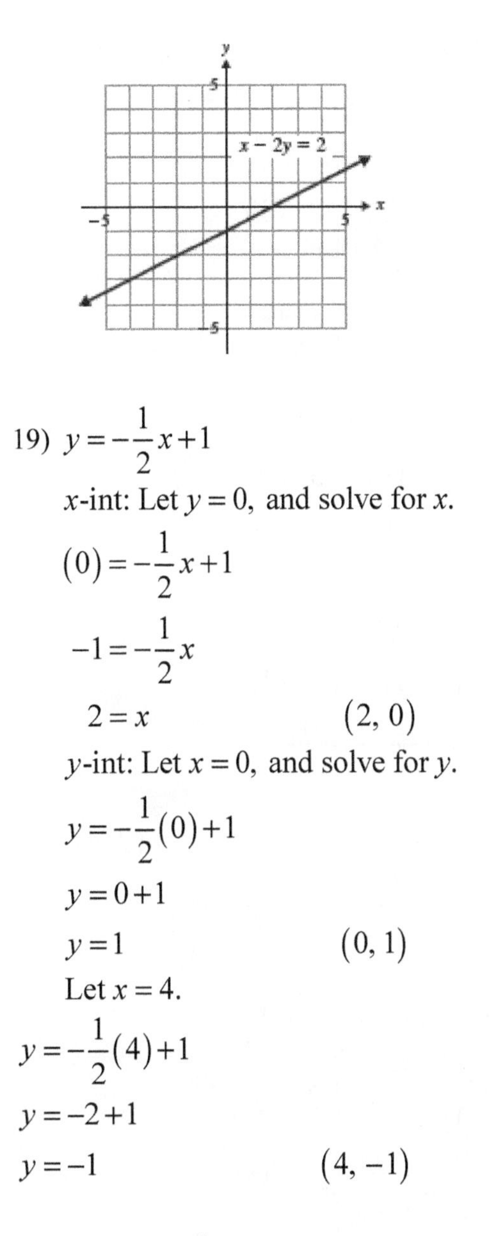

19) $y=-\frac{1}{2}x+1$

x-int: Let $y=0$, and solve for x.

$(0)=-\frac{1}{2}x+1$
$-1=-\frac{1}{2}x$
$2=x$ $\quad$ (2, 0)

y-int: Let $x=0$, and solve for y.

$y=-\frac{1}{2}(0)+1$
$y=0+1$
$y=1$ $\quad$ (0, 1)

Let $x=4$.

$y=-\frac{1}{2}(4)+1$
$y=-2+1$
$y=-1$ $\quad$ (4, −1)

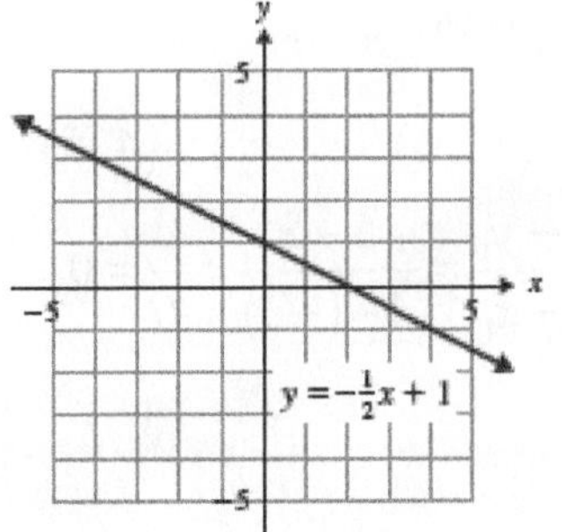

21) $y = 4$

$(0, 4), (2, 4), (-1, 4)$.

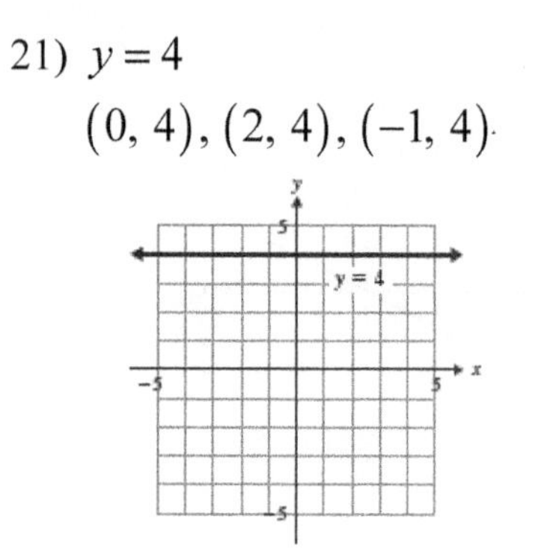

23) $m = \dfrac{\text{rise}}{\text{run}} = \dfrac{3}{2}$

25) $(x_1, y_1) = (5, 8)$

$(x_2, y_2) = (1, -12)$

$m = \dfrac{y_2 - y_1}{x_2 - x_1} = \dfrac{-12-8}{1-5} = \dfrac{-20}{-4} = 5$

27) $(x_1, y_1) = (-7, -2)$ $(x_2, y_2) = (2, 4)$

$m = \dfrac{y_2 - y_1}{x_2 - x_1} = \dfrac{4-(-2)}{2-(-7)} = \dfrac{6}{9} = \dfrac{2}{3}$

29) $(x_1, y_1) = \left(-\dfrac{1}{4}, 1\right)$

$(x_2, y_2) = \left(\dfrac{3}{4}, -6\right)$

$m = \dfrac{y_2 - y_1}{x_2 - x_1} = \dfrac{-6-1}{\frac{3}{4} - \left(-\frac{1}{4}\right)} = \dfrac{-7}{1} = -7$

31) $(x_1, y_1) = (-2, 5)$ $(x_2, y_2) = (4, 5)$

$m = \dfrac{y_2 - y_1}{x_2 - x_1} = \dfrac{5-5}{4-(-2)} = \dfrac{0}{6} = 0$

Slope $= 0$

33) a)

In 1975, Christine paid $4.00 for the album.

b) The slope is positive, so the value of the album is increasing over time.

c)

$m = \dfrac{34-4}{2005-1975} = \dfrac{30}{30} = 1$. The value of the album is increasing by $1.00 per year.

35)

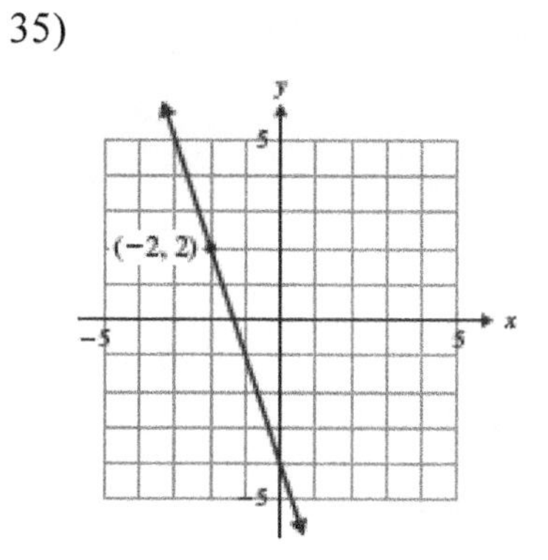

37)

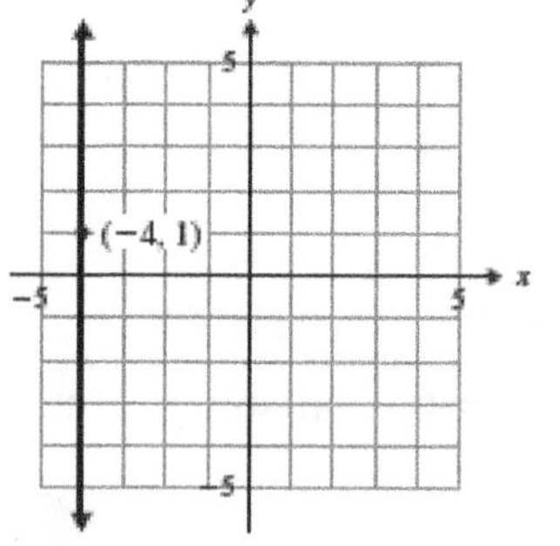

39) $m=-1$, y-int $=(0, 5)$

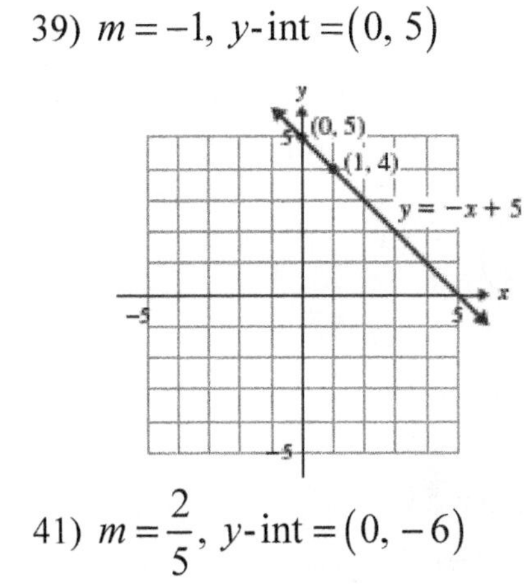

41) $m=\frac{2}{5}$, y-int $=(0, -6)$

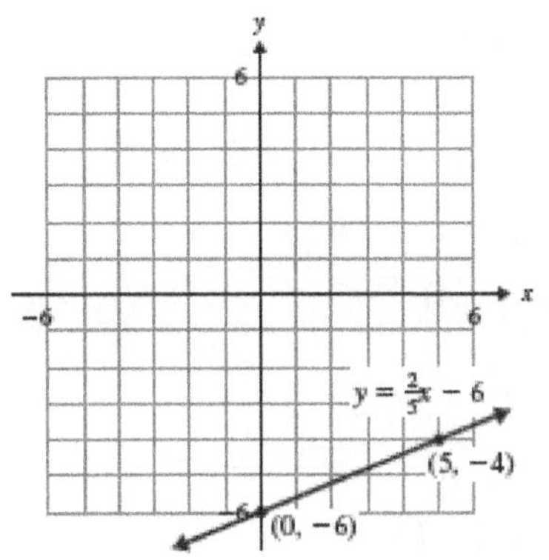

43) $x+3y=-6$

$3y=-x-6$

$y=-\frac{1}{3}x-2$

$m=-\frac{1}{3}$, y-int $=(0,-2)$

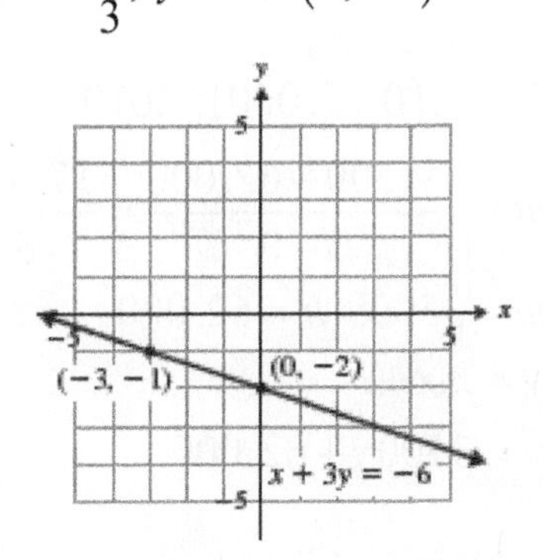

45) $x+y=0$

$y=-x$

$m=-1$, y-int :$(0, 0)$

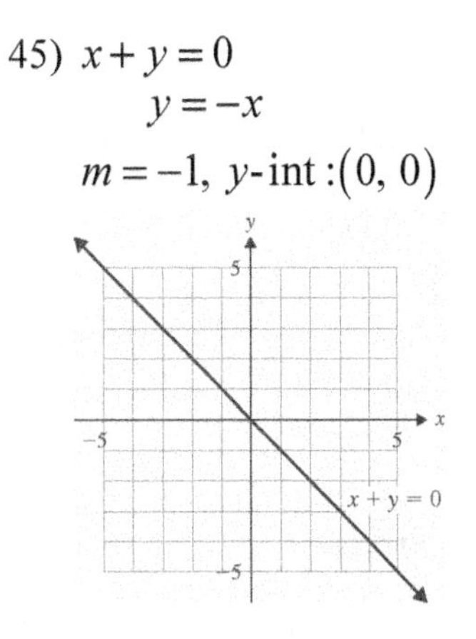

47) a) $(0, 197.6)$; In 2003 the squash crop was worth about \$197.6 million.

b) It has been increasing by \$7.9 million per year.

c) Estimate: \$213 million.

$y=7.9x+197.6$

$y=7.9(2)+197.6$

$y=15.8+197.6$

$y=213.4$

\$213.4 million

49) $x-4y=20$

$-4y=-x+20$

$y=\frac{1}{4}x-5$

$m=\frac{1}{4}$

$-x+4y=6$

$4y=x+6$

$y=\frac{1}{4}x+\frac{3}{2}$

$m=\frac{1}{4}$

parallel

51) $x=7$

$m=$ undefined

$y=-3$

$m=0$

perpendicular

53) $(x_1, y_1) = (-1, 4);\ m = 6$
$$y - y_1 = m(x - x_1)$$
$$y - 4 = 6(x - (-1))$$
$$y - 4 = 6x + 6$$
$$y = 6x + 10$$

55) $y = mx + b \quad m = -\frac{3}{4};\ b = 7$
$$y = -\frac{3}{4}x + 7$$

57) $m = \frac{-3-1}{6-4} = \frac{-4}{2} = -2$
$$(x_1, y_1) = (4, 1)$$
$$y - y_1 = m(x - x_1)$$
$$y - 1 = -2(x - 4)$$
$$y - 1 = -2x + 8$$
$$y = -2x + 9$$

59) Horizontal Line
$$(c, d) = (3, 7)$$
$$y = d$$
$$y = 7$$

61) $m = \frac{-10-5}{-1-4} = \frac{-15}{-5} = 3$
$$(x_1, y_1) = (4, 5)$$
$$y - y_1 = m(x - x_1)$$
$$y - 5 = 3(x - 4)$$
$$y - 5 = 3x - 12$$
$$y = 3x - 7$$
$$-3x + y = -7$$
$$3x - y = 7$$

63) $(x_1, y_1) = \left(1, -\frac{3}{2}\right);\ m = \frac{5}{2}$
$$y - y_1 = m(x - x_1)$$
$$y - \left(-\frac{3}{2}\right) = \frac{5}{2}(x - 1)$$
$$y + \frac{3}{2} = \frac{5}{2}x - \frac{5}{2}$$
$$2y + 3 = 5x - 5$$
$$2y = 5x - 8$$
$$2y - 5x = -8$$
$$5x - 2y = 8$$

65) $y = mx + b \quad m = -4;\ b = 0$
$$y = (-4)x - 0$$
$$4x + y = 0$$

67) $m = \frac{5-1}{2-6} = \frac{4}{-4} = -1$
$$(x_1, y_1) = (6, 1)$$
$$y - y_1 = m(x - x_1)$$
$$y - 1 = -1(x - 6)$$
$$y - 1 = -x + 6$$
$$y = -x + 7$$
$$x + y = 7$$

69) a) $2005: (0, 62{,}000);\ 2010: (5, 79{,}500)$
$$m = \frac{79{,}500 - 62{,}000}{5 - 0} = \frac{17{,}500}{5}$$
$$m = 3500;\quad b = 62{,}000$$
$$y = mx + b$$
$$y = 3500x + 62{,}000$$

b) Mr. Romanski's salary is increasing by $3500 per year.

c) $y = 3500x + 62{,}000$

In 2008, $x = 3$

$y = 3500(3) + 62000$

$y = 10,500 + 62,000$

$y = 72,500$

He earned \$72,500.

d) $y = 3500x + 62000$

$93,500 = 3500x + 62,000$

$31,500 = 3500x$

$x = 9$

He expects to earn \$93,500 in 2014.

71) $(x_1, y_1) = (-1, 14);\ m = -8$

$y - y_1 = m(x - x_1)$

$y - 14 = -8(x - (-1))$

$y - 14 = -8x - 8$

$y = -8x + 6$

73) Determine the slope.

$x - 2y = 6$

$-2y = -x + 6$

$y = \frac{1}{2}x - 3$

$(x_1, y_1) = (4, 11);\ m = \frac{1}{2}$

$y - y_1 = m(x - x_1)$

$y - 11 = \frac{1}{2}(x - 4)$

$2(y - 11) = 2\left(\frac{1}{2}x - 2\right)$

$2y - 22 = x - 4$

$-x + 2y = 18$

$x - 2y = -18$

75) Determine the slope.

$x + 5y = 10$

$5y = -x + 10$

$y = -\frac{1}{5}x + 2$

$(x_1, y_1) = (15, 7);\ m = -\frac{1}{5}$

$y - y_1 = m(x - x_1)$

$y - 7 = -\frac{1}{5}(x - 15)$

$y - 7 = -\frac{1}{5}x + 3$

$y = -\frac{1}{5}x + 10$

77) $(x_1, y_1) = (3, -9);\ m_{perp} = 1$

$y - y_1 = m(x - x_1)$

$y - (-9) = 1(x - 3)$

$y + 9 = x - 3$

$y = x - 12$

79) Determine the slope.

$2x + 3y = -3$

$3y = -2x - 3$

$y = -\frac{2}{3}x - 1$

$(x_1, y_1) = (-4, -4);\ m_{perp} = \frac{3}{2}$

$y - y_1 = m(x - x_1)$

$y - (-4) = \frac{3}{2}(x - (-4))$

$y + 4 = \frac{3}{2}x + 6$

$y = \frac{3}{2}x + 2$

81) $y = 4$

83) Domain: $\{-6, 5, 8, 10\}$
Range: $\{0, 1, 4\}$
function

85) Domain: {CD, DVD, Blueray Disc}
Range: {Music, Movie, PS3 }
not a function

87) Domain: $(-\infty, \infty)$
Range: $(-\infty, \infty)$
Function

89) $(-\infty, \infty)$

91) $x \neq 0$
Domain: $(-\infty, 0) \cup (0, \infty)$

93) $(-\infty, \infty)$

95) $f(2) = 7;\ \ f(-1) = -8$

97) a) $f(3) = -3(3) + 8$
$= -9 + 8 = -1$
$f(3) = -1$

b) $f(-5) = -3(-5) + 8$
$= 15 + 8 = 23$
$f(-5) = 23$

c) $g(4) = (4)^2 - 9(4) + 4$
$= 16 - 36 + 4 = -16$
$g(4) = -16$

d) $g(0) = (0)^2 - 9(0) + 4$
$= 0 - 0 + 4 = 4$
$g(0) = 4$

99) $m = \frac{1}{4}$, y-int $= (0, -2)$

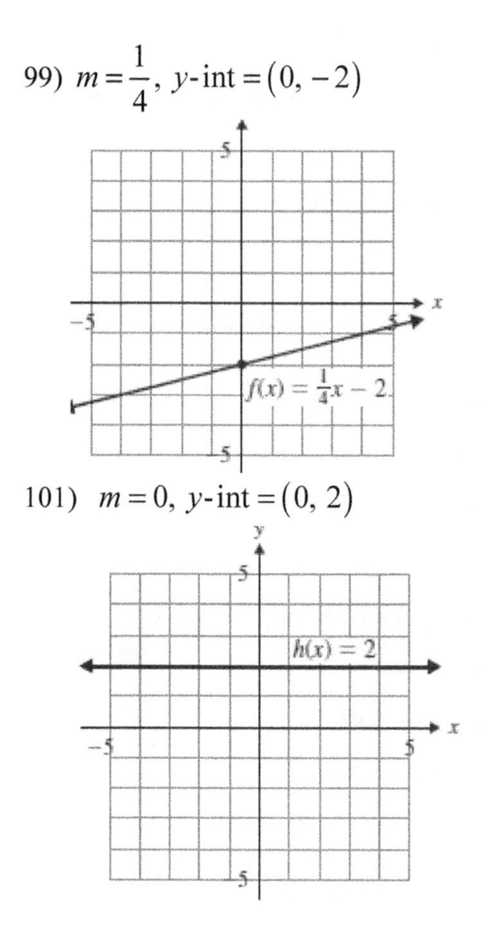

101) $m = 0$, y-int $= (0, 2)$

Chapter 4 Test

1) Yes
$$2x - 7y = 8$$
$$2(-3) - 7(-2) = 8$$
$$-6 + 14 = 8$$
$$8 = 8$$

3) Positive; Negative.

5)

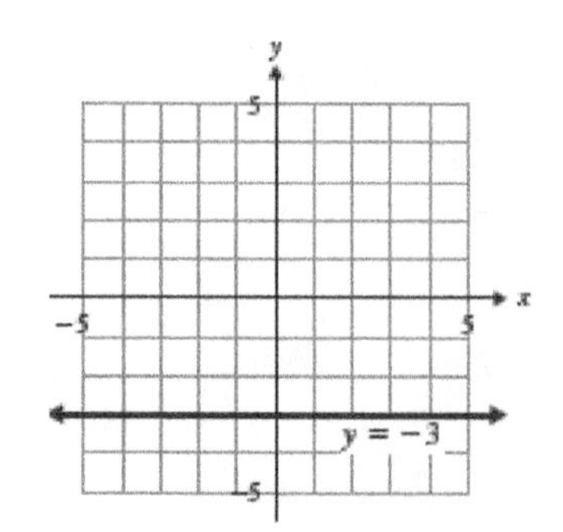

7) a) $(x_1, y_1) = (3, -1)$

$(x_2, y_2) = (-5, 9)$

$m = \frac{y_2 - y_1}{x_2 - x_1} = \frac{9-(-1)}{-5-3}$

$= \frac{10}{-8} = -\frac{5}{4}$

b) $(x_1, y_1) = (8, 6)$

$(x_2, y_2) = (11, 6)$

$m = \frac{y_2 - y_1}{x_2 - x_1} = \frac{6-6}{11-8}$

$= \frac{0}{3} = 0$

9)

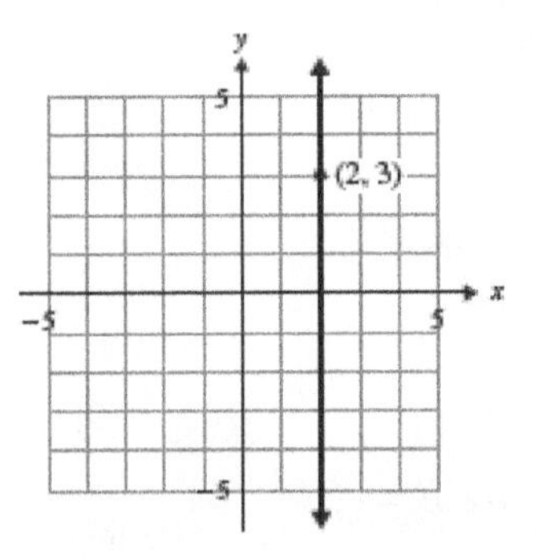

11) $y = mx + b \quad m = 7;\ b = -10$

$y = 7x - 10$

13) $4x + 18y = 9 \qquad 9x - 2y = -6$

$18y = -4x + 9$

$y = -\frac{4}{18}x + \frac{9}{18}$

$y = -\frac{2}{9}x + \frac{1}{2}$

$-2y = -9x - 6$

$y = \frac{-9}{-2}x - \frac{6}{-2}$

$y = \frac{9}{2}x + 3$

$m = -\frac{2}{9}$

$m = \frac{9}{2}$

perpendicular

15) a) $2007: (2, 399)$

The number in 2007 was 399.

b) $2005: (0, 419); \quad 2010: (5, 374)$

$m = \frac{374 - 419}{5 - 0} = \frac{-45}{5} = -9; \ b = 419$

$y = mx + b$

$y = -9x + 419$

c) $y = -9x + 419$

$y = -9(2) + 419$

$= -18 + 419$

$= 401$

According to the equation, 401 students attended the school in 2007.

The actual number was 399.

d)

$m = -9$

The school is losing 9 students per year.

e)

y-int:$(0, 419)$

In 2005, 419 students attended this school.

f) $y = -9x + 419;\ 2013: x = 8$

$y = -9(8) + 419$

$= -72 + 419$

$= 347$

347 children are expected to attend in 2013.

17) Domain: $\{0, 1, 16\}$

Range: $\{-2, -1, 0, 2\}$

not a function

19) $(-\infty, \infty)$

21) $f(-3) = 4$

23) $f(5) = -6(5) + 11 = -30 + 11 = -19$

25) $g(-2) = (-2)^2 - 5(-2) - 3 = 4 + 10 - 3 = 11$

27) $m = 2,\ y\text{-int} = (0, 1)$

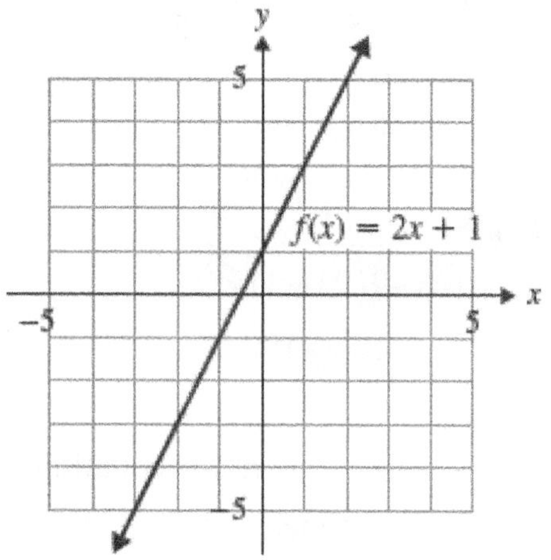

y
5
$f(x) = 2x + 1$
x
−5
5
−5

Cumulative Review: Chapters 1 - 4

1) $\dfrac{336}{792} = \dfrac{336 \div 24}{792 \div 24} = \dfrac{14}{33}$

3) $-3^4 = -81$

5) $\dfrac{3}{8} - 2 = \dfrac{3}{8} - \dfrac{16}{8} = -\dfrac{13}{8}$

7) $2(17) - 9;\ 34 - 9;\ 25$

9) $\left(\dfrac{30w^5}{15w^{-3}}\right)^{-4} = \left(\dfrac{15w^{-3}}{30w^5}\right)^4 = \left(\dfrac{1}{2w^8}\right)^4 = \dfrac{1}{16w^{32}}$

11) $\dfrac{3}{2}(7c-5) - 1 = \dfrac{2}{3}(2c+1)$

$$6 \cdot \frac{3}{2}(7c-5) - 6 \cdot 1 = 6 \cdot \frac{2}{3}(2c+1)$$
$$9(7c-5) - 6 = 4(2c+1)$$
$$63c - 45 - 6 = 8c + 4$$
$$63c - 51 = 8c + 4$$
$$55c = 55$$
$$c = 1$$

The solution set is $\{1\}$.

13) $3x + 14 \le 7x + 4$

$$3x \le 7x - 10$$
$$-4x \le -10$$
$$x \ge \frac{-10}{-4}$$
$$x \ge \frac{5}{2}$$
$$\left[\frac{5}{2}, \infty\right)$$

15) $m\angle A = x^\circ \quad m\angle B = (5x - 14)^\circ$

$$m\angle C = 20^\circ$$
$$m\angle A + m\angle B + m\angle C = 180$$
$$x + 5x - 14 + 20 = 180$$
$$6x + 6 = 180$$
$$6x = 174$$
$$x = 29$$
$$m\angle A = x^\circ = 29^\circ$$
$$m\angle B = (5x-14)^\circ = [5(29) - 14]^\circ = 131^\circ$$

17) $(x_1, y_1) = (4, -11)$

$$(x_2, y_2) = (10, 5)$$
$$m = \frac{y_2 - y_1}{x_2 - x_1} = \frac{5 - (-11)}{10 - 4} = \frac{16}{6} = \frac{8}{3}$$

19) $(x_1, y_1) = (16, 5);\ m = \frac{3}{8}$

$$y - y_1 = m(x - x_1)$$
$$y - 5 = \frac{3}{8}(x - 16)$$
$$8(y - 5) = 8\left(\frac{3}{8}x - 6\right)$$
$$8y - 40 = 3x - 48$$
$$-3x + 8y = -8$$
$$3x - 8y = 8$$

21) $(-\infty, \infty)$

23) $f(0) = 2(0) + 8 = 8$

25) $m = -\frac{1}{3},\ y\text{-int} = (0, 2)$

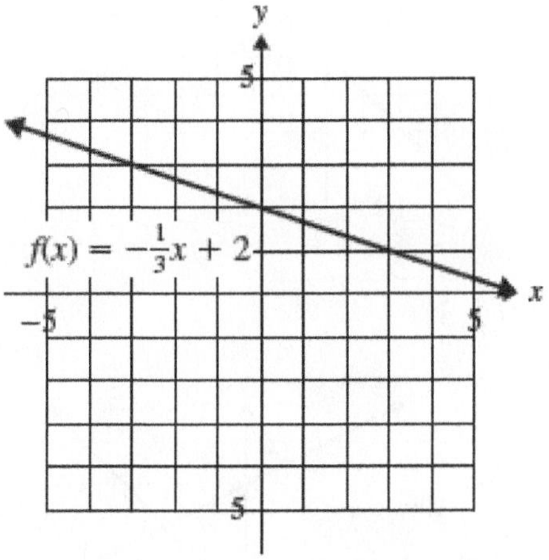

Section 5.1 Exercises

1) Yes

$$x+2y=-6$$
$$8+2(-7)=-6$$
$$8-14=-6$$
$$-6=-6$$

$$-x-3y=13$$
$$-8-3(-7)=13$$
$$-8+21=13$$
$$13=13$$

3) No

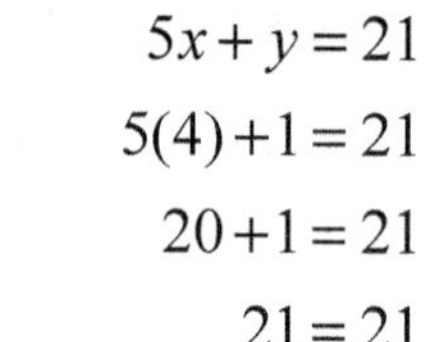

$$5x+y=21$$
$$5(4)+1=21$$
$$20+1=21$$
$$21=21$$

$$2x-3y=11$$
$$2(4)-3(1)\overset{?}{=}21$$
$$8-3\overset{?}{=}21$$
$$5\neq 21$$

5) Yes

$$5y-4x=-5$$
$$5(-3)-4\left(-\frac{5}{2}\right)=-5$$
$$-15+10=-5$$
$$-5=-5$$

$$6x+2y=-21$$
$$6\left(-\frac{5}{2}\right)+2(-3)=-21$$
$$-15-6=-21$$
$$-21=-21$$

7) No

$$y=-x+11$$
$$9\overset{?}{=}-0+11$$
$$9\neq 11$$

$$x=5y-2$$
$$0\overset{?}{=}5(9)-2$$
$$0\overset{?}{=}45-2$$
$$0\neq 43$$

9) The lines are parallel.

11) $(3,1)$

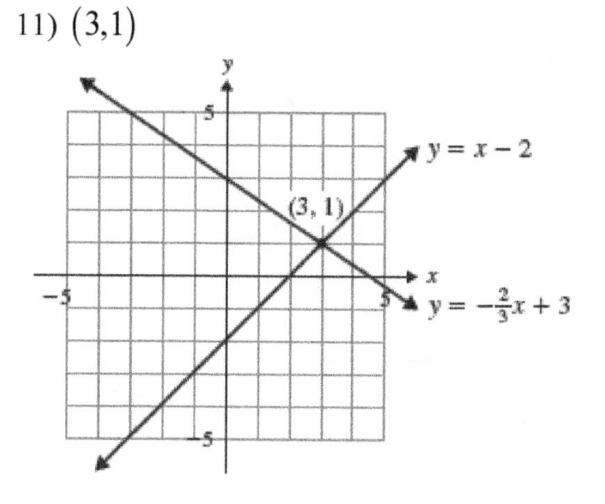

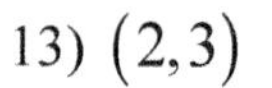

13) $(2,3)$

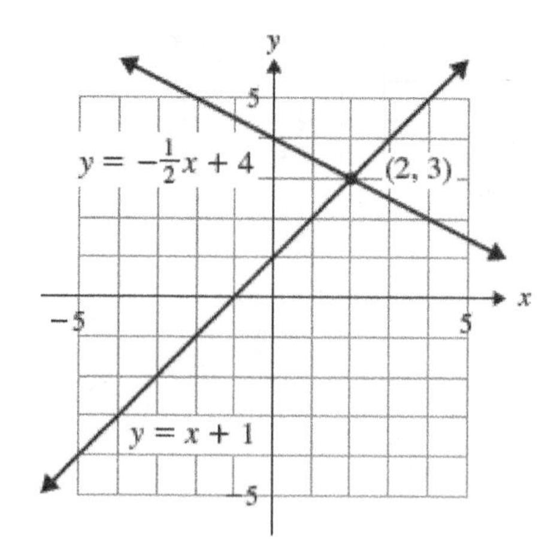

15) $(4, -5)$

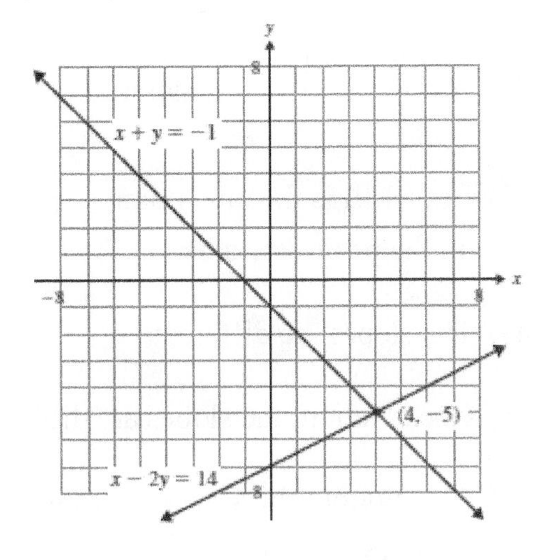

17) $(-1, -4)$

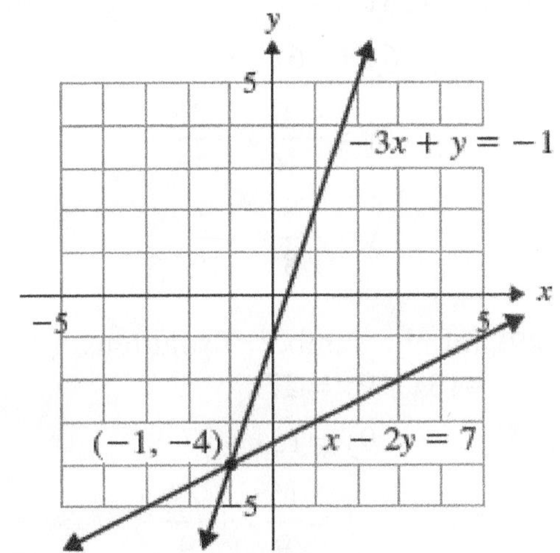

19) $\varnothing$; inconsistent system

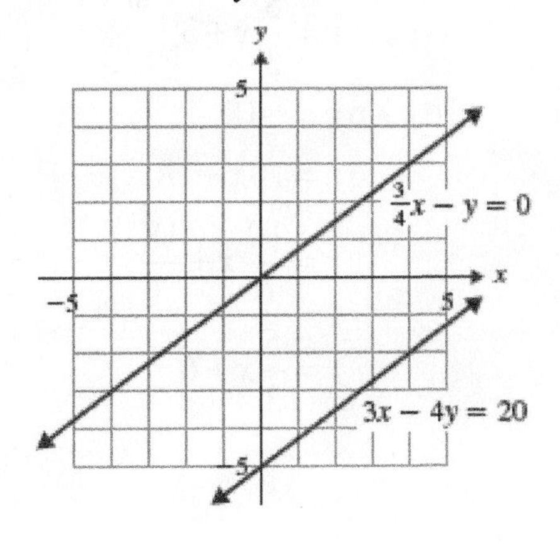

21) There is one solution set of this form

$\{(x, y) \mid y = \frac{1}{3}x - 2\}$

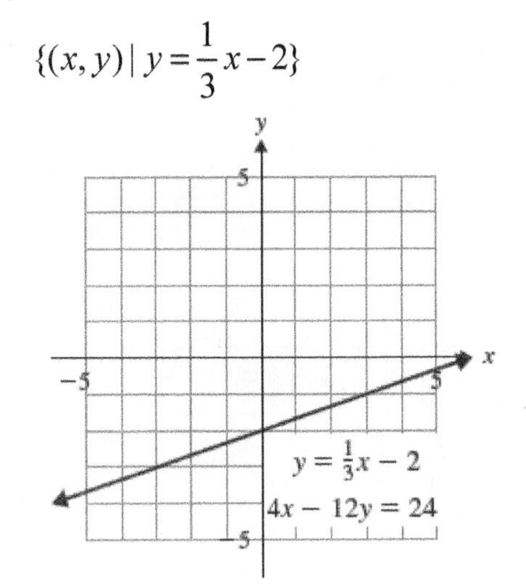

23) $(0, 2)$

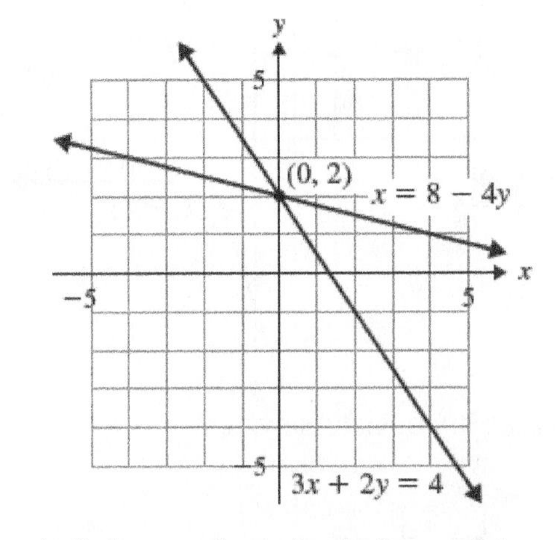

25) infinite number of solutions of the form $\{(x, y) \mid y = -3x + 1\}$; dependent system

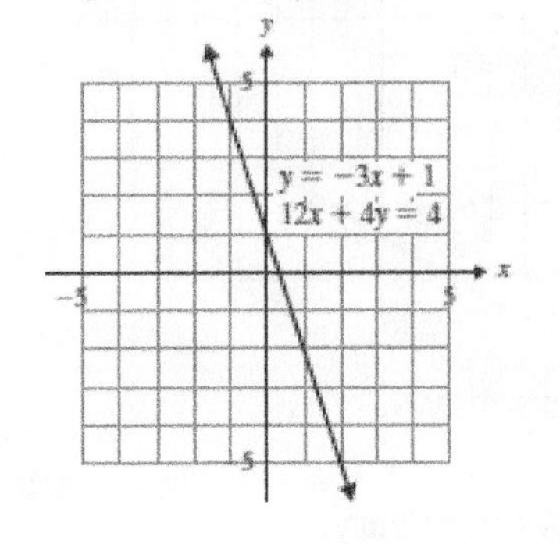

27) $(-2,2)$

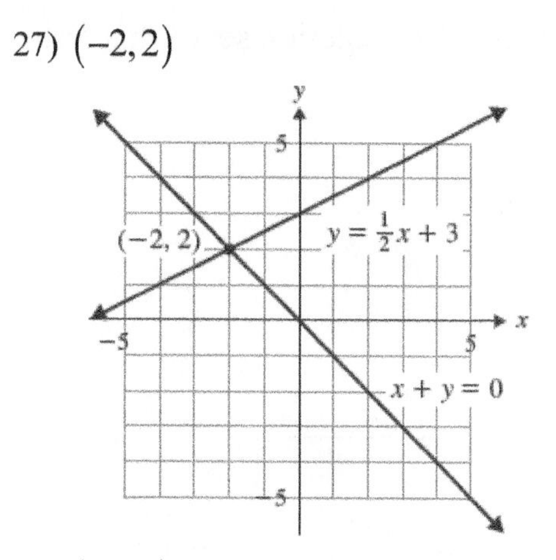

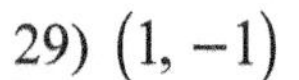

29) $(1,-1)$

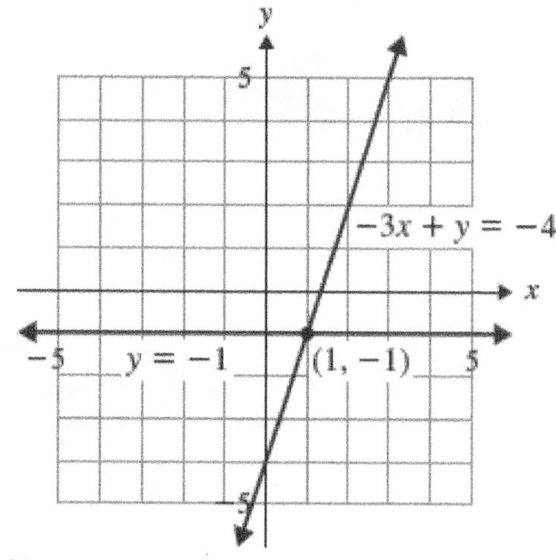

31) $\varnothing$; inconsistent system

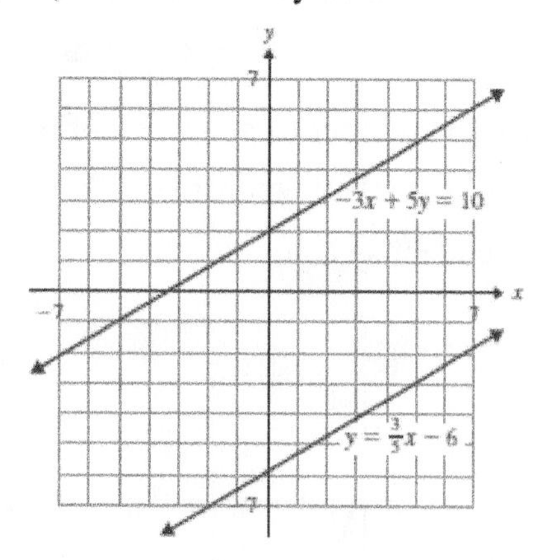

33) Answers may vary.

35) Answers may vary.

37) Answers may vary.

39) C. $(-3,\ 4)$ is in quadrant II.

40) A. $\left(\dfrac{7}{2},-\dfrac{1}{2}\right)$ is in quadrant IV.

41) B; (4.1,0) is the only point on the positive x-axis.

43) The slopes are different.

45) One solution; the slopes are different.

47) No solution; the slopes are same; the lines are parallel.

$$y=-\frac{3}{8}x+1$$
$$6x+16y=-9$$
$$16y=-6x-9$$
$$y=-\frac{6}{16}x-\frac{9}{16}$$
$$y=-\frac{3}{8}x-\frac{9}{16}$$

49) Infinite number of solutions

$$-15x+9y=27$$
$$9y=15x+27$$
$$y=\frac{15}{9}x+\frac{27}{9}$$
$$y=\frac{5}{3}x+3$$
$$10x-6y=-18$$
$$6y=10x+18$$
$$y=\frac{10}{6}x+\frac{18}{3}$$
$$y=\frac{5}{3}x+6$$

51) One solution; the lines have different slopes.

$$3x+12y=9$$
$$12y=-3x+9$$
$$y=-\frac{3}{12}x+\frac{9}{12}$$
$$y=-\frac{1}{4}x+\frac{3}{4}$$
$$x-4y=3$$
$$4y=x-3$$
$$y=\frac{1}{4}x-\frac{3}{4}$$

53) No solution
The lines are parallel.

55) a) 1985-2000

b) (2000,1.4); in the year 2000, 1.4% of foreign students were from Hong Kong and 1.4% were from Malaysia.

c) 1985-1990 and 2000-2005

d) 1985-1990; this line segment has the most negative slope.

57) $y=-2x+2$ $\quad$ $(3,-4)$
$y=x-7$

59) $x-y=3$ $\quad$ $(4,1)$
$x+4y=8$

61) $4x+5y=-17$ $\quad$ $(-2.25,-1.6)$
$3x-4y=4.45$

Section 5.2 Exercises

1) It is the only variable with a coefficient of 1.

3) The variables are eliminated, and you get a false statement.

5) Substitute $y=4x-3$ into
$$5x+y=15$$
$$5x+(4x-3)=15$$
$$9x-3=15$$
$$9x=18$$
$$x=2$$
Substitute $x=2$ into
$$y=4x-3$$
$$y=4(2)-3$$
$$y=8-3$$
$$y=5$$
$(2,5)$

7) Substitute $x=7y+11$ into
$$4x-5y=-2$$
$$4(7y+11)-5y=-2$$
$$28y+44-5y=-2$$
$$23y+44=-2$$
$$23y=-2-44$$
$$23y=-46$$
$$y=-2$$
Substitute $y=-2$ into
$$x=7y+11$$
$$x=7(-2)+11$$
$$x=-14+11$$
$$x=-3$$
$(-3,-2)$

9) Substitute $x=-2y-3$ into

$$\begin{aligned}4x+5y&=-6\\4(-2y-3)+5y&=-6\\-8y-12+5y&=-6\\-3y-12&=-6\\-3y&=-6+12\\-3y&=6\\y&=-2\end{aligned}$$

Substitute $y=-2$ into

$$\begin{aligned}x+2y&=-3\\x+2(-2)&=-3\\x-4&=-3\\x&=-3+4\\x&=1\end{aligned}$$

$(1,-2)$

11) Substitute $y=4x-7$ into

$$\begin{aligned}2y-7x&=-14\\2(4x-7)-7x&=-14\\8x-14-7x&=-14\\x-14&=-14\\x&=-14+14\\x&=0\end{aligned}$$

Substitute $x=0$ into

$$\begin{aligned}4x-y&=7\\4(0)-y&=7\\0-y&=7\\-y&=7\\y&=-7\end{aligned}$$

$(0,-7)$

13) Substitute $y=2x-3$ into

$$\begin{aligned}9y-18x&=5\\9(2x-3)-18x&=5\\18x-27-18x&=5\\-27&\neq 5\end{aligned}$$

There is no solution.

15) Substitute $x=2y+10$ into

$$\begin{aligned}3x-6y&=30\\3(2y+10)-6y&=30\\6y+30-6y&=30\\30&=30\end{aligned}$$

Infinite number of solutions of the form

$\{(x,y)\,|\,x-2y=10\}$

17) Substitute $y=-10x-5$ into

$$\begin{aligned}-5x+2y&=10\\-5x+2(-10x-5)&=10\\-5x-20x-10&=10\\-25x-10&=10\\-25x&=20\\x&=-\frac{4}{5}\end{aligned}$$

Substitute $x=-\frac{4}{5}$ into

$$\begin{aligned}10x+y&=-5\\10\left(-\frac{4}{5}\right)+y&=-5\\-8+y&=-5\\y&=-5+8\\y&=3\end{aligned}$$

$\left(-\frac{4}{5},3\right)$

19) Substitute $x=-\frac{3}{5}y+7$ into

$$x+4y=24$$
$$\left(-\frac{3}{5}y+7\right)+4y=24$$
$$-3y+35+20y=120$$
$$17y+35=120$$
$$17y=120-35$$
$$17y=85$$
$$y=5$$

Substitute $y=5$ into

$$x=-\frac{3}{5}y+7$$
$$x=\left(-\frac{3}{5}\right)5+7$$
$$x=-3+7$$
$$x=4 \qquad (4,5)$$

21) Substitute $x=2y-2$ into

$$4y=2x+4$$
$$4y=2(2y-2)+4$$
$$4y=4y-4+4$$
$$4y=4y$$

Infinite number of solutions of the form $\{(x,y)\,|\,-x+2y=2\}$

23) $2x+3y=6$

$$2x=6-3y$$
$$x=\frac{1}{2}(6-3y)$$

Substitute $x=\frac{1}{2}(6-3y)$ into

$$5x+2y=-7$$
$$5\left[\frac{1}{2}(6-3y)\right]+2y=-7$$
$$15-\frac{15}{2}y+2y=-7$$
$$30-15y+4y=-14$$
$$-11y+30=-14$$
$$-11y=-44$$
$$y=4$$

Substitute $y=4$ into

$$2x+3y=6$$
$$2x+3(4)=6$$
$$2x+12=6$$
$$2x=6-12$$
$$2x=-6$$
$$x=-3 \qquad (-3,4)$$

25) $9x-2y=11$
$-2y=11-9x$
$y=\frac{9}{2}x-\frac{11}{2}$
Substitute $y=\frac{9}{2}x-\frac{11}{2}$ into
$6x-7y=-4$
$6x-7\left(\frac{9}{2}x-\frac{11}{2}\right)=-4$
$6x-\frac{63}{2}x+\frac{77}{2}=-4$
$12x-63x+77=-8$
$-51x+77=-8$
$-51x=-8-77$
$-51x=-85$
$x=\frac{85}{51}$
$x=\frac{5}{3}$

Substitute $x=\frac{5}{3}$ into
$6x-7y=-4$
$6\left(\frac{5}{3}\right)-7y=-4$
$10-7y=-4$
$-7y=-4-10$
$-7y=-14$
$y=2$ $\left(\frac{5}{3},2\right)$

27) $18x+6y=-66$
$6y=-66-18x$
$y=-11-3x$
Substitute $y=-11-3x$ into
$12x+4y=-19$
$12x+4(-11-3x)=-19$
$12x-44-12x=-19$
$-44\neq-19$
There is no solution. $\varnothing$

29) Multiply the equation by the LCD of the fractions to eliminate the fractions.

31) $\frac{1}{4}x-\frac{1}{2}y=1$
$4\left(\frac{1}{4}x-\frac{1}{2}y\right)=4\cdot1$
$x-2y=4$
$x=2y+4$

$\frac{2}{3}x+\frac{1}{6}y=\frac{25}{6}$
$6\left(\frac{2}{3}x+\frac{1}{6}y\right)=6\cdot\frac{25}{6}$
$4x+y=25$
Substitute $x=2y+4$ into
$4x+y=25$
$4(2y+4)+y=25$
$8y+16+y=25$
$9y+16=25$
$9y=9$
$y=1$
Substitute $y=1$ into
$x=2y+4$
$x=2(1)+4$
$x=2+4$
$x=6$ (6, 1)

33) $\frac{1}{6}x+\frac{4}{3}y=\frac{13}{3}$

$x+8y=26$

$x=-8y+26$

Substitute $x=-8y+26$ into

$\frac{2}{5}x+\frac{3}{2}y=\frac{18}{5}$

$4x+15y=36$

$4(-8y+26)+15y=36$

$-32y+104+15y=36$

$-17y=36-104$

$-17y=-68$

$y=4$

Substitute $y=4$ into

$x=-8y+26$

$x=-8(4)+26$

$x=-32+26$

$x=-6$ $(-6, 4)$

35) $\frac{x}{10}-\frac{y}{2}=\frac{13}{10}$

$x-5y=13$

$x=5y+13$

$\frac{x}{3}+\frac{5}{4}y=-\frac{3}{2}$

$4x+15y=-18$

Substitute $x=5y+13$ into

$4x+15y=-18$

$4(5y+13)+15y=-18$

$20y+52+15y=-18$

$35y+52=-18$

$35y=-18-52$

$35y=-70$

$y=-2$

Substitute $y=-2$ into

$x=5y+13$

$x=5(-2)+13$

$x=-10+13$

$x=3$ $(3, -2)$

37) $y-\frac{5}{2}x=-2$

$y=\frac{5}{2}x-2$

Substitute $y=\frac{5}{2}x-2$ into

$\frac{3}{4}x-\frac{3}{10}y=\frac{3}{5}$

$\frac{3}{4}x-\frac{3}{10}\left(\frac{5}{2}x-2\right)=\frac{3}{5}$

$\frac{3}{4}x-\frac{3}{4}x+\frac{6}{10}=\frac{3}{5}$

$\frac{6}{10}=\frac{3}{5}$

$\frac{3}{5}=\frac{3}{5}$

There are infinite number of solutions of the form $\{(x,y)\mid y-\frac{5}{2}x=-2\}$

39) $\frac{3}{4}x+\frac{1}{2}y=6$

$3x+2y=24$

$x=3y+8$

Substitute $x=3y+8$ into

$3x+2y=24$

$3(3y+8)+2y=24$

$9y+24+2y=24$

$11y+24=24$

$11y=0$

$y=0$

Substitute $y=0$ into

$x=3y+8$

$x=3(0)+8$

$x=0+8$

$x=8$ $(8, 0)$

41) $0.2x-0.1y=0.1$

$10(0.2x-0.1y)=10(0.1)$

$2x-y=1$

$y=2x-1$

$0.01x+0.04y=0.23$

$100(0.01x+0.04y)=100(0.23)$

$x+4y=23$

Substitute $y=2x-1$ into

$x+4y=23$

$x+4(2x-1)=23$

$x+8x-4=23$

$9x-4=23$

$9x=27$

$x=3$

Substitute $x=3$ into

$y=2x-1$

$y=2(3)-1$

$y=6-1$

$x=5$ $(3, 5)$

43) $0.6x-0.1y=1$

$10(0.6x-0.1y)=10(1)$

$6x-y=10$

$y=6x-10$

$-0.4x+0.5y=-1.1$

$10(-0.4x+0.5y)=10(-1.1)$

$-4x+5y=-11$

Substitute $y=6x-10$ into

$-4x+5y=-11$

$-4x+5(6x-10)=-11$

$-4x+30x-50=-11$

$26x=-11+50$

$26x=39$

$x=1.5$

Substitute $x=1.5$ into

$y=6x-10$

$y=6(1.5)-10$

$y=9-10$

$y=-1$ $(1.5, -1)$

45) $0.02x+0.01y=-0.44$
$100(0.02x+0.01y)=100(-0.44)$
$2x+y=-44$
$y=-2x-44$
$-0.1x-0.2y=4$
$10(-0.1x-0.2y)=10(4)$
$-x-2y=40$
Substitute $y=-2x-44$ into
$-x-2y=40$
$-x-2(-2x-44)=40$
$-x+4x+88=40$
$3x+88=40$
$3x=-48$
$x=-16$
Substitute $x=-16$ into
$y=-2x-44$
$y=-2(-16)-44$
$y=32-44$
$y=-12$ $(-16,-12)$

47) $2.8x+0.7y=0.1$
$10(2.8x+0.7y)=10(0.1)$
$28x+7y=1$

$0.04x+0.01y=-0.06$
$100(0.04x+0.01y)=100(-0.06)$
$4x+y=-6$
$y=-6-4x$
Substitute y $=-6-4x$ into
$28x+7y=1$
$28x+7(-6-4x)=1$
$28x-42-28x=1$
$-42\neq 1$
There is no solution. $\varnothing$

49) $8+2(3x-5)-7x+6y=16$
$8+6x-10-7x+6y=16$
$-x-2+6y=16$
$-x+6y=18$
$x=6y-18$
$9(y-2)+5x-13y=-4$
$9y-18+5x-13y=-4$
$5x-4y-18=-4$
$5x-4y=-4+18$
$5x-4y=14$
Substitute $x=6y-18$ into
$5x-4y=14$
$5(6y-18)-4y=14$
$30y-90-4y=14$
$26y=14+90$
$26y=104$
$y=4$
Substitute $y=4$ into
$x=6y-18$
$x=6(4)-18$
$x=24-18$
$x=6$ $(6,4)$

51) $10(x+3)-7(y+4)=2(4x-3y)+3$
$10x+30-7y-28=8x-6y+3$
$10x-7y+2=8x-6y+3$
$10x-7y-8x+6y=3-2$
$2x-y=1$
$10-3(2x-1)+5y=3y-7x-9$
$10-6x+3+5y=3y-7x-9$
$13-6x+5y=3y-7x-9$
$-6x+5y-3y+7x=-9-13$
$x+2y=-22$
$x=-22-2y$
Substitute $x=-22-2y$ into
$2x-y=1$
$2(-22-2y)-y=1$
$-44-4y-y=1$
$-44-5y=1$
$-5y=1+44$
$-5y=45$
$y=-9$
Substitute $y=-9$ into
$2x-y=1$
$2x-(-9)=1$
$2x+9=1$
$2x=1-9$
$2x=-8$
$x=-4$ $(-4,-9)$

53) $-(y+3)=5(2x+1)-7x$
$-y-3=10x+5-7x$
$-y-3=3x+5$
$-y-3x=5+3$
$-y-3x=8$
$y=-3x-8$
$x+12-8(y+2)=6(2-y)$
$x+12-8y-16=12-6y$
$x-8y-4=12-6y$
$x-8y+6y=12+4$
$x-2y=16$
Substitute $y=-3x-8$ into
$x-2y=16$
$x-2(-3x-8)=16$
$x+4x+16=16$
$5x+16=16$
$5x=16-16$
$5x=0$
$x=0$
Substitute $x=0$ into
$y=-3x-8$
$y=-3(0)-8$
$y=0-8$
$y=-8$ $(0,-8)$

55) A+ Rental: $y = 0.60x$
Rock Bottom Rental: $y = 0.25x + 70$

a) How much would it cost to drive 160 miles?

A+: $y = 0.60(160)$
$y = 96.00$ $96.00

Rock: $y = 0.25x + 70$
$y = 0.25(160) + 70$
$y = 40 + 70$
$y = 110$ $110.00

b) How much would it cost to drive 300 miles?

A+: $y = 0.60x$
$y = 0.60(300)$
$y = 180$ $180.00

Rock Bottom: $y = 0.25x + 70$
$y = 0.25(300) + 70$
$y = 75 + 70$
$y = 145$ $145.00

c) Requirement: A+Rental and Rock Bottom Rental to be the same Break Even Condition

$$y = 0.60x$$
$$y = 0.25x + 70$$
$$0.60x = 0.25x + 70$$
$$100(0.60x) = 100(0.25x + 70)$$
$$60x = 25x + 7000$$
$$60x - 25x = 7000$$
$$35x = 7000$$

If the cargo trailer is driven 200 miles, the cost would be the same from each company: $120.00
(200, 120)

d) If the cargo trailer is driven less than 200 miles, it is cheaper to rent from A+ if the distance driven is more than 200 miles, then, it is cheaper to rent from Rock Bottom Rentals. If the trailer is driven exactly 200 miles, then the cost would be the same from each company.

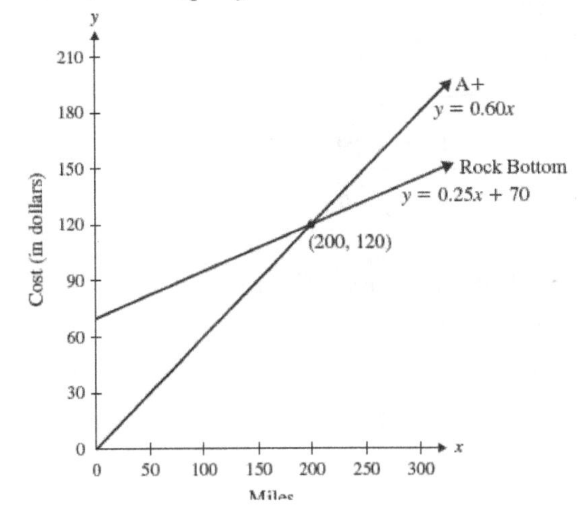

Section 5.3 Exercises

1) Add the equations

3) Add the equations.

$$x - y = -3$$
$$\underline{2x + y = 18}$$
$$3x = 15$$
$$x = 5$$

Substitute $x = 5$ into
$x - y = -3$
$5 - y = -3$
$-y = -3 - 5$
$y = 8$ $(5, 8)$

5) Add the equations

$$-x + 2y = 2$$
$$\underline{x - 7y = 8}$$

$-5y = 10$

$y = -2$

Substitute $y = -2$ into

$-x + 2y = 2$

$-x + 2(-2) = 2$

$-x - 4 = 2$

$-x = 2 + 4$

$x = -6$ $\qquad (-6, -2)$

7) Add the equations

$x + 4y = 1$

$\underline{3x - 4y = -29}$

$4x = -28$

$x = -7$

Substitute $x = -7$ into

$x + 4y = 1$

$-7 + 4y = 1$

$4y = 1 + 7$

$4y = 8$

$y = 2$ $\qquad (-7, 2)$

9) Multiply the second equation by 2.

$4x - 7y = 8$

$8x - 14y = 16$

Add the equations

$8x - 14y = 16$

$\underline{-8x + \ \ 5y = -16}$

$-11y = 0$

$y = 0$

Substitute $y = 0$ into

$4x - 7y = 8$

$4x - 7(0) = 8$

$4x - 0 = 8$

$4x = 8$

$x = 2$ $\qquad (2, 0)$

11) Multiply the second equation by -3

$3x + 5y = 16$

$-9x - 15y = 48$

Add the equations

$4x + 15y = 13$

$\underline{-9x - 15y = -48}$

$-5x = -35$

$x = 7$

Substitute $x = 7$ into

$3x + 5y = 16$

$3(7) + 5y = 16$

$21 + 5y = 16$

$5y = -5$

$y = -1$ $\qquad (7, -1)$

13) Multiply the first equation by 3.

$9x - 7y = -14$

$27x - 21y = -42$

Multiply the second equation by 7

$4x + 3y = 6$

$28x + 21y = 42$

Add the equations

$28x + 21y = \ \ 42$

$\underline{27x - 21y = -42}$

$x \qquad = 0$

Substitute $x = 0$ into

$9x - 7y = -14$

$9(0) - 7y = -14$

$0 - 7y = -14$

$-7y = -14$

$y = 2$ $\qquad (0, 2)$

15) Multiply the first equation by 3.

$-9x+2y=-4$

$-27x+6y=-12$

Multiply the second equation by 2.

$6x-3y=11$

$12x-6y=22$

Add the equations

$$\begin{array}{r} -27x+6y=-12 \\ \underline{12x-6y=22} \\ -15x=10 \end{array}$$

$x=-\frac{2}{3}$

Substitute $x=-\frac{2}{3}$ into

$-9x+2y=-4$

$-9\left(-\frac{2}{3}\right)+2y=-4$

$6+2y=-4$

$2y=-10$

$y=-5$ $\left(-\frac{2}{3},-5\right)$

17) Multiply the first equation by -2.

$9x-y=2$

$-18x+2y=-4$

Add the equations.

$$\begin{array}{r} 18x-2y=4 \\ \underline{-18x+2y=-4} \\ 0=0 \end{array}$$

There are infinite number of solutions of the form $\{(x,y)\,|\,9x-y=2\}$

19) Multiply the first equation by -2

$x=12-4y$

$-2x=-24+8y$

$-2x-8y=-24$

$2x-7=9y$

$2x-9y=7$

Add the equations.

$$\begin{array}{r} -2x-8y=-24 \\ \underline{2x-9y=7} \\ -17y=-17 \end{array}$$

$y=1$

Substitute $y=1$ into

$x=12-4y$

$x=12-4(1)$

$x=12-4$

$x=8$ $(8, 1)$

21) Multiply the first equation by -3, then write it in standard form.

$4y=9-3x$

$-12y=-27+9x$

$-9x-12y=-27$

Multiply the second equation by 2, then write it in the standard form.

$5x-16=-6y$

$10x-32=-12y$

$10x+12y=32$

Add the equations.

$$\begin{array}{r} -9x-12y=-27 \\ \underline{10x+12y=32} \\ x=5 \end{array}$$

Substitute $x=5$ into

$4y=9-3x$

$4y=9-3(5)$

$4y=9-15$

$4y=-6$

$y=-\frac{3}{2}$ $\left(5,-\frac{3}{2}\right)$

23) Multiply the first equation by 5, then write it in standard form.

$$2x-9=8y$$
$$10x-45=40y$$
$$10x-40y=45$$

Multiply the second equation by 2.

$$20y-5x=6$$
$$40y-10x=12$$
$$-10x+40y=12$$

Add the equations.

$$\begin{array}{r} 10x-40y=45 \\ \underline{-10x+40y=12} \\ 0\neq 57 \end{array}$$

There is no solution. $\varnothing$

25) Multiply the first equation by 7.

$$6x-11y=-1$$
$$42x-77y=-7$$

Multiply the second equation by 6.

$$-7x+13y=2$$
$$-42x+78y=12$$

Add the equations.

$$\begin{array}{r} 42x-77y=-7 \\ \underline{-42x+78y=12} \\ y=5 \end{array}$$

Substitute $y=5$ into

$$6x-11y=-1$$
$$6x-11(5)=-1$$
$$6x-55=-1$$
$$6x=-1+55$$
$$6x=54$$
$$x=9$$

$(9, 5)$

27) Multiply the first equation by 2.

$$9x+6y=-2$$
$$18x+12y=-4$$

Multiply the second equation by 3.

$$-6x-4y=11$$
$$-18x-12y=33$$

Add the equations.

$$\begin{array}{r} -18x-12y=33 \\ \underline{18x+12y=-4} \\ 0\neq 29 \end{array}$$

There is no solution. $\varnothing$

29) Eliminate the fractions. Multiply the first equation by 4 and multiply the second equation by 24.

31) Multiply the first equation by -10.

$$\frac{4}{5}x-\frac{1}{2}y=-\frac{3}{2}$$
$$-8x+5y=15$$

Multiply the second equation by 4

$$2x-\frac{1}{4}y=\frac{1}{4}$$
$$8x-y=1$$

Add the equations.

$$\begin{array}{r} -8x+5y=15 \\ \underline{8x-\ y=\ 1} \\ 4y=16 \\ y=4 \end{array}$$

Substitute $y=5$ into

$$8x-y=1$$
$$8x-4=1$$
$$8x=1+4$$
$$8x=5$$
$$x=\frac{5}{8}$$

$\left(\frac{5}{8},4\right)$

33) Multiply the first equation by -16.

$$\frac{5}{4}x-\frac{1}{2}y=\frac{7}{8}$$

$$-20x+8y=-14$$

Multiply the second equation by 50.

$$\frac{2}{5}x-\frac{1}{10}y=-\frac{1}{2}$$

$$20x-5y=-25$$

Add the equations.

$$\begin{aligned} -20x+8y&=-14 \\ 20x-5y&=-25 \\ \hline 3y&=-39 \\ y&=-13 \end{aligned}$$

Substitute $y=-13$ into

$$\begin{aligned} 20x-5y&=-25 \\ 20x-5(-13)&=-25 \\ 20x+65&=-25 \\ 20x&=-25-65 \\ 20x&=-90 \\ x&=-\frac{9}{2} \end{aligned} \qquad \left(-\frac{9}{2},-13\right)$$

35) Multiply the first equation by 36.

$$\frac{x}{4}+\frac{y}{2}=-1$$

$$9x+18y=-36$$

Multiply the second equation by -24.

$$\frac{3}{8}x+\frac{5}{3}y=-\frac{7}{12}$$

$$-9x-40y=14$$

Add the equations.

$$\begin{aligned} 9x+18y&=-36 \\ -9x-40y&=14 \\ \hline -22y&=-22 \\ y&=1 \end{aligned}$$

Substitute $y=1$ into

$$\begin{aligned} 9x+18y&=-36 \\ 9x+18(1)&=-36 \\ 9x+18&=-36 \\ 9x&=-36-18 \\ 9x&=-54 \\ x&=-6 \end{aligned} \qquad (-6,1)$$

37) Multiply the first equation by 24.

$$\frac{x}{12}-\frac{y}{8}=\frac{7}{8}$$

$$2x-3y=21$$

Multiply the second equation by 3.

$$y=\frac{2}{3}x-7$$

$$3y=2x-21$$

$$-2x+3y=-21$$

Add the equations.

$$\begin{aligned} 2x-3y&=21 \\ -2x+3y&=-21 \\ \hline 0&=0 \end{aligned}$$

There are infinite solutions of the form

$$\left\{(x,y)\mid y=\frac{2}{3}x-7\right\}.$$

39) Multiply the first equation by 8.

$-\frac{1}{2}x+\frac{5}{4}y=\frac{3}{4}$

$-4x+10y=6$

Multiply the second equation by 10.

$\frac{2}{5}x-\frac{1}{2}y=-\frac{1}{10}$

$4x-5y=-1$

Add the equations.

$$\begin{aligned} -4x+10y&=6 \\ 4x-5y&=-1 \\ \hline 5y&=5 \\ y&=1 \end{aligned}$$

Substitute $y=1$ into

$$\begin{aligned} 4x-5y&=-1 \\ 4x-5(1)&=-1 \\ 4x-5&=-1 \\ 4x&=-1+5 \\ 4x&=4 \\ x&=1 \end{aligned}$$

$(1,1)$

41) Multiply the first equation by 400.

$0.08x+0.07y=-0.84$

$32x+28y=-336$

Multiply the second equation by -100.

$0.32x-0.06y=-2$

$-32x+6y=200$

Add the equations

$$\begin{aligned} 32x+28y&=-336 \\ -32x+6y&=\quad 200 \\ \hline 34y&=-136 \\ y&=-4 \end{aligned}$$

Substitute $y=-4$

$$\begin{aligned} -32x+6y&=200 \\ -32x+6(-4)&=200 \\ -32x-24&=200 \\ -32x&=200+24 \\ -32x&=224 \\ x&=-7 \end{aligned}$$

$(-7,-4)$

43) Multiply the first equation by 30.

$0.1x+2y=-0.8$

$3x+60y=-24$

Multiply the second equation by -100.

$0.03x+0.10y=0.26$

$-3x-10y=-26$

Add the equations

$$\begin{aligned} 3x+60y&=-24 \\ -3x-10y&=-26 \\ \hline 50y&=-50 \\ y&=-1 \end{aligned}$$

Substitute $y=-1$ into

$$\begin{aligned} 3x+60y&=-24 \\ 3x+60(-1)&=-24 \\ 3x-60&=-24 \\ 3x&=-24+60 \\ 3x&=36 \\ x&=12 \end{aligned}$$

$(12,-1)$

45) Multiply the first equation by 30.

$-0.4x+0.2y=0.1$

$-12x+6y=3$

Multiply the second equation by 20.

$0.6x-0.3y=1.5$

$12x-6y=30$

Add the equations

$-12x+6y=3$

$12x-6y=30$

$0 \neq 33$

There is no solution. $\varnothing$

47) Multiply the first equation by 300.

$0.04x+0.03y=0.16$

$12x+9y=48$

Multiply the second equation by -20.

$0.6x+0.2y=1.15$

$-12x-4y=-23$

Add the equations

$-12x-4y=-23$

$12x+9y=48$

$5y=25$

$y=5$

Substitute $y=5$ into

$12x+9y=48$

$12x+9(5)=48$

$12x+45=48$

$12x=48-45$

$12x=3$

$x=0.25$ $\quad (0.25,5)$

49) $17x-16(y+1)=4(x-y)$

$17x-16y-16=4x-4y$

$13x-12y=16$

$19-10(x+2)=-4(x+6)-y+2$

$19-10x-20=-4x-24-y+2$

$-6x+y=-21$

Multiply the second equation by 12.

$-6x+y=-21$

$-72x+12y=-252$

Add the new equations

$-72x+12y=-252$

$13x-12y=16$

$-59x=-236$

$x=4$

Substitute $x=4$ into

$13x-12y=16$

$13(4)-12y=16$

$52-12y=16$

$-12y=-36$

$y=3$ $\quad (4,3)$

51) $5-3y=6(3x+4)-8(x+2)$

$5-3y=18x+24-8x-16$

$-10x-3y=3$

Multiply the above equation by 4.

$-40x-12y=12$

$6x-2(5y+2)=-7(2y-1)-4$

$6x-10y-4=-14y+7-4$

$6x+4y=7$

Multiply the above equation by 3.

$18x+12y=21$

Add the new equations

$18x+12y=21$

$-40x-12y=12$

$-22x=33$

$x=-\frac{3}{2}$

Substitute $x=-\frac{3}{2}$ into

$$6x+4y=7$$
$$6\left(-\frac{3}{2}\right)+4y=7$$
$$-9+4y=7$$
$$4y=16$$
$$y=4 \qquad \left(-\frac{3}{2},4\right)$$

53) $6(x-3)+x-4y=1+2(x-9)$

$$6x-18+x-4y=1+2x-18$$
$$5x-4y=1$$

Multiply the above equation by 2.

$$10x-8y=2$$
$$4(2y-3)+10x=5(x+1)-4$$
$$8y-12+10x=5x+5-4$$
$$5x+8y=13$$

Add the new equations.

$$\begin{array}{r} 10x-8y=2 \\ \underline{5x+8y=13} \\ 15x \qquad =15 \end{array}$$
$$x=1$$

Substitute $x=1$ into

$$5x+8y=13$$
$$5(1)+8y=13$$
$$5+8y=13$$
$$8y=8$$
$$y=1 \qquad (1,1)$$

55) Multiply the first equation by 3.

$$4x+5y=-6$$
$$12x+15y=-18$$

Multiply the second equation by -4.

$$3x+8y=15$$
$$-12x-32y=-60$$

Add the equations.

$$12x+15y=-18$$
$$-12x-32y=-60$$
$$-17y=-78$$
$$y=\frac{78}{17}$$

Multiply the second equation by 8.

$$4x+5y=-6$$
$$32x+40y=-48$$

Multiply the original second equation by -5.

$$3x+8y=15$$
$$-15x-40y=-75$$

Add the equations.

$$-15x-40y=-75$$
$$32x+40y=-48$$
$$17x=-123$$
$$x=-\frac{123}{17} \qquad \left(-\frac{123}{17},\frac{78}{17}\right)$$

57) Multiply the equation by 3

$$4x+9y=7$$
$$12x+27y=21$$

Multiply the second equation by -2.

$$6x+11y=-14$$
$$-12x-22y=28$$

Add the equations.

$$\begin{aligned} 12x+27y&=21 \\ -12x-22y&=28 \\ \hline -5y&=49 \\ y&=-\frac{49}{5} \end{aligned}$$

Multiply the original first equation by 11

$$4x+9y=7$$
$$44x+99y=77$$

Multiply the original second equation by -9

$$6x+11y=-14$$
$$-54x-99y=126$$

Add the equations.

$$\begin{aligned} -54x-99y&=126 \\ 44x+99y&=\ 77 \\ \hline -10x\qquad&=203 \\ x&=-\frac{203}{10} \end{aligned} \qquad \left(-\frac{203}{10},-\frac{49}{13}\right)$$

59) Ensure $(5,4)$ is a solution; substitute for (x,y) in

$$\begin{aligned} 2x-3y&=-2 \\ 2(5)-3(4)&=-2 \\ 10-12&=-2 \\ -2&=-2 \end{aligned}$$

To obtain the value of k, substitute $(5,4)$ in

$$\begin{aligned} x+ky&=17 \\ 5+k(4)&=17 \\ 4k&=17-5 \\ 4k&=12 \\ k&=3 \end{aligned}$$

61) Ensure that $(-7,3)$ is a solution; substitute for (x,y) in

$$\begin{aligned} 3x+4y&=-9 \\ 3(-7)+4(3)&=-9 \\ -21+12&=-9 \\ -9&=-9 \end{aligned}$$

To calculate k, substitute $(-7,3)$ in

$$\begin{aligned} kx-5y&=41 \\ k(-7)-5(3)&=41 \\ -7k-15&=41 \\ -7k&=41+15 \\ -7k&=56 \\ k&=-8 \end{aligned}$$

63) a) To have infinite solutions, the variables need to be eliminated

Multiply the equation by -1

$x-y=5$

$-x+y=-5$

Add to the second equation

$x-y=c$

$-x+y=-5$

$0=c-5$

To satisfy the above equation, c needs to be 5.

b) Any value for c other than 5 will produce no solution.

65) a) To have infinite solutions, the variables need to be eliminated

Multiply the equation by -3

$ax+4y=-5$

$-3ax-12y=15$

Add to the first equation

$9x+12y=-15$

$-3ax-12y=15$

$x(9-3a)=0$

$9-3a=0$

$a=3$

To satisfy the above equation, a needs to be 3

b) Any value for a other than 3 will produce exactly one solution

67) Add the equations

$-5x+4by=6$

$5x+3by=8$

$7by=14$

$y=\frac{14}{7b}=\frac{2}{b}$

Substitute $y=\frac{2}{b}$ in

$5x+3by=8$

$5x+3b\left(\frac{2}{b}\right)=8$

$5x+6=8$

$5x=8-6$

$x=\frac{2}{5}$ $\left(\frac{2}{5},\frac{2}{b}\right)$

69) Add the equations.

$3ax+by=4$

$ax-by=-5$

$4ax \quad =-1$

$x=-\frac{1}{4a}$

Substitute $x=-\frac{1}{4a}$ in

$ax-by=-5$

$a\left(-\frac{1}{4a}\right)-by=-5$

$-\frac{1}{4}-by=-5$

$-by=-5+\frac{1}{4}$

$-by=-\frac{19}{4}$

$y=\frac{19}{4b}$ $\left(-\frac{1}{4a},\frac{19}{4b}\right)$

Chapter 5: Putting It All Together

1) Elimination method; none of the coefficients is 1 or -1

$2x-3y=-8$

$-4(2x-3y)=-4(-8)$

$-8x+12y=32$

Add the equations.

$$\begin{array}{r} 8x-5y=10 \\ + \ -8x+12y=32 \\ \hline 7y=42 \end{array}$$

$$y=6$$

Substitute $y=6$ into

$$2x-3y=-8$$
$$2x-3(6)=-8$$
$$2x-18=-8$$
$$2x=10$$
$$x=5 \qquad (5,\ 6)$$

3) Since the coefficient of y in the second equation is 1, you can solve for y and use substitution. Or, multiply the second equation by -5 and use the elimination method. Either method will work well.

$$8x+y=-1$$
$$y=-1-8x$$

Substitute $y=-8x-1$ into

$$12x-5y=18$$
$$12x-5(-8x-1)=18$$
$$12x+40x+5=18$$
$$52x+5=18$$
$$52x=13$$
$$x=\frac{1}{4}$$

Substitute $x=\frac{1}{4}$ into

$$8x+y=-1$$
$$8\left(\frac{1}{4}\right)+y=-1$$
$$2=y=-1$$
$$y=-3 \qquad \left(\frac{1}{4},\ -3\right)$$

5) Substitution; the second equation is solved for x and does not contain any fractions.

Substitute $x=y+8$ into

$$y-4x=-11$$
$$y-4(y+8)=-11$$
$$y-4y-32=-11$$
$$-3y-32=-11$$
$$-3y=21$$
$$y=-7$$

Substitute $y=-7$ into

$$x=y+8$$
$$x=(-7)+8$$
$$x=1 \qquad (1,\ -7)$$

7) Substitute $x=3y+6$ into

$$4x+5y=24$$
$$4(3y+6)+5y=24$$
$$12y+24+5y=24$$
$$17y+24=24$$
$$17y=0$$
$$y=0$$

Substitute $y=0$ into

$$x=3y+6$$
$$x=3(0)+6$$
$$x=0+6$$
$$x=6 \qquad (6,\ 0)$$

9)
$$6x+15y=-1$$
$$2(6x+15y)=2(-1)$$
$$12x+30y=-2$$
$$9x=10y-8$$
$$3(9x)=3(10y-8)$$
$$27x=30y-24$$
$$27x-30y=-24$$

Add the equations.

$$\begin{aligned}27x-30y&=-24\\12x+30y&=-2\\\hline 39x&=-26\end{aligned}$$

$$x=-\frac{2}{3}$$

Substitute $x=-\frac{2}{3}$ into

$$6x+15y=-1$$
$$6\left(-\frac{2}{3}\right)+15y=-1$$
$$-4+15y=-1$$
$$15y=3$$
$$y=\frac{1}{5} \qquad \left(-\frac{2}{3},\frac{1}{5}\right)$$

11) $10x+4y=7$ $\qquad$ $15x+6y=-2$

$$3(10x+4y)=3(7) \qquad -2(15x+6y)=-2(-2)$$
$$30x+12y=21 \qquad -30x-12y=4$$

Add the equations.

$$\begin{aligned}30x+12y&=21\\-30x-12y&=4\\\hline 0&\neq 25\end{aligned}$$

$\varnothing$

13) Substitute $x=-\frac{1}{2}$ into

$$10x+9y=4$$
$$10\left(-\frac{1}{2}\right)+9y=4$$
$$-5+9y=4$$
$$9y=9$$
$$y=1 \qquad \left(-\frac{1}{2},1\right)$$

15) $7y-2x=13$

$$3(7y-2x)=3(13)$$
$$21y-6x=39$$
$$-6x+21y=39$$
$$3x-2y=6$$
$$2(3x-2y)=2(6)$$
$$6x-4y=12$$

Add the equations.

$$\begin{aligned}-6x+21y&=39\\6x-4y&=12\\\hline 17y&=51\end{aligned}$$
$$y=3$$

Substitute $y=3$ into

$$3x-2y=6$$
$$3x-2(3)=6$$
$$3x-6=6$$
$$3x=12$$
$$x=4 \qquad (4,3)$$

17) $\frac{2}{5}x+\frac{4}{5}y=-2$

$$-5\left(\frac{2}{5}x+\frac{4}{5}y\right)=-5(-2)$$
$$-2x-4y=10$$
$$\frac{1}{6}x+\frac{1}{6}y=\frac{1}{3}$$
$$12\left(\frac{1}{6}x+\frac{1}{6}y\right)=12\left(\frac{1}{3}\right)$$
$$2x+2y=4$$

Add the equations.

$$\begin{aligned}-2x-4y&=10\\2x+2y&=4\\\hline -2y&=14\end{aligned}$$
$$y=-7$$

Substitute $y=-7$ into

$$2x+2y=4$$
$$2x+2(-7)=4$$
$$2x-14=4$$
$$2x=18$$
$$x=9 \qquad (9,-7)$$

19) $$-0.3x+0.1y=0.4$$
$$10(-0.3x+0.1y)=10(0.4)$$
$$-3x+y=4$$
$$0.01x+0.05y=0.2$$
$$300(0.01x+0.05y)=300(0.2)$$
$$3x+15y=60$$

Add the equations.

$$\begin{array}{r} -3x+y=4 \\ \underline{3x+15y=60} \\ 16y=64 \end{array}$$
$$y=4$$

Substitute $y=4$ into

$$-3x+y=4$$
$$-3x+4=4$$
$$-3x=0$$
$$x=0 \qquad (0,4)$$

21) $$-6x+2y=-10$$
$$7(-6x+2y)=7(-10)$$
$$-42x+14y=-70$$
$$21x-7y=35$$
$$2(21x-7y)=2(35)$$
$$42x-14y=70$$

Add the equations.

$$\begin{array}{r} -42x+14y=-70 \\ \underline{42x-14y=70} \\ 0=0 \end{array}$$

There are infinite solutions of the form $\{(x,y)\,|\,-6x+2y=-10\}$.

23) $$y=\frac{3}{2}x-\frac{1}{2}$$
$$10(y)=10\left(\frac{3}{2}x-\frac{1}{2}\right)$$
$$10y=15x-5$$
$$-15x+10y=-5$$
$$2=5y-8x$$
$$-2(2)=-2(5y-8x)$$
$$-4=-10y+16x$$
$$16x-10y=-4$$

Add the equations.

$$\begin{array}{r} -15x+10y=-5 \\ \underline{16x-10y=-4} \\ x=-9 \end{array}$$

Substitute $x=-9$ into

$$16x-10y=-4$$
$$16(-9)-10y=-4$$
$$-144-10y=-4$$
$$-10y=140$$
$$y=-14 \qquad (-9,-14)$$

Substitute $y=\frac{4}{3}$ into

$$10x-9y=8$$
$$10x-9\left(\frac{4}{3}\right)=8$$
$$10x-12=8$$
$$10x=20$$
$$x=2 \qquad \left(2,\frac{4}{3}\right)$$

25) $2x-3y=-8$

$10(2x-3y)=10(-8)$

$20x-30y=-80$

$7x+10y=4$

$3(7x+10y)=3(4)$

$21x+30y=12$

Add the equations.

$$\begin{array}{l} 21x+30y=\ \ 12 \\ \underline{20x-30y=-80} \\ 41x \qquad\ \ =-68 \end{array}$$

$x=-\dfrac{68}{41}$

$7(2x-3y)=7(-8)$

$14x-21y=-56$

$-2(7x+10y)=-2(4)$

$-14x-20y=-8$

Add the equations.

$$\begin{array}{l} 14x-21y=-56 \\ -14x-20y=-8 \\ \qquad -41y=-64 \end{array}$$

$y=\dfrac{64}{41}$ $\left(-\dfrac{68}{41},\dfrac{64}{41}\right)$

27) $6(2x-3)=y+4(x-3)$

$12x-18=y+4x-12$

$8x-y=6$

$7(8x-y)=7(6)$

$56x-7y=42$

$5(3x+4)+4y=11-3y+27x$

$15x+20+4y=11-3y+27x$

$-12x+7y=-9$

Add the equations

$$\begin{array}{l} 56x-7y=42 \\ \underline{-12x+7y=-9} \\ 44x \qquad =33 \end{array}$$

$x=\dfrac{3}{4}$

Substitute $x=\dfrac{3}{4}$ into

$8x-y=6$

$8\left(\dfrac{3}{4}\right)-y=6$

$6-y=6$

$y=0$ $\left(\dfrac{3}{4},0\right)$

29) $2y-2(3x+4)=-5(y-2)-17$

$$2y-6x-8=-5y+10-17$$
$$-6x+7y=1$$
$$4(-6x+7y)=4(1)$$
$$-24x+28y=4$$
$$4(2x+3)=10+5(y+1)$$
$$8x+12=10+5y+5$$
$$8x-5y=3$$
$$3(8x-5y)=3(3)$$
$$24x-15y=9$$

Add the equations

$$-24x+28y=4$$
$$24x-15y=9$$
$$13y=13$$
$$y=1$$

Substitute $y=1$ into

$$8x-5y=3$$
$$8x-5(1)=3$$
$$8x-5=3$$
$$8x=8$$
$$x=1 \qquad (1,1)$$

31) Substitute $y=-4x$ into

$$10x+2y=-5$$
$$10x+2(-4x)=-5$$
$$10x-8x=-5$$
$$2x=-5$$
$$x=-\frac{5}{2}$$

Substitute $x=-\frac{5}{2}$ into

$$y=-4x$$
$$y=-4\left(-\frac{5}{2}\right)$$
$$y=10 \qquad \left(-\frac{5}{2},10\right)$$

33) $(2,2)$

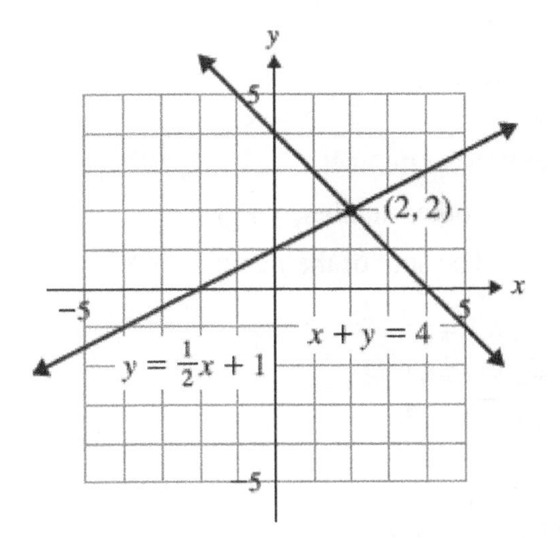

35) $(-4,-4)$

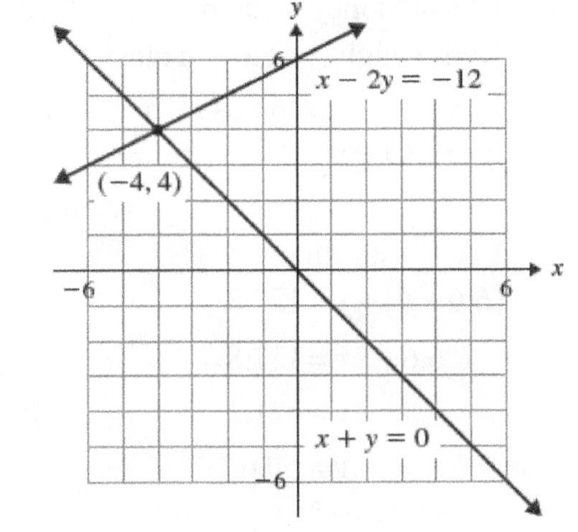

37) $\varnothing$

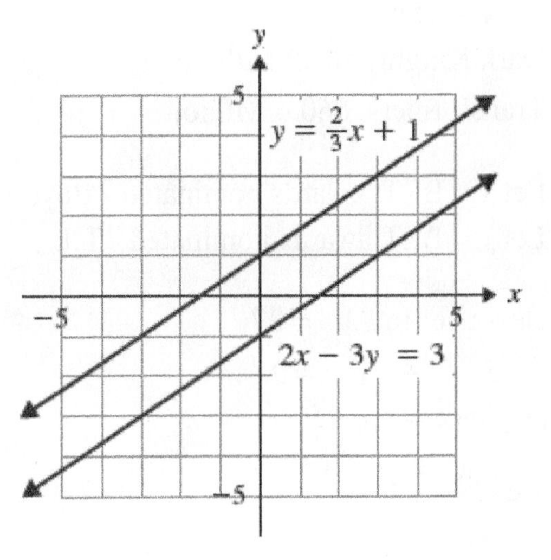

39) Using a graphing calculator, the solution set is $(-1.25,\ -0.5)$

Section 5.4 Exercises

1) Let $x =$ one number
$y =$ other number, $x+11$
The sum of the numbers is 87

$$x+y=87$$
$$x+x+11=87$$
$$2x+11=87$$
$$2x=76$$
$$x=38$$

The numbers are 38, 49

3) Let $x =$ Money made by Dark Knight
let $y =$ Money made by Transformers
Dark Knight = 6.6 + Transformers
$x = 6.6 + y$
Dark Knight + Transformers = 127.8
$x + y = 127.8$
Use substitution.
$$(6.6+y)+y=127.8$$
$$6.6+2y=127.8$$
$$2y=121.2$$
$$y=60.6$$
$$x=6.6+y;\ y=60.6$$
$$x=60.6+6.6$$
$$x=67.2$$
Dark Knight, \$67.2 Million;
Transformers, \$60.6 Million

5) Let $x =$ BET Awards nominated - Beyonce
Let $y =$ BET awards nominated - T.I.

Beyonce + T.I. = 27
$x + y = 27$
Beyonce = $T.I. + 5$
$x = y + 5$

Use substitution.
$$(y+5)+y=27$$
$$2y+5=27$$
$$2y=22$$
$$y=11$$
$$x=y+5;\ y=11$$
$$x=11+5$$
$$x=16$$
Beyonce – 16; T.I. – 11

7) Let $x =$ number who speak Urdu
Let $y =$ number who speak Polish.

Who speak Urdu = $\frac{1}{2}$(who speak Polish)
$$x=\frac{1}{2}y$$
who speak Urdu + who speak Polish = 975,000
$x + y = 975,000$

Use substitution.
$$y+\frac{1}{2}(y)=975,000$$
$$2y+y=1,950,000$$
$$3y=1,950,000$$
$$y=650,000$$
$$x=\frac{1}{2}y;\ y=650,000$$
$$x=\frac{1}{2}(650,000)$$
$$x=325,000$$
Polish Speaking: 650,000;
Urdu Speaking: 325,000

9) Let $x =$ minutes by Yuri
let $y =$ minutes by Sheperd.
Mins. by Yuri = Mins. by Sheperd + 93
$x = y + 93$
Mins. by Yuri + Mins. by Sheperd = 123
$x + y = 123$

Use substitution.

$$y+(y+93)=123$$
$$2y+93=123$$
$$2y=30$$
$$y=15$$

$$x=y+93;\ y=15$$
$$x=93+15$$
$$x=108$$

Minutes in space by Yuri = 108;
by Sheperd = 15.

11) Let w = the width, and
let h = the height.
height = width − 50
$h \quad = w-50$
$2h+2w=220$
Use substitution.
$$2(h-50)+2h=220$$
$$4h-100=220$$
$$4h=320$$
$$h=80$$
$$w=h-50;\ h=80$$
$$w=h-50$$
$$w=30$$
Height: 80 in. width: 30 in.

13) Let w = the width, and
let l = the length
$2l+2w=343.6$
Length = width + 48.2
$l=w+48.2$
Use substitution.
$$2(w+48.2)+2w=343.6$$
$$2w+96.4+2w=343.6$$
$$4w+96.4=343.6$$
$$4w=247.2$$
$$w=61.8$$
$$l=w+48.2;\ w=61.8$$
$$l=61.8+48.2$$
$$l=110.0$$
Length: 110 mm; width: 61.8 mm

15) Let w = the width, and
let l = the length
$l+2w=119$
length = 1.5(width)
$l \quad =1.5(w)$
Use substitution.
$$1.5w+2w=119$$
$$3.5w=119$$
$$7w=238$$
$$w=34$$
$$l=1.5w;\ w=34$$
$$l=1.5(34)$$
$$l=51$$
length: 51 ft; width: 34 ft.

17) $x^\circ=\frac{3}{5}y^\circ$
supplementary angles
$x^\circ+y^\circ=180^\circ$
Use substitution.
$$\frac{3}{5}y+y=180$$
$$\frac{8}{5}y=180$$
$$y=112.5$$
$$x=\frac{3}{5}y;\ y=112.5$$
$$x=\frac{3}{5}(112.5)$$
$$x=67.5$$
$m\angle x=67.5^\circ;\ m\angle y=112.5^\circ$

19) Let x = cost of a t-shirt
and let y = cost of a souvenir puck.
Kenny's Purchase:
t-shirt+2puck = 36
$x+ \quad 2y \quad =36$
Kyle's Purchase:
2 t-shirts + 3 pucks = 64
$2x \quad + \quad 3y \quad =64$

$$x+2y=36$$
$$-2(x+2y)=-2(36)$$
$$-2x-4y=-72$$
Add the equations.

$$\begin{array}{r} -2x-4y=-72 \\ 2x+3y=\ \ 64 \\ \hline -y=-8 \end{array}$$

$y=8$

Substitute $y=8$ into

$x+2y=36$

$x+2(8)=36$

$x+16=36$

$x=20$

t-shirt: \$20; Souvenir Puck: \$8

21) Let $x=$ cost of a Bobblehead
let $y=$ cost of a mug

3(bobblehead)+4(mug) $=105$

$3x \quad + \quad 4y \quad =105$

2(bobblehead)+3(mug) $=74$

$2x \quad + \quad 3y \quad =74$

$2(3x+4y)=2(105)$

$6x+8y=210$

$-3(2x+3y)=-3(74)$

$-6x-9y=-222$

Add the equations

$$\begin{array}{r} -6x-9y=-222 \\ 6x+8y=\ \ 210 \\ \hline -y=-12 \end{array}$$

$y=12$

Substitute $y=12$ into

$2x+3y=74$

$2x+3(12)=74$

$2x+36=74$

$2x=38$

$x=19$

Bobblehead:\$19.00; Mug:\$12.00

23) Let $x=$ cost of hamburger
let $y=$ cost of fries.

$5\cdot$ hamburgers $+1\cdot$ fries $=4.44$

$5x \quad + \quad y \quad =4.44$

$4\cdot$ hamburgers $+2\cdot$ fries $=5.22$

$4x \quad + \quad 2y \quad =5.22$

$5x+y=4.44$

$-2(5x+y)=-2(4.44)$

$-10x-2y=-8.88$

Add the equations.

$$\begin{array}{r} -10x-2y=-8.88 \\ 4x+\ 2y=\ \ 5.22 \\ \hline -6x=-3.66 \end{array}$$

$x=0.61$

Substitute $x=0.61$ into

$5x+y=4.44$

$5(0.61)+y=4.44$

$3.05+y=4.44$

$y=1.39$

hamburger : \$0.61; fries: \$1.39

25) Let $x=$ wrapping paper
let $y=$ gift bag

$4\cdot$ wrapping paper $+3\cdot$ gift bag $=52$

$4x \quad + \quad 3y \quad =52$

$3\cdot$ wrapping paper $+1\cdot$ gift bag $=29$

$3x \quad + \quad y \quad =29$

$3x+y=29$

$-3(3x+y)=-3(29)$

$-9x-3y=-87$

Add the equations

$$\begin{array}{r} -9x-3y=-87 \\ 4x+3y=\ \ 52 \\ \hline -5x\quad\ =-35 \end{array}$$

$x=7$

Substitute $x=7$ into

$4x+3y=52$

$4(7)+3y=52$

$28+3y=52$

$3y=24$

$y=8$

Wrapping Paper: \$7.00;
Gift Bag: \$8.00

27) $x =$ number of ounces of 9% solution
$y =$ number of ounces of 17% solution

Soln	Concn	No of Oz of soln	No of Oz of alcohol in the soln
9%	0.09	x	$0.09x$
17%	0.17	y	$0.17y$
15%	0.15	12	$0.15(12)$

$$x+y=12 \qquad 0.09x+0.17y=0.15(12)$$
$$y=12-x \quad 100(0.09x+0.17y)=100[0.15(12)]$$

$$9x+17y=15(12)$$

Use substitution.

$$9x+17(12-x)=15(12)$$
$$9x+204-17x=180$$
$$-8x+204=180$$
$$-8x=-24$$
$$x=3$$

$$y=12-x;\ x=3$$
$$y=12-3$$
$$y=9$$

9% : 3 oz; 17% : 9 oz

29) $x =$ amount of 25% acid solution
$y =$ amount of 100% acid solution

Solution.	*Amount*	*Concn.*	*Amount in Solution*
25%	x	0.25	$0.25x$
100%	y	1.00	y
40%	10	0.40	$0.40(10)$

$$x+y=10 \qquad 0.25x+y=0.4(10)$$
$$y=10-x \quad 100(0.25x+y)=100[0.4(10)]$$

$$25x+100y=400$$

Use substitution.

$$25x+100(10-x)=400$$
$$25x+1000-100x=400 \quad y=10-x;\ x=8$$
$$-75x+1000=400 \quad y=10-8$$
$$-75x=-600 \quad y=2$$
$$x=8$$

Pure acid: 2 Liters; 25% Solution: 8 Liters

31) $x =$ tea that sells for \$7.50/oz
$y =$ tea that sells for \$5.00/oz

$$x+y=60$$
$$x(7.5)+y(5.0)=60(6.0)$$
$$7.5x+5.0y=360$$

Use substitution.

$$x=60-y$$
$$7.5(60-y)+5y=360$$
$$450-7.5y+5y=360$$
$$-2.5y=-90$$
$$y=36$$
$$x=60-y;\ y=36$$
$$x=60-36$$
$$x=24$$

Asian Treasure: 24 oz; Pearadise: 36 oz.

33) $x =$ sodium in chalupa; $y =$ sodium in crunchy taco

$$x + 3y = 1640$$
$$2x + 2y = 1960$$

Use substituion.

$$x = 1640 - 3y$$
$$2(1640 - 3y) + 2y = 1960$$
$$3280 - 6y + 2y = 1960$$
$$3280 - 4y = 1960$$
$$-4y = -1320$$
$$y = 330$$
$$x = 1640 - 3y; y = 330$$
$$x = 1640 - 3(330)$$
$$x = 1640 - 990$$
$$x = 650$$

Sodium in chicken chalupa: 650 mg;

Sodium in crunchy taco: 330 mg

35) $x =$ Amount invested at 2%
$y =$ Amount invested at 4%

$$x + y = 6000 \qquad x(0.02) + y(0.04) = 190$$
$$y = 6000 - x \qquad 100(0.02x + 0.04y) = 100(190)$$
$$2x + 4y = 19000$$

Use substitution.

$$2x + 4(6000 - x) = 19,000$$
$$2x + 24000 - 4x = 19,000$$
$$24000 - 2x = 19,000$$
$$-2x = -5000$$
$$x = 2500$$
$$y = 6000 - x; x = 2500$$
$$y = 6000 - 2500$$
$$y = 3500$$

37) $x =$ Number of \$0.44 stamps;
$y =$ Number of \$0.28 stamps

$$x + y = 16 \qquad 0.44x + 0.28y = 6.40$$
$$x = 16 - y \qquad 100(0.44x + 0.28y) = 100(6.40)$$
$$44x + 28y = 640$$

Use substitution.

$$44(16 - y) + 28y = 640$$
$$704 - 44y + 28y = 640$$
$$704 - 16y = 640$$
$$-16y = -64$$
$$y = 4$$
$$x = 16 - y; y = 4$$
$$x = 16 - 4$$
$$x = 12$$

\$0.44 Stamps: 12; \$0.28 Stamps: 4

39) $x =$ Michael's Speed; $y =$ Jan's Speed

$$x = y + 1$$

Distance Traveled in 3 Hrs.

Michael: $3x = 3(y + 1)$

Jan: $3y$

Since Jan and Michael traveled in opposite directions,

$$3(y + 1) + 3y = 51$$
$$3y + 3 + 3y = 51$$
$$6y + 3 = 51$$
$$6y = 48$$
$$y = 8$$
$$x = y + 1$$
$$x = 8 + 1$$
$$x = 9$$

Michael: 9 mph; Jan: 8 mph

41) $x =$ Speed of small plane;
$y =$ Speed of Jet

$$x = y - 160$$
$$2x + 2y = 1280$$

Use substitution.

$$2(y-160)+2y=1280$$
$$2y-320+2y=1280$$
$$4y-320=1280$$
$$4y=1600$$
$$y=400$$
$$x=y-160; y=400$$
$$x=400-160$$
$$x=240$$

Speed of small plane: 240 mph;
Speed of Jet: 400 mph

42) $x =$ Tyreese's jogging Speed;
$y =$ Justine's jogging Speed

$$x=y+3 \qquad \frac{1}{2}x+\frac{1}{2}y=6.5$$
$$2(\frac{1}{2}x+\frac{1}{2}y)=2(6.5)$$
$$x+y=13$$

Use substitution.

$$(y+3)+y=13$$
$$2y+3=13$$
$$2y=10$$
$$y=5$$

$$x=y+3; y=5$$
$$x=5+3$$
$$x=8$$

Tyreese's jogging speed: 8 mph;
Justine's jogging speed: 5 mph

43) $x =$ Pam's Speed; $y =$ Jim's Speed

$$x=y-2 \qquad \frac{1}{2}x+\frac{1}{2}y=9$$
$$2\left(\frac{1}{2}x+\frac{1}{2}y\right)=2(9)$$
$$x+y=18$$

Use substitution

$$(y-2)+y=18$$
$$y-2+y=18$$
$$2y-2=18$$
$$2y=20$$
$$y=10$$

$$x=y-2; y=10$$
$$x=10-2$$
$$x=8$$

Pam's speed: 8 mph;
Jim's speed: 10 mph

45)
$$2(x+y)=14$$
$$2x+2y=14$$
$$2(x-y)=10$$
$$2x-2y=10$$

Add the equations

$$2x+2y=14$$
$$\underline{2x-2y=10}$$
$$4x \quad =24$$
$$x=6$$

Substitute $x=6$ into

$$2x+2y=14$$
$$2(6)+2y=14$$
$$12+2y=14$$
$$2y=2$$
$$y=1$$

Speed of Boat in still water: 6 mph;
Speed of current: 1 mph

47) $x =$ Speed of Boat;
$y =$ Speed of current

$$5x+5y=80$$
$$8(5x+5y)=8(80)$$
$$40x+40y=640$$
$$8x-8y=80$$
$$5(8x-8y)=5(80)$$
$$40x-40y=400$$

Add the equations.

$$40x-40y=400$$
$$\underline{40x+40y=640}$$
$$80x=1040$$
$$x=13$$

Substitute $x = 13$ into

$$5x + 5y = 80$$
$$5(13) + 5y = 80$$
$$65 + 5y = 80$$
$$5y = 15$$
$$y = 3$$

Speed of Boat: 13 mph;

Speed of Current: 3 mph

9)

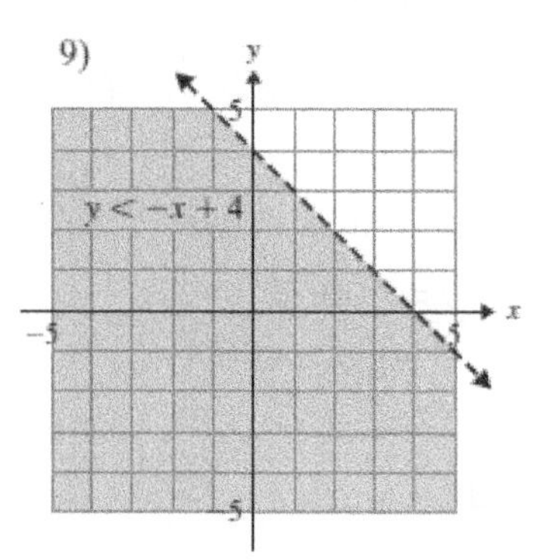

49) x = Speed of Jet;

y = Speed of Wind

Add the equations

$$2.5x - 2.5y = 1000$$
$$2.5x + 2.5y = 1250$$
$$5x = 2250$$
$$x = 450$$

Substitute $x = 450$ into

$$2.5x + 2.5y = 1250$$
$$2.5(450) + 2.5y = 1250$$
$$1125 + 2.5y = 1250$$
$$2.5y = 125$$
$$y = 50$$

Speed of Jet: 450 mph;

Speed of Wind: 50 mph

x = Speed of Plane;

y = Speed of Wind

11)

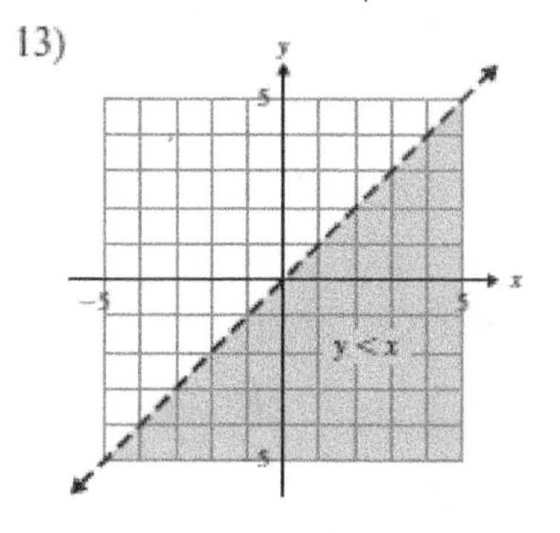

13)

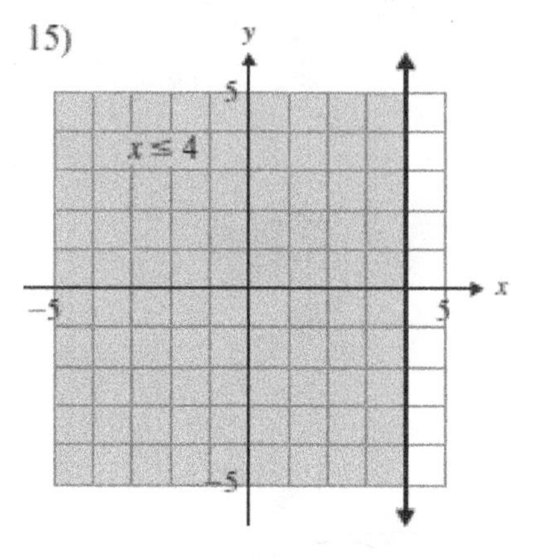

Section 5.5 Exercises

1-5) Answers may vary.

7)

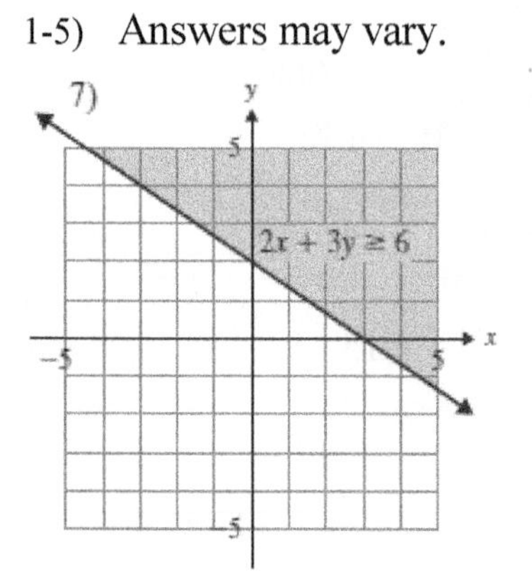

15)

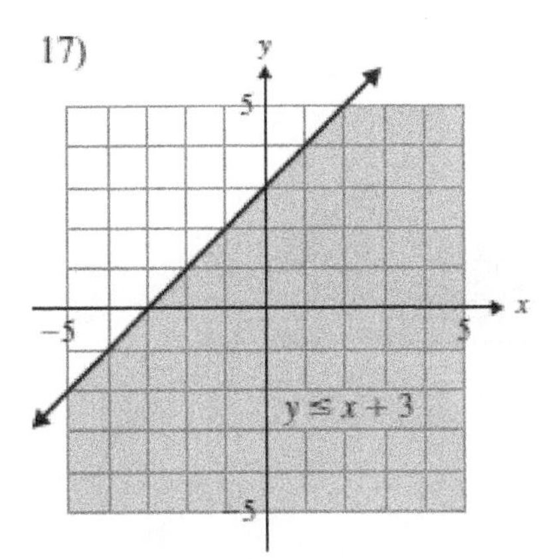

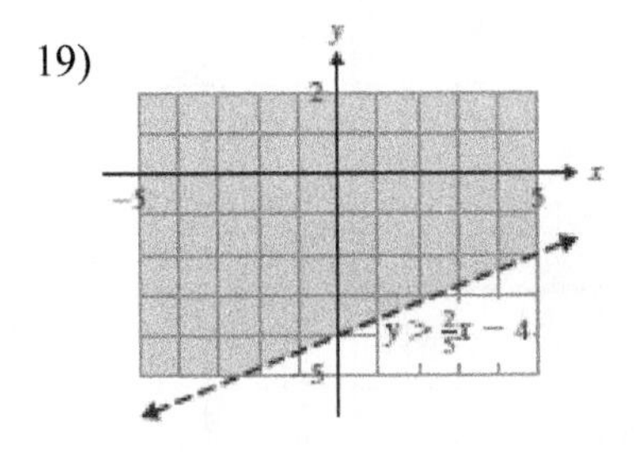

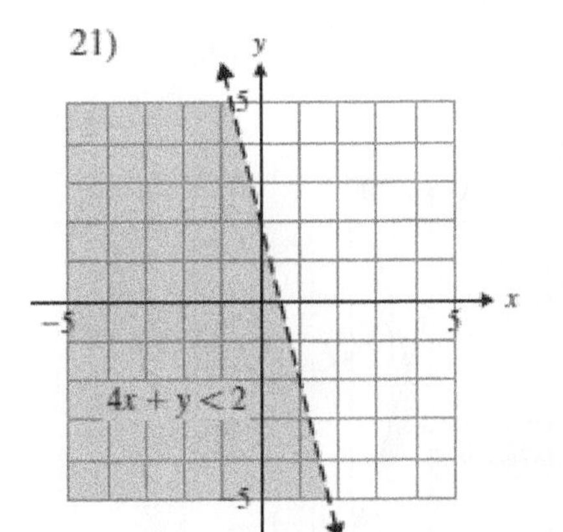

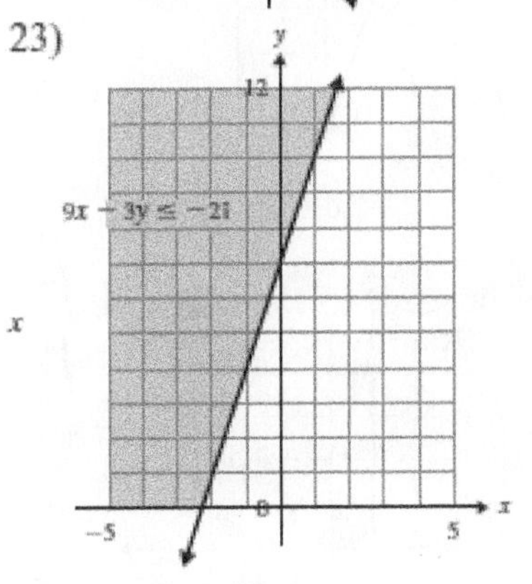

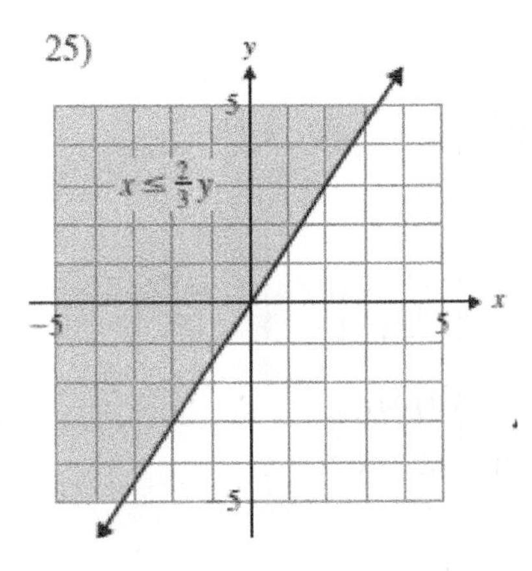

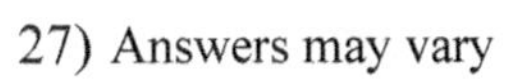
27) Answers may vary

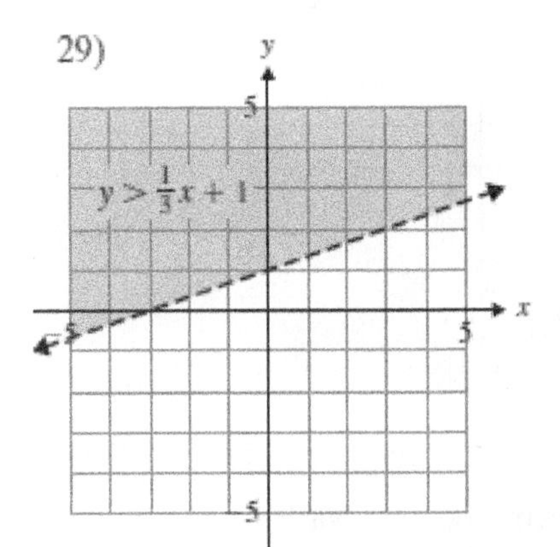

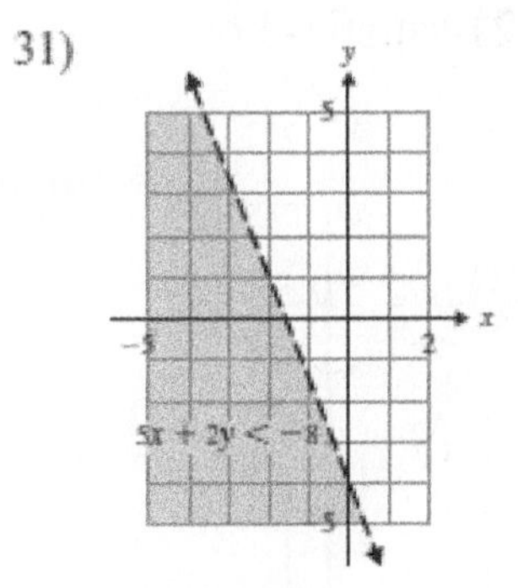

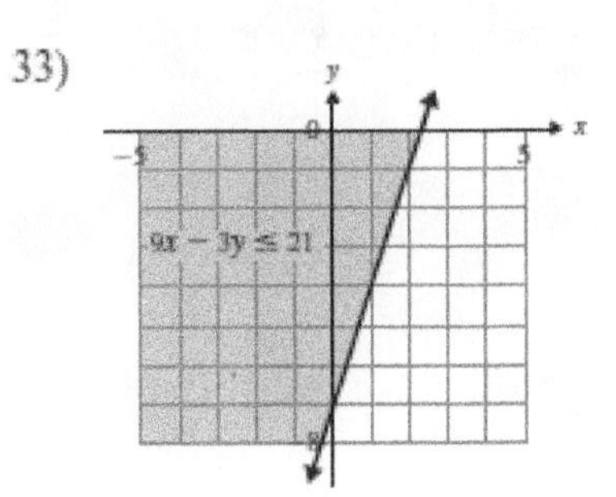

35)

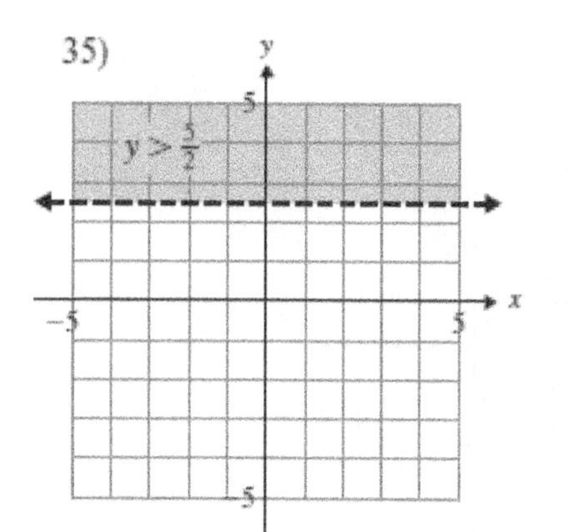

37)

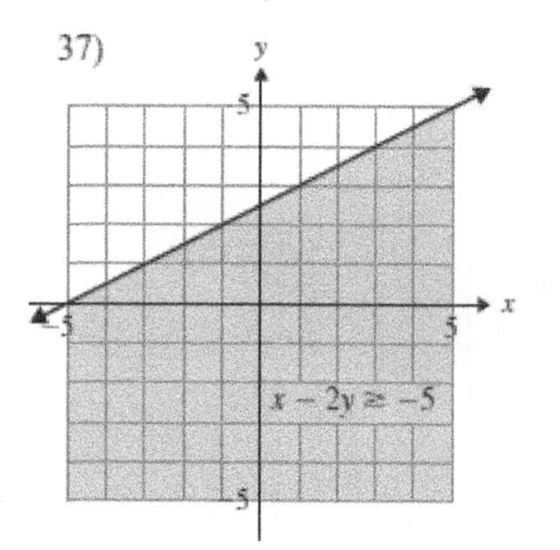

39-42) Answers may vary.

43) Yes; $(9,-2)$ satisfies both inequalities.

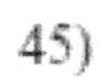

45)

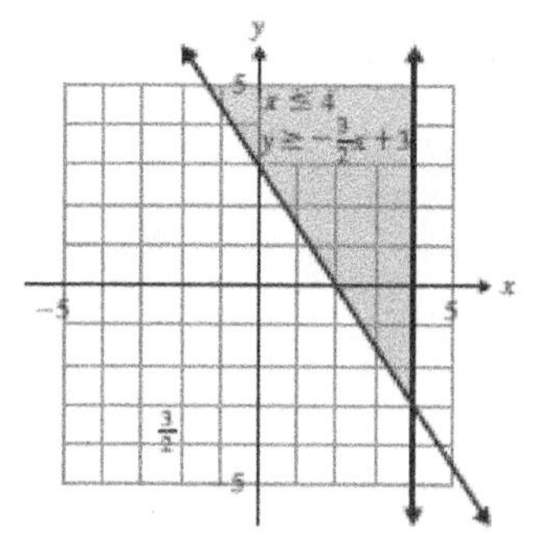

47)

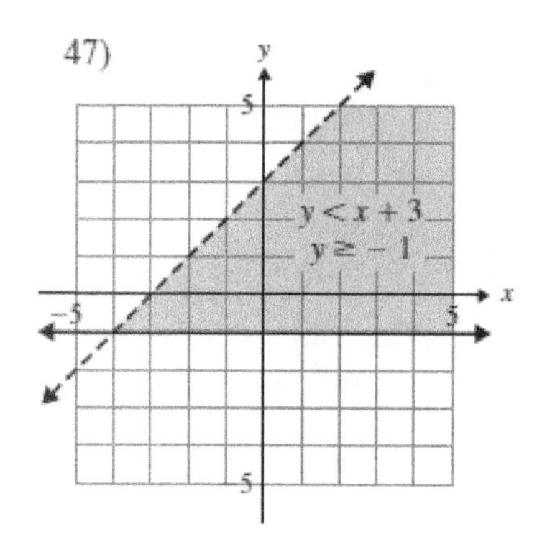

49)

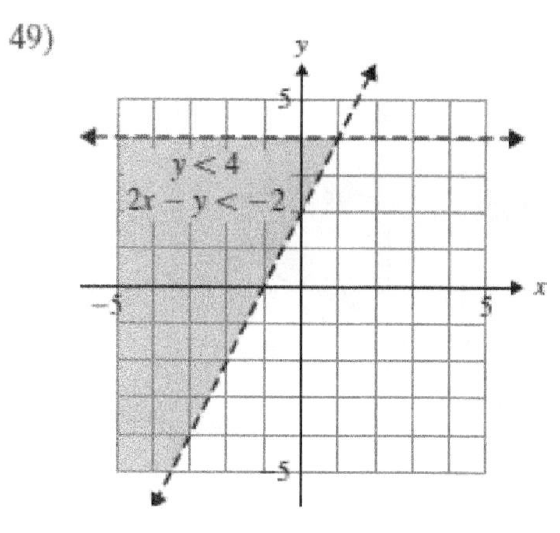

51)

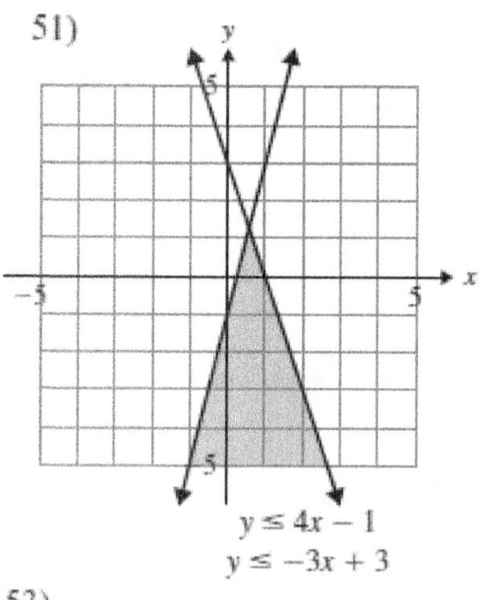

53)

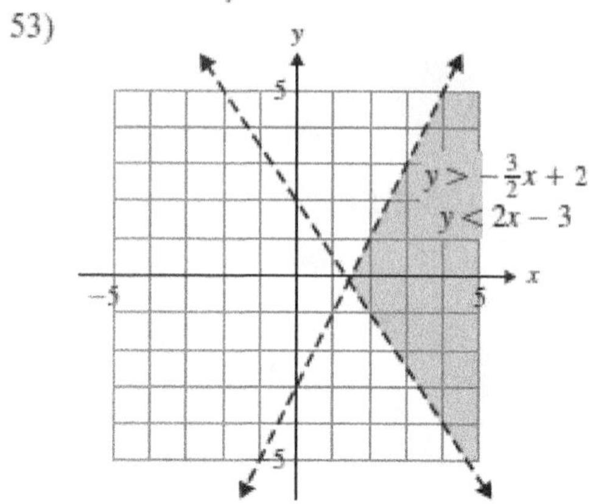

55)

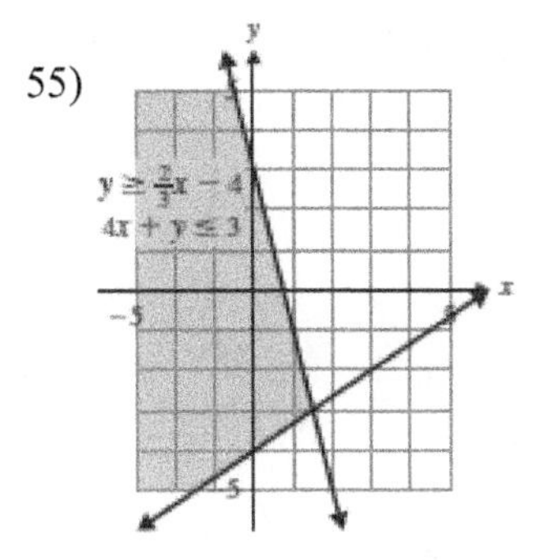

57)

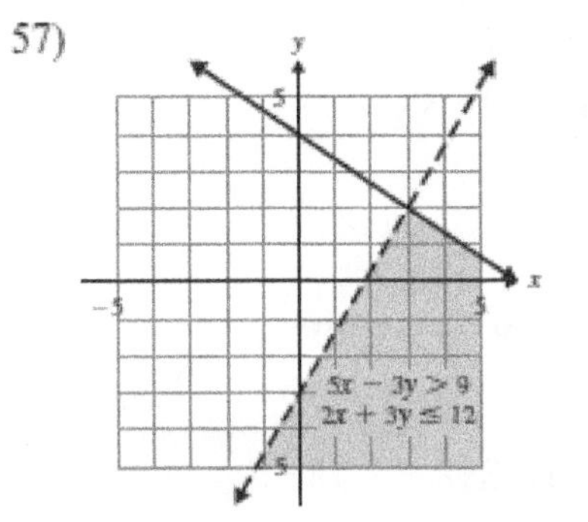

59)

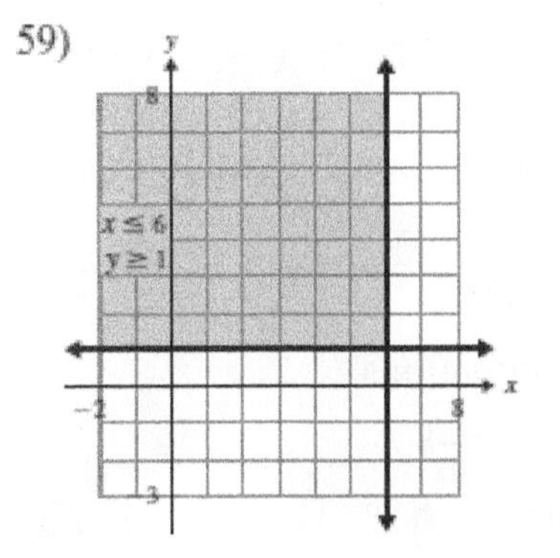

61)

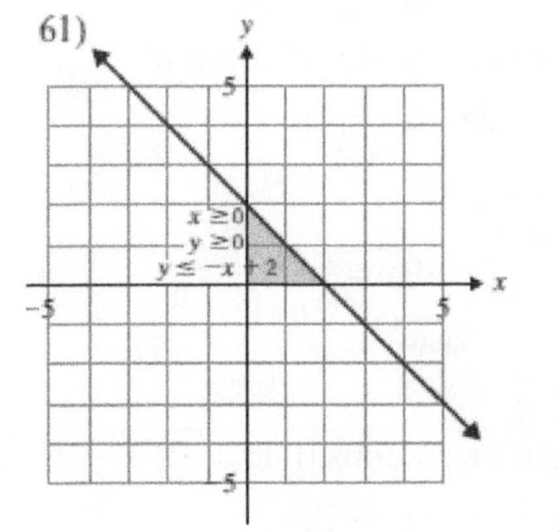

63)

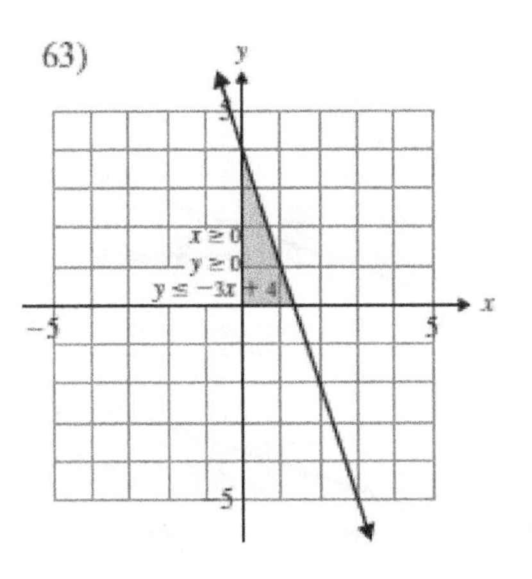

65)

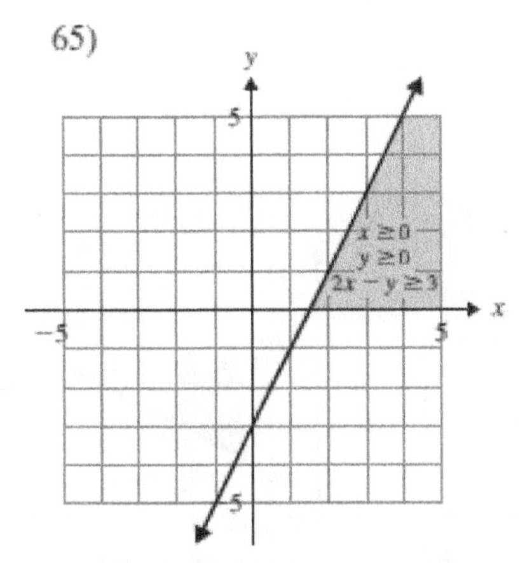

Chapter 5 Review

1) No

$$x-5y=13$$
$$-4-5(-5)=13$$
$$-4+25=13$$
$$21\neq 13$$

3) The lines are parallel

5) $\varnothing$

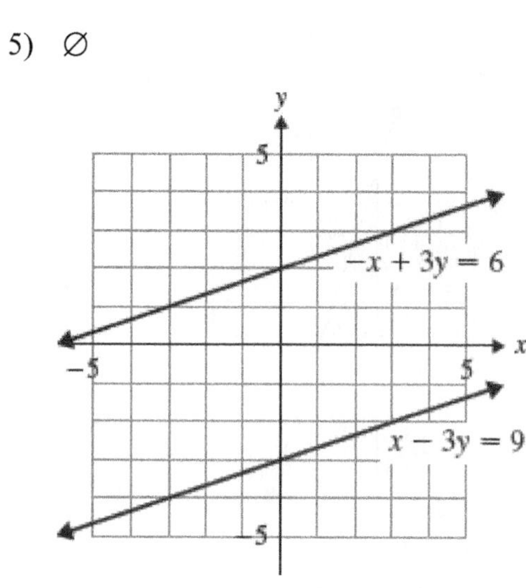

7) $(-3,-1)$

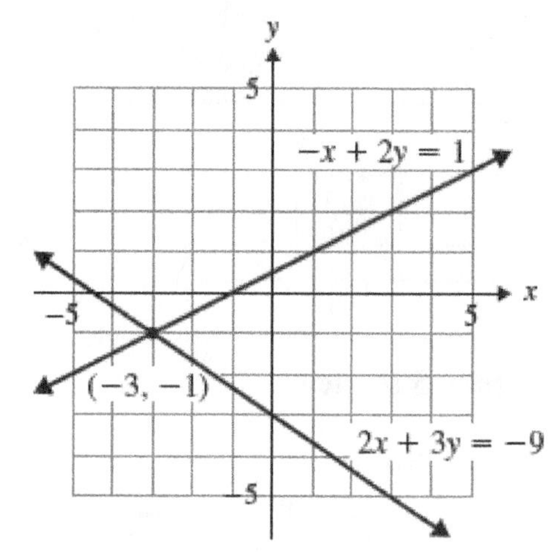

9) Infinite number of solutions

11) $9x-2y=8$

$y=2x+1$

$9x-2(2x+1)=8$

$9x-4x-2=8$

$5x=10$

$x=2$

$y=2x+1$

$y=2(2)+1$

$y=5$ $\qquad (2,5)$

13) $-x+8y=19$

$x=8y-19$

$4x-3y=11$

$4(8y-19)-3y=11$

$32y-3y-76=11$

$29y=87$

$y=3$

$x=8(3)-19$

$x=5$ $\qquad (5,3)$

15) Add the equations

$$\begin{array}{r} x-7y=3 \\ -x+5y=-1 \\ \hline -2y=2 \end{array}$$

$y=-1$

Substitute $y=-1$ into

$x-7y=3$

$x-7(-1)=3$

$x+7=3$

$x=-4$ $\qquad (-4,-1)$

17) Multiply the second equation by 2

$5x-2y=4$

$10x-4y=8$

Add the equations

$$\begin{array}{r} -10x+4y=-8 \\ 10x-8y=8 \\ \hline 0=0 \end{array}$$

There are infinite solutions of

$\{(x,y)\,|\,5x-2y=4\}$

19) $\boxed{\text{I}}$ $\qquad 2x+9y=-6$

$\boxed{\text{II}}$ $\qquad 5x+y=3$

Add equations $\boxed{\text{I}}$ and $\boxed{\text{II}}\cdot(-9)$

$$\begin{array}{r} -45x-9y=-27 \\ 2x+9y=-6 \\ \hline -43x=-33 \end{array}$$

$$x=\frac{33}{43}$$

Add equations $\text{I}\cdot(5)$ and $\text{II}\cdot(-2)$.

$$\begin{array}{r} 10x+45y=-30 \\ \underline{-10x-2y=-6} \\ 43y=-36 \\ y=-\dfrac{36}{43} \end{array} \qquad \left(\frac{33}{43},-\frac{36}{43}\right)$$

21) when one of the variables has a coefficient of 1 or -1

23) $6x+y=-8$

$y=-6x-8$

Substitute $y=-6x-8$ into

$$9x+7y=-1$$
$$9x+7(-6x-8)=-1$$
$$9x-42x-56=-1$$
$$-33x=55$$
$$x=-\frac{5}{3}$$

Substitute $x=-\dfrac{5}{3}$ into

$$9x+7y=-1$$
$$9\left(-\frac{5}{3}\right)+7y=-1$$
$$-15+7y=-1$$
$$7y=14$$
$$y=2 \qquad \left(-\frac{5}{3},2\right)$$

25)

$\boxed{\text{I}}$ $\quad \frac{1}{3}x-\frac{2}{9}y=-\frac{2}{3};\ 3x-2y=-6$

$\boxed{\text{II}}$ $\quad \frac{5}{12}x+\frac{1}{3}y=1;\ 5x+4y=12$

Add equations $\boxed{\text{I}}\cdot(36)$ and $\boxed{\text{II}}\cdot(24)$

$$\begin{array}{r} 12x-8y=-24 \\ \underline{10x+8y=\ \ 24} \\ 22x=0 \\ x=0 \end{array}$$

Substitute $x=0$ into

$$10x+8y=24$$
$$0+8y=24$$
$$8y=24$$
$$y=3 \qquad (0,3)$$

27) $6(2x-3)=y+4(x-3)$

$12x-18=y+4x-12$

$\boxed{\text{I}}$ $\quad 8x-y=6$

$5(3x+4)+4y=11-3y+27x$

$15x+20+4y=11-3y+27x$

$\boxed{\text{II}}$ $\ -12x+y=-9$

Add equations $\boxed{\text{I}}$ and $\boxed{\text{II}}$

$$\begin{array}{r} 8x-y=\ \ 6 \\ \underline{-12x+y=-9} \\ -4x=-3 \\ x=\dfrac{3}{4} \end{array}$$

Substitute $x=\dfrac{3}{4}$ into $\boxed{\text{I}}$

$$8\left(\frac{3}{4}\right)-y=6$$
$$6-y=6$$
$$y=0 \qquad \left(\frac{3}{4},0\right)$$

29) $\boxed{\text{I}}$ $\frac{3}{4}x-\frac{5}{4}y=\frac{7}{8}$

$4-2(x+5)-y=3(1-2y)+x$

$4-2x-10-y=3-6y+x$

$\boxed{\text{II}}$ $-3x+5y=9$

Add equations $\boxed{\text{I}}\cdot(8)$ and $\boxed{\text{II}}\cdot(2)$

$6x-10y=7$

$-6x+10y=18$

$0\neq 25$

$\varnothing$; There is no solution

31) $x=$ white milk; $y=$ chocolate milk

$\boxed{\text{I}}$ $x=2y$

$\boxed{\text{II}}$ $x+y=141$

Substitute $x=2y$ into $\boxed{\text{II}}$

$2y+y=141$

$3y=141$

$y=47$

Substitute $y=47$ into $\boxed{\text{I}}$

$x=2(47)$

$x=94$

white milk: 94; chocolate milk: 47

33) $x=$ Edwin's speed; $y=$ Camille's speed

$\boxed{\text{I}}$ $x=y+2$

$\boxed{\text{II}}$ $0.5x+0.5y=7$

Substitute $x=y+2$ into $\boxed{\text{II}}$

$0.5(y+2)+0.5y=7$

$y+1=7$

$y=6$

Substitute $y=6$ into $\boxed{\text{I}}$

$x=6+2$

$x=8$

Edwin's speed: 8 mph;

Camille's speed: 6 mph

35) $x=$ width; $y=$ length

$\boxed{\text{I}}$ $x=y-5$

$\boxed{\text{II}}$ $2x+2y=38$

Substitute $x=y-5$ into $\boxed{\text{II}}$

$2(y-5)+2y=38$

$4y-10=38$

$4y=48$

$y=12$

Substitute $y=12$ into $\boxed{\text{I}}$

$x=12-5$

$x=7$

Width: 7 cm; Length: 12 cm

37) $x=$ quarters; $y=$ dimes

$\boxed{\text{I}}$ $x+y=63$

$0.25x+0.10y=11.55$

$100(0.25x+0.10y)=100(11.55)$

$\boxed{\text{II}}$ $25x+10y=1155$

Add equations $\boxed{\text{I}}\cdot(-10)$ and $\boxed{\text{II}}$

$-10x-10y=-630$

$25x+10y=1155$

$15x=525$

$x=35$

Substitute $x=35$ into $\boxed{\text{I}}$

$35+y=63$

$y=28$

Quarters: 35; Dimes: 28

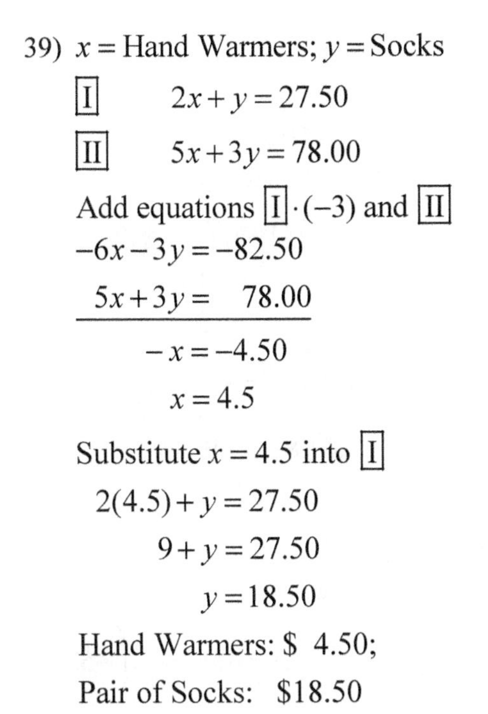

39) $x = \text{Hand Warmers}$; $y = \text{Socks}$

$\boxed{\text{I}} \quad 2x + y = 27.50$

$\boxed{\text{II}} \quad 5x + 3y = 78.00$

Add equations $\boxed{\text{I}} \cdot (-3)$ and $\boxed{\text{II}}$

$$\begin{array}{r} -6x - 3y = -82.50 \\ 5x + 3y = 78.00 \\ \hline -x = -4.50 \\ x = 4.5 \end{array}$$

Substitute $x = 4.5$ into $\boxed{\text{I}}$

$$2(4.5) + y = 27.50$$
$$9 + y = 27.50$$
$$y = 18.50$$

Hand Warmers: \$ 4.50;

Pair of Socks: \$18.50

41) Answers may vary.

43)

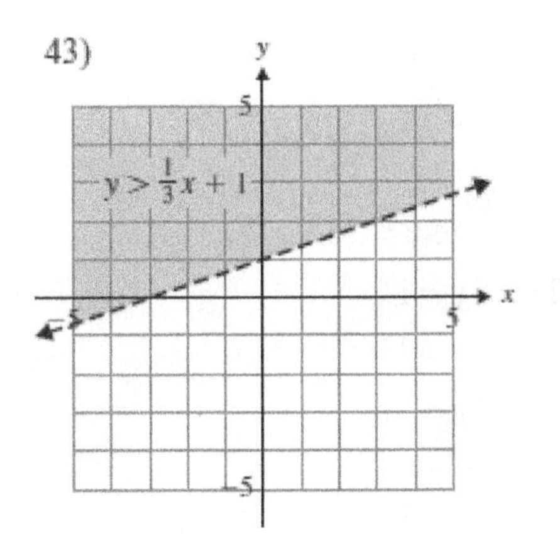

45)

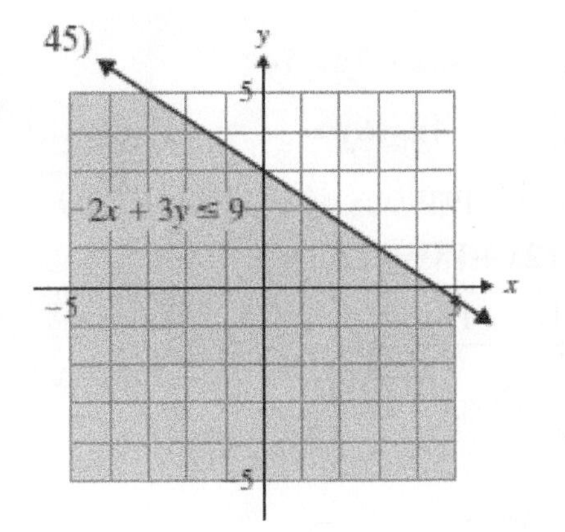

47) 49) 51) 53)

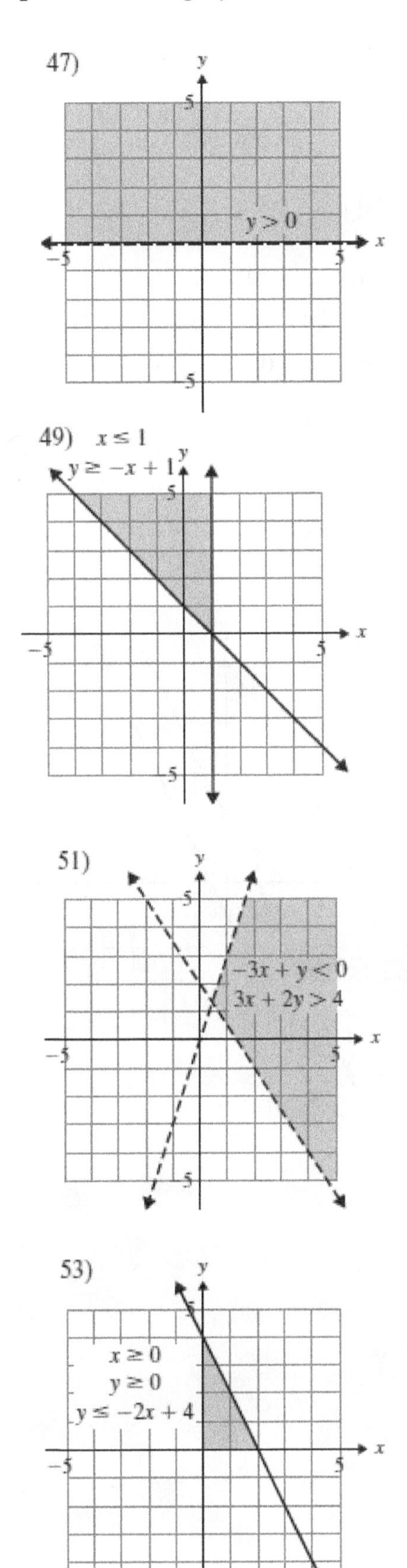

Chapter 5 Test

1) Yes

$$9x+5y=14$$
$$9\left(-\frac{2}{3}\right)+5(4)=14$$
$$-6+20=14$$
$$14=14$$
$$-6x-y=0$$
$$-6\left(-\frac{2}{3}\right)-4=0$$
$$4-4=0$$
$$0=0$$

3) $\varnothing$

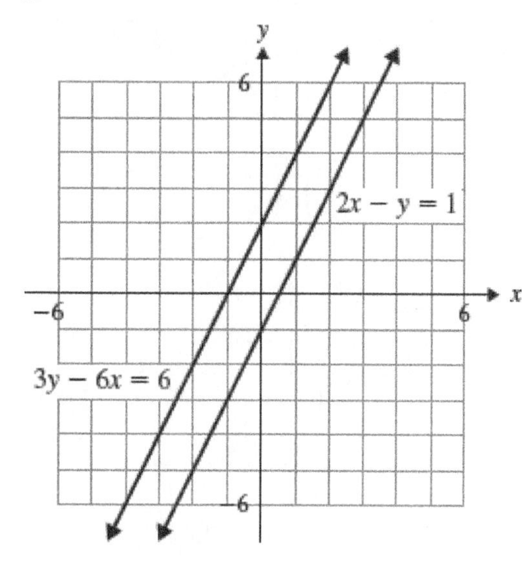

5) Substitute $x=-9-8y$ into

$$3x-10y=-10$$
$$3(-9-8y)-10y=-10$$
$$-27-24y-10y=-10$$
$$-34y=17$$
$$y=-\frac{1}{2}$$

Substitute $y=-\frac{1}{2}$ into

$$x+8y=-9$$
$$x+8\left(-\frac{1}{2}\right)=-9$$
$$x-4=-9$$
$$x=-5 \qquad \left(-5,-\frac{1}{2}\right)$$

7) Add the equations

$$2x+5y=11$$
$$\underline{7x-5y=16}$$
$$9x=27$$
$$x=3$$

Substitute $x=3$ into

$$2x+5y=11$$
$$2(3)+5y=11$$
$$6+5y=11$$
$$5y=5$$
$$y=1 \qquad (3,1)$$

9) $\boxed{\text{I}}$ $\quad -6x+9y=14$

$\boxed{\text{II}}$ $\quad 4x-6y=5$

Add Equations $\boxed{\text{I}}\cdot(2)$ and $\boxed{\text{II}}\cdot(3)$

$$-12x+18y=28$$
$$\underline{12x-18y=15}$$
$$0\neq 43$$

$\varnothing$

11)

$\boxed{\text{I}}$ $\frac{5}{8}x+\frac{1}{4}y=\frac{1}{4}$

$\boxed{\text{II}}$ $\frac{1}{3}x+\frac{1}{2}y=-\frac{4}{3}$

Add Equations $\boxed{\text{I}}\cdot(24)$ and $\boxed{\text{II}}\cdot(-12)$

$15x+6y=6$

$\underline{-4x-6y=16}$

$11x=22$

$x=2$

Substitute $x=2$ into $\boxed{\text{II}}\cdot(-12)$

$-4(2)-6y=16$

$-8-6y=16$

$-6y=24$

$y=-4$ $(2,-4)$

13) Answers may vary.

Solution is $(5,-1)$

Substitute $x=5$ and $y=-1$ into

$x+y$

$5-1=4$

$x+y=4$

$x-y$

$5-(-1)=5+1=6$

$x-y=6$

The system of equations:

$\boxed{\text{I}}$ $x+y=4$

$\boxed{\text{II}}$ $x-y=6$

15) $x=$ Adult Ticket; $y=$ Children's Ticket

$\boxed{\text{I}}$ $x+2y=85$

$\boxed{\text{II}}$ $2x+3y=150$

Add Equations $\boxed{\text{I}}\cdot(2)$ and $\boxed{\text{II}}\cdot(-1)$

$2x+4y=170$

$\underline{-2x-3y=-150}$

$y=20$

Substitute $y=20$ into I's

$x+2(20)=85$

$x=45$

Adult Ticket: \$45.00; Children Ticket: \$20.00

17) $x=$ 12% Solution; $y=$ 30% Solution

$\boxed{\text{I}}$ $x+y=72$

$0.12x+0.30y=0.2(70)$

$100(0.12x+0.30y)=(100)(0.2)(72)$

$\boxed{\text{II}}$ $12x+30y=1440$

Add Equations $\boxed{\text{I}}\cdot(12)$ and $\boxed{\text{II}}\cdot(-1)$

$12x+12y=864$

$-12x-30y=-1440$

$-18y=-576$

$y=32$

Substitute $y=32$ into $\boxed{\text{I}}$

$x+32=72$

$x=40$

12% Solution: 40 ml;

30% Solution: 32 ml

19)

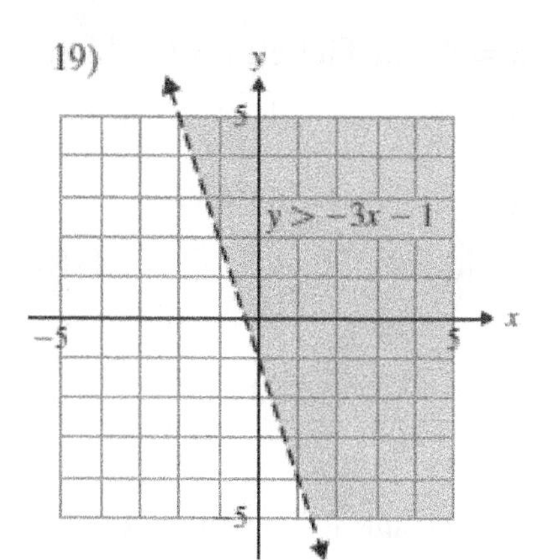

21)

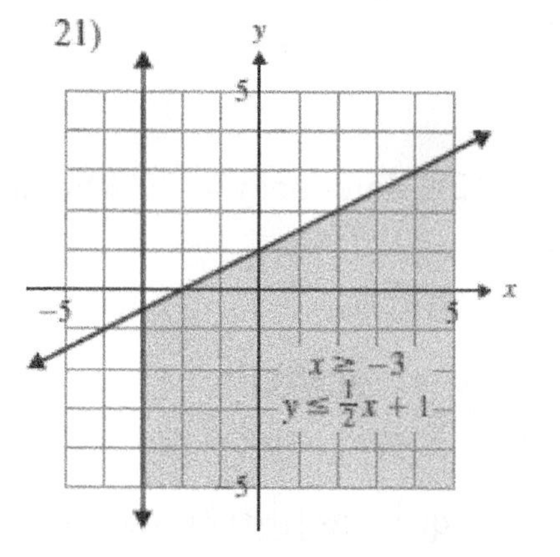

Cumulative Review: Chapters 1-5

1) $\dfrac{7}{15}+\dfrac{9}{10}=\dfrac{14}{30}+\dfrac{27}{30}=\dfrac{41}{30}$

3) $3(5-7)^3+18\div 6-8$

$=3(-2)^3+18\div 6-8$

$=3(-8)+18\div 6-8$

$=-24+3-8$

$=-29$

5) $-3(4x^2+5x-1)$

$=-12x^2-15x+3$

7) $9x^2\cdot 7x^{-6}$

$=63x^{-4}=\dfrac{63}{x^4}$

9) $0.0007319=7.319\cdot 10^{-4}$

11) $11-3(2k-1)=2(6-k)$

$11-6k+3=12-2k$

$-4k=-2$

$k=\dfrac{1}{2}$

13) $x=$ Gas Milege for the old car

$y=$ Gas Milege for the new car

$y=1.61x$

Substitute the value for y as 25.4

$x=\dfrac{25.4}{1.61}$

$x\approx 15.8$ mpg

15)

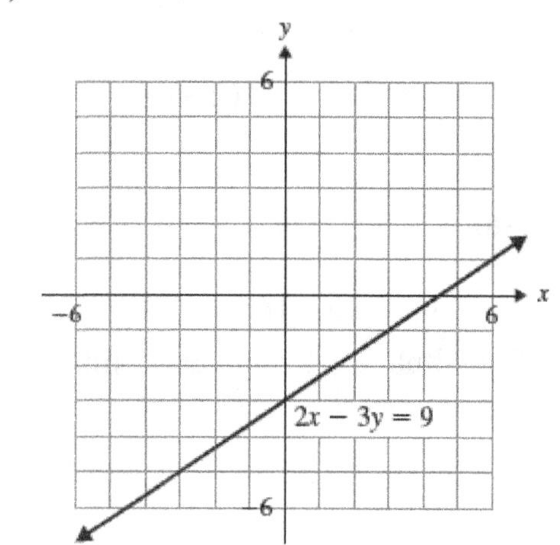

17) Slope, $m=\dfrac{-1-2}{-9-3}$

$m=\dfrac{-3}{-12}$

$m=\dfrac{1}{4}$

$y-y_1=m(x-x_1)$

$y-2=\dfrac{1}{4}(x-3)$

$y=\dfrac{1}{4}x-\dfrac{3}{4}+2$

$y=\dfrac{1}{4}x+\dfrac{5}{4}$

19)
$9x-3y=6$

$\boxed{I}$ $y=3x-2$

$\boxed{II}$ $3x-2y=-8$

Substitute $\boxed{I}$ into $\boxed{II}$

$3x-2(3x-2)=-8$

$3x-6x+4=-8$

$-3x=-12$

$x=4$

$y=3(4)-2$

$y=12-2$

$y=10$ (4,10)

21) I $x+2y=4$

II $-3x-6y=6$

Add Equations $I\cdot(3)$ and II

$3x+6y=12$

$\underline{-3x-6y=6}$

$0\neq 18$

$\varnothing$

23) $\boxed{I}$ $y=4x+1$

$\boxed{II}$ $2x-y=3$

Substitute Equation $\boxed{I}$ into $\boxed{II}$

$2x-(4x+1)=3$

$2x-4x-1=3$

$-2x-1=3$

$-2x=4$

$x=-2$

Substitute $x=-2$ into $\boxed{I}$

$y=4(-2)+1$

$y=-7$ (−2,−7)

25)

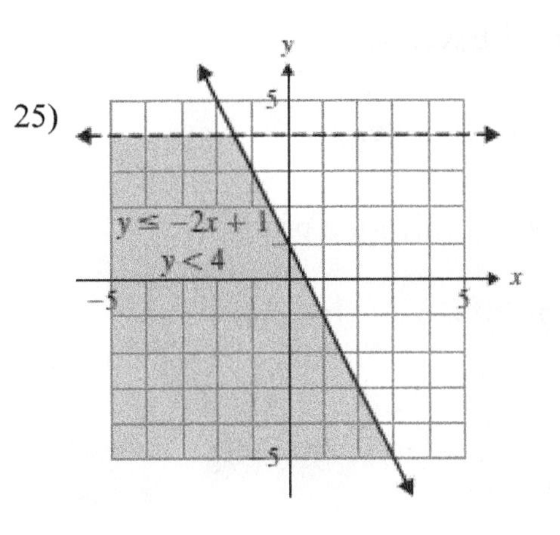

Chapter 6: Polynomials

Section 6.1 Exercises

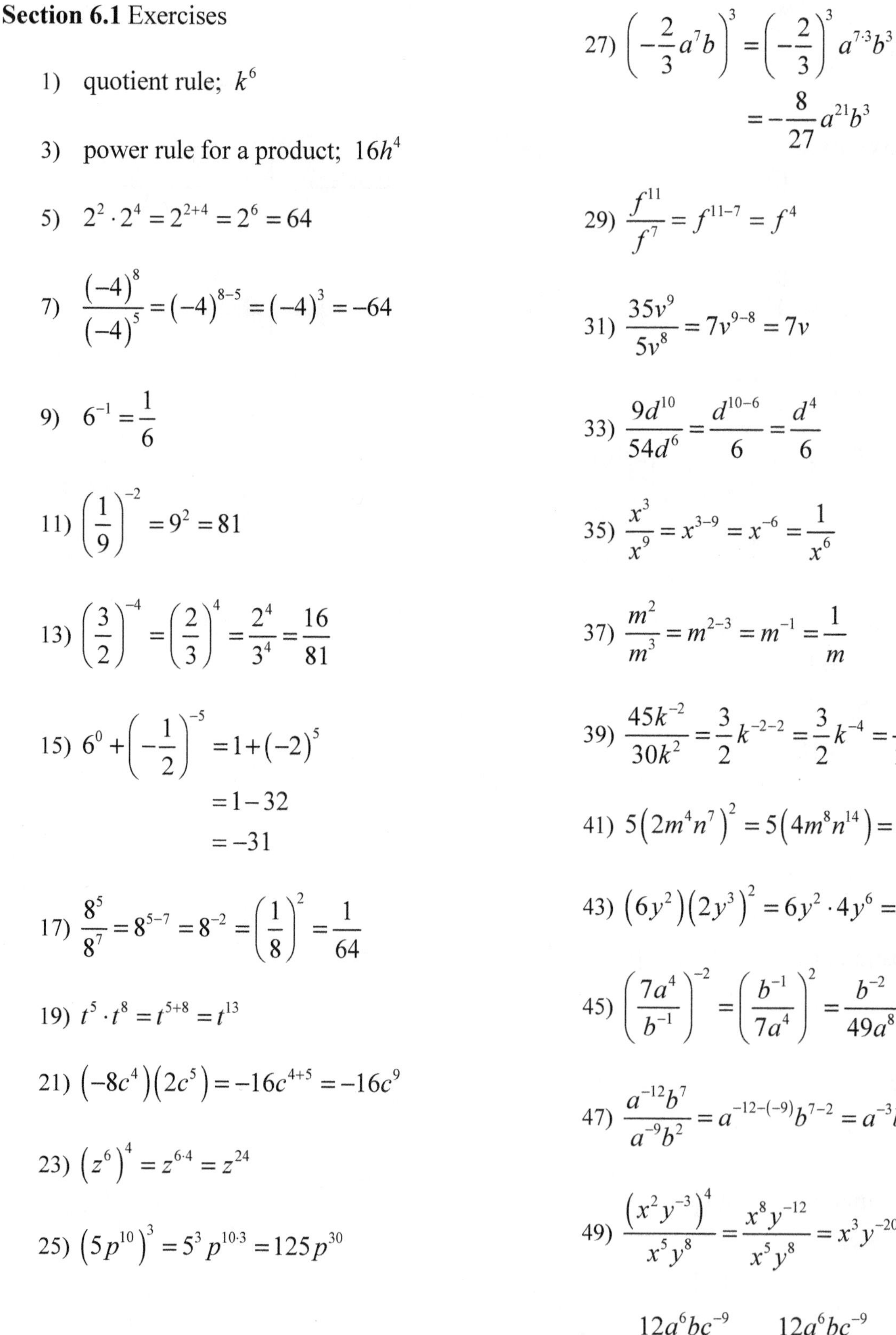

1) quotient rule; k^6

3) power rule for a product; $16h^4$

5) $2^2 \cdot 2^4 = 2^{2+4} = 2^6 = 64$

7) $\dfrac{(-4)^8}{(-4)^5} = (-4)^{8-5} = (-4)^3 = -64$

9) $6^{-1} = \dfrac{1}{6}$

11) $\left(\dfrac{1}{9}\right)^{-2} = 9^2 = 81$

13) $\left(\dfrac{3}{2}\right)^{-4} = \left(\dfrac{2}{3}\right)^4 = \dfrac{2^4}{3^4} = \dfrac{16}{81}$

15) $6^0 + \left(-\dfrac{1}{2}\right)^{-5} = 1 + (-2)^5$
$= 1 - 32$
$= -31$

17) $\dfrac{8^5}{8^7} = 8^{5-7} = 8^{-2} = \left(\dfrac{1}{8}\right)^2 = \dfrac{1}{64}$

19) $t^5 \cdot t^8 = t^{5+8} = t^{13}$

21) $(-8c^4)(2c^5) = -16c^{4+5} = -16c^9$

23) $(z^6)^4 = z^{6\cdot 4} = z^{24}$

25) $(5p^{10})^3 = 5^3 p^{10\cdot 3} = 125p^{30}$

27) $\left(-\dfrac{2}{3}a^7 b\right)^3 = \left(-\dfrac{2}{3}\right)^3 a^{7\cdot 3}b^3$
$= -\dfrac{8}{27}a^{21}b^3$

29) $\dfrac{f^{11}}{f^7} = f^{11-7} = f^4$

31) $\dfrac{35v^9}{5v^8} = 7v^{9-8} = 7v$

33) $\dfrac{9d^{10}}{54d^6} = \dfrac{d^{10-6}}{6} = \dfrac{d^4}{6}$

35) $\dfrac{x^3}{x^9} = x^{3-9} = x^{-6} = \dfrac{1}{x^6}$

37) $\dfrac{m^2}{m^3} = m^{2-3} = m^{-1} = \dfrac{1}{m}$

39) $\dfrac{45k^{-2}}{30k^2} = \dfrac{3}{2}k^{-2-2} = \dfrac{3}{2}k^{-4} = \dfrac{3}{2k^4}$

41) $5(2m^4n^7)^2 = 5(4m^8n^{14}) = 20m^8n^{14}$

43) $(6y^2)(2y^3)^2 = 6y^2 \cdot 4y^6 = 24y^8$

45) $\left(\dfrac{7a^4}{b^{-1}}\right)^{-2} = \left(\dfrac{b^{-1}}{7a^4}\right)^2 = \dfrac{b^{-2}}{49a^8} = \dfrac{1}{49a^8b^2}$

47) $\dfrac{a^{-12}b^7}{a^{-9}b^2} = a^{-12-(-9)}b^{7-2} = a^{-3}b^5 = \dfrac{b^5}{a^3}$

49) $\dfrac{(x^2y^{-3})^4}{x^5y^8} = \dfrac{x^8y^{-12}}{x^5y^8} = x^3y^{-20} = \dfrac{x^3}{y^{20}}$

51) $\dfrac{12a^6bc^{-9}}{(3a^2b^{-7}c^4)^2} = \dfrac{12a^6bc^{-9}}{9a^4b^{-14}c^8} = \dfrac{4a^2b^{15}c^{-17}}{3}$
$= \dfrac{4a^2b^{15}}{3c^{17}}$

53) $\left(xy^{-3}\right)^{-5} = x^{-5}y^{15} = \dfrac{y^{15}}{x^5}$

55) $\left(\dfrac{a^2b}{4c^2}\right)^{-3} = \left(\dfrac{4c^2}{a^2b}\right)^3 = \dfrac{64c^6}{a^6b^3}$

57) $\left(\dfrac{7h^{-1}k^9}{21h^{-5}k^5}\right)^{-2} = \left(\dfrac{21h^{-5}k^5}{7h^{-1}k^9}\right)^2$

$= \left(3h^{-5-(-1)}k^{5-9}\right)^2$

$= \left(3h^{-4}k^{-4}\right)^2$

$= 9h^{-8}k^{-8}$

$= \dfrac{9}{h^8k^8}$

59) $\left(\dfrac{15cd^{-4}}{5c^3d^{-10}}\right)^{-3} = \left(\dfrac{5c^3d^{-10}}{15cd^{-4}}\right)^3$

$= \left(\dfrac{1}{3}c^{3-1}d^{-10-(-4)}\right)^3$

$= \left(\dfrac{1}{3}c^2d^{-6}\right)^3$

$= \dfrac{1}{27}c^6d^{-18} = \dfrac{c^6}{27d^{18}}$

61) $\dfrac{\left(2u^{-5}v^2w^4\right)^{-5}}{\left(u^6v^{-7}w^{-10}\right)^2}$

$= \dfrac{1}{\left(u^6v^{-7}w^{-10}\right)^2\left(2u^{-5}v^2w^4\right)^5}$

$= \dfrac{1}{u^{12}v^{-14}w^{-20}2^5u^{-25}v^{10}w^{20}}$

$= \dfrac{1}{2^5u^{-13}v^{-4}w^0}$

$= \dfrac{u^{13}v^4}{32}$

63) $A = 5x \cdot 2x = 10x^2$ sq. units

$P = 2(5x) + 2(2x) = 10x + 4x$

$= 14x$ units

65) $A = \dfrac{1}{4}p \cdot \dfrac{3}{4}p = \dfrac{3}{16}p^2$ sq. units

$P = 2\left(\dfrac{1}{4}p\right) + 2\left(\dfrac{3}{4}p\right)$

$= \dfrac{1}{2}p + \dfrac{3}{2}p$

$= 2p$ units

67) $k^{4a} \cdot k^{2a} = k^{4a+2a} = k^{6a}$

69) $\left(g^{2x}\right)^4 = g^{2x\cdot 4} = g^{8x}$

71) $\dfrac{x^{7b}}{x^{4b}} = x^{7b-4b} = x^{3b}$

73) $\left(2r^{6m}\right)^{-3} = \dfrac{1}{\left(2r^{6m}\right)^3} = \dfrac{1}{8r^{18m}}$

Section 6.2 Exercises

1) Yes; The coefficients are real numbers and the exponents are whole numbers.

3) No; One of the exponents is a negative number.

5) No; Two of the exponents are fractions.

7) binomial

9) trinomial

11) monomial

13) It is the same as the degree of the term in the polynomial with the highest degree.

15) Add the exponents on the variables.

17) Degree of polynomial = 4

Term	Coeff	Degree
$3y^4$	3	4
$7y^3$	7	3
$-2y$	−2	1
8	8	0

19) Degree of polynomial = 5

Term	Coeff	Degree
$-4x^2y^3$	−4	5
$-x^2y^2$	−1	4
$\frac{2}{3}xy$	$\frac{2}{3}$	2
$5y$	5	1

21) a) $2r^2-7r+4,\ r=3$

$2(3)^2-7(3)+4=18-21+4=1$

b) $2r^2-7r+4,\ r=-1$

$2(-1)^2-7(-1)+4=2+7+4=13$

23) $9x+4y,\ x=5; y=-2$

$9(5)+4(-2)=45-8=37$

25) $x^2y^2-5xy+2y,\ x=5; y=-2$

$(5)^2(-2)^2-5(5)(-2)+2(-2)$

$=25\cdot4+40-4$

$=100+50+10$

$=146$

27) $\frac{1}{2}xy-4x-y,\ x=5; y=-2$

$\frac{1}{2}(5)(-2)-4(5)-(-2)$

$=-5-20+2$

$=-23$

29) a) $y=60x+380,\ x=5$

$y=60(5)+380=300+380=680$

If he rents the equipment for 5 hours, the cost of building will be \$680.00

b) $y=60x+380,\ x=9$

$y=60(9)+380=540+380=920$

If he keeps the equipment for 9 hours, the cost of building will be \$920.00

c) $y=60x+380,\ y=860.00$

$860=60x+380$

$480=60x$

$8=x$

If the cost of road is \$860.00, then he needed to rent the equipment for 8 hours.

31) $-6z+8z+11z=13z$

33) $5c^2+9c-16c^2+c-3c=-11c^2+7c$

35) $6.7t^2-9.1t^6-2.5t^2+4.8t^6=-4.3t^6+4.2t^2$

37) $7a^2b^2+4ab^2-16ab^2-a^2b^2+5ab^2$

$=6a^2b^2-7ab^2$

39) $$\begin{array}{r} 5n-8 \\ +\ \underline{4n+3} \\ 9n-5 \end{array}$$

41) $$\begin{array}{r} -7a^3+11a \\ +\ \underline{2a^3-4a} \\ -5a^3+7a \end{array}$$

43) $$\begin{array}{r} 9r^2+16r+2 \\ +\ \underline{3r^2-10r+9} \\ 12r^2+6r+11 \end{array}$$

45) $$\begin{array}{r} b^2-8b-14 \\ +\ \underline{3b^2+8b+11} \\ 4b^2-3 \end{array}$$

47) $$\begin{array}{r} \frac{5}{6}w^4-\frac{2}{3}w^2\qquad\quad+\frac{1}{2} \\ +\ \underline{-\frac{4}{9}w^4+\frac{1}{6}w^2-\frac{3}{8}w-2} \\ \frac{7}{18}w^4-\frac{1}{2}w^2-\frac{3}{8}w-\frac{3}{2} \end{array}$$

49) $(6m^2-5m+10)+(-4m^2+8m+9)$
$=2m^2+3m+19$

51) $\left(-2c^4-\frac{7}{10}c^3+\frac{3}{4}c-\frac{2}{9}\right)+\left(12c^4+\frac{1}{2}c^3-c+3\right)$

$=10c^4-\frac{1}{5}c^3-\frac{1}{4}c+\frac{25}{9}$

53) $(2.7d^3+5.6d^2-7d+3.1)$
$+(-1.5d^3+2.1d^2-4.3d-2.5)$
$=1.2d^3+7.7d^2-11.3d+0.6$

55) $$\begin{array}{r} 15w+7 \\ -\ \underline{3w+11} \end{array} \qquad \begin{array}{r} 15w+7 \\ +\ \underline{-3w-11} \\ 12w-4 \end{array}$$

57) $$\begin{array}{r} y-6 \\ -\ \underline{2y-8} \end{array} \qquad \begin{array}{r} y-6 \\ +\ \underline{-2y+8} \\ -y+2 \end{array}$$

59) $$\begin{array}{r} 3b^2-8b+12 \\ -\ \underline{5b^2+2b-7} \end{array}$$

$$\begin{array}{r} 3b^2-8b+12 \\ +\ \underline{-5b^2-2b+7} \\ -2b^2-10b+19 \end{array}$$

61) $$\begin{array}{r} f^4-6f^3+5f^2-8f+13 \\ -\ \underline{-3f^4+8f^3-f^2\qquad+4} \end{array}$$

$$\begin{array}{r} f^4-6f^3+5f^2-8f+13 \\ +\ \underline{3f^4-8f^3+f^2\qquad-4} \\ 4f^4-14f^3+6f^2-8f+9 \end{array}$$

63) $$\begin{array}{r} 10.7r^2+1.2r+9 \\ -\ \underline{4.9r^2-5.3r-2.8} \end{array}$$

$$\begin{array}{r} 10.7r^2+1.2r+9 \\ +\ \underline{-4.9r^2+5.3r+2.8} \\ 5.8r^2+6.5r+11.8 \end{array}$$

65) $(j^2+16j)-(-6j^2+7j+5)$
$=(j^2+16j)+(6j^2-7j-5)$
$=7j^2+9j-5$

67) $(17s^5-12s^2)-(9s^5+4s^4-8s^2-1)$
$=(17s^5-12s^2)+(-9s^5-4s^4+8s^2+1)$
$=8s^5-4s^4-4s^2+1$

69) $\left(-\frac{3}{8}r^2+\frac{2}{9}r+\frac{1}{3}\right)-\left(-\frac{7}{16}r^2-\frac{5}{9}r+\frac{7}{6}\right)=\left(-\frac{6}{16}r^2+\frac{2}{9}r+\frac{2}{6}\right)+\left(\frac{7}{16}r^2+\frac{5}{9}r-\frac{7}{6}\right)=\frac{1}{16}r^2+\frac{7}{9}r-\frac{5}{6}$

71) Answers may vary.

73) No. If the coefficients of the like terms are opposite in sign, their sum will be zero.
Example: $(3x^2+4x+5)+(2x^2-4x+1)=5x^2+6$

75) $(8a^4-9a^2+17)-(15a^4+3a^2+3)=(8a^4-9a^2+17)+(-15a^4-3a^2-3)=-7a^4-12a^2+14$

77) $(-11n^2-8n+21)+(4n^2+15n-3)+(7n^2-10)=7n+8$

79) $(w^3+5w^2+3)-(6w^3-2w^2+w+12)+(9w^3+7)=(w^3+5w^2+3)+(-6w^3+2w^2-w-12)+(9w^3+7)$
$=4w^3+7w^2-w-2$

81) $\left(y^3-\frac{3}{4}y^2-5y+\frac{3}{7}\right)+\left(\frac{1}{3}y^3-y^2+8y-\frac{1}{2}\right)=\left(\frac{3}{3}y^3-\frac{3}{4}y^2-5y+\frac{6}{14}\right)+\left(\frac{1}{3}y^3-\frac{4}{4}y^2+8y-\frac{7}{14}\right)$
$=\frac{4}{3}y^3-\frac{7}{4}y^2+3y-\frac{1}{14}$

83) $(3m^3-5m^2+m+12)-[(7m^3+4m^2-m+11)+(-5m^3-2m^2+6m+8)]$
$=(3m^3-5m^2+m+12)-(2m^3+2m^2+5m+19)=(3m^3-5m^2+m+12)+(-2m^3-2m^2-5m-19)$
$=m^3-7m^2-4m-7$

85) $(p^2-7)+(8p^2+2p-1)=9p^2+2p-8$

87) $(6z^6+z^2+9)-(z^6-8z^2+13)=(6z^6+z^2+9)+(-z^6+8z^2-13)=5z^6+9z^2-4$

89) $(2p^2+p+5)-[(6p^2+1)+(3p^2-8p+5)]=(2p^2+p+5)-[9p^2-8p+5]$
$$=(2p^2+p+5)+(-9p^2+8p-5)$$
$$=-7p^2+9p$$

91) $(5w+17z)-(w+3z)=(5w+17z)+(-w-3z)=4w+14z$

93) $(ac+8a+6c)+(-6ac+4a-c)=-5ac+12a+5c$

95) $(-6u^2v^2+11uv+14)-(-10u^2v^2-20uv+18)=(-6u^2v^2+11uv+14)+(10u^2v^2+20uv-18)$
$$=4u^2v^2+31uv-4$$

97) $(12x^3y^2-5x^2y^2+9x^2y-17)+(5x^3y^2+x^2y-1)-(6x^2y^2+10x^2y+2)$
$$=(17x^3y^2-5x^2y^2+10x^2y-18)+(-6x^2y^2-10x^2y-2)$$
$$=17x^3y^2-11x^2y^2-20$$

99) $P=2(2x+7)+2(x-4)$
$$=4x+14+2x-8$$
$$=6x+6 \text{ units}$$

101) $P=2(5p^2-2p+3)+2(p-6)=10p^2-4p+6+2p-12=10p^2-2p-6$ units

103) a) $f(-3)=5(-3)^2+7(-3)-8$
$f(-3)=5(9)+7(-3)-8$
$f(-3)=45-21-8$
$f(-3)=16$

b) $f(1)=5(1)^2+7(1)-8$
$f(1)=5(1)+7(1)-8$
$f(1)=5+7-8$
$f(1)=4$

105) a) $P(3)=(3)^3-2(3)^2+5(3)+8$
$P(3)=27-2(9)+5(3)+8$
$P(3)=27-18+15+8$
$P(3)=32$

b) $P(0)=(0)^3-2(0)^2+5(0)+8$
$P(0)=0-2(0)+5(0)+8$
$P(0)=0-0+0+8$
$P(0)=8$

107)
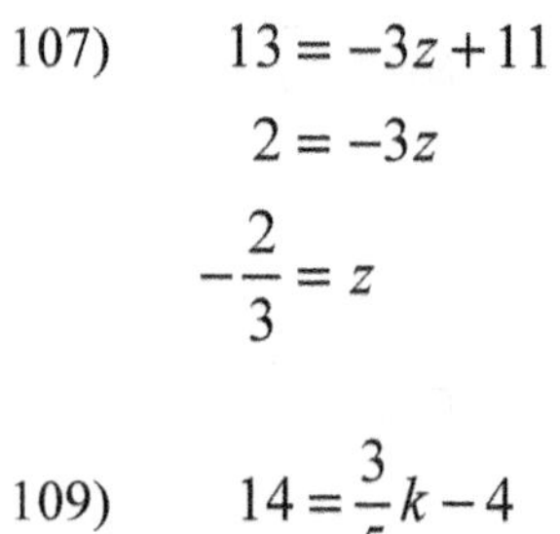

$$13 = -3z + 11$$
$$2 = -3z$$
$$-\frac{2}{3} = z$$

109)
$$14 = \frac{3}{5}k - 4$$
$$18 = \frac{3}{5}k$$
$$\frac{90}{3} = k$$
$$30 = k$$

111)
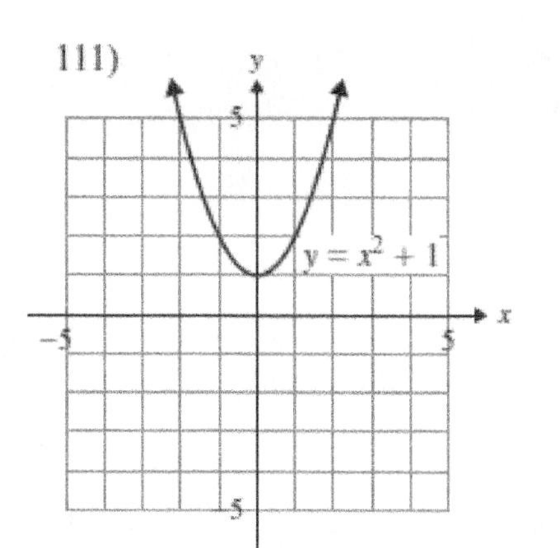

113)
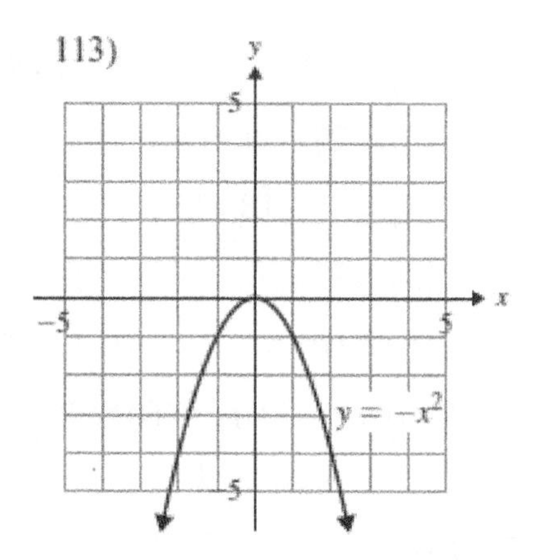

115)
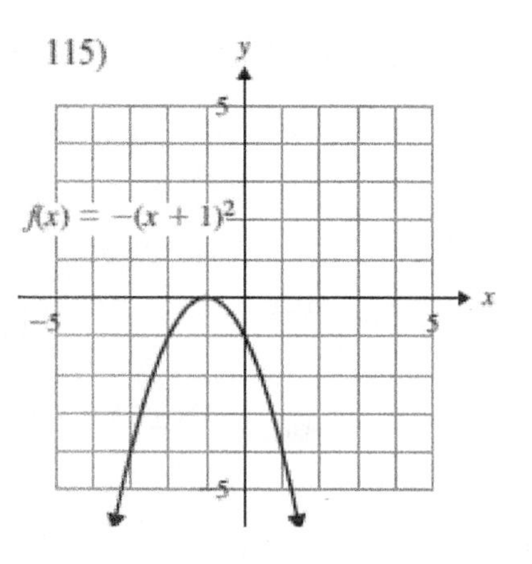

117)
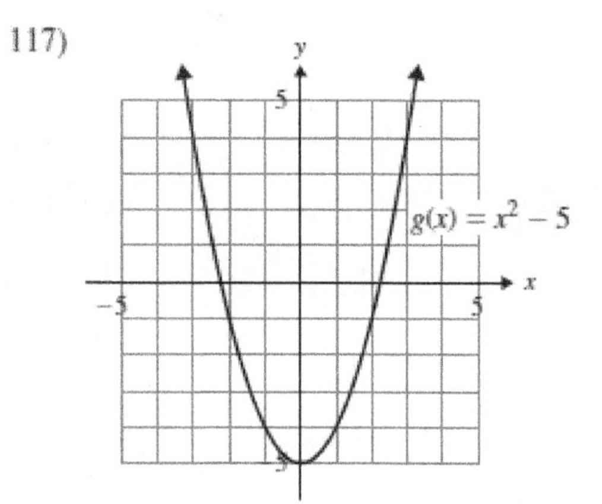

119)
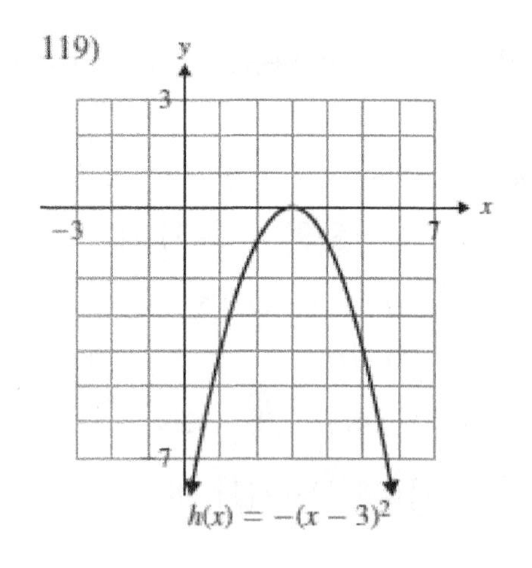

Section 6.3 Exercises

1) Answers may vary.

3) $\left(3m^5\right)\left(8m^3\right) = 24m^8$

5) $(-8c)(4c^5) = -32c^6$

7) $5a(2a-7) = (5a)(2a) + (5a)(-7) = 10a^2 - 35a$

9) $-6c(7c+2) = (-6c)(7c) + (-6c)(2) = -42c^2 - 12c$

11) $6v^3(v^2 - 4v - 2) = (6v^3)(v^2) + (6v^3)(-4v) + (6v^3)(-2) = 6v^5 - 24v^4 - 12v^3$

13) $-9b^2(4b^3 - 2b^2 - 6b - 9) = (-9b^2)(4b^3) + (-9b^2)(-2b^2) + (-9b^2)(-6b) + (-9b^2)(-9)$
$= -36b^5 + 18b^4 + 54b^3 + 81b^2$

15) $3a^2b(ab^2 + 6ab - 13b + 7) = (3a^2b)(ab^2) + (3a^2b)(6ab) + (3a^2b)(-13b) + (3a^2b)(7)$
$= 3a^3b^3 + 18a^3b^2 - 39a^2b^2 + 21a^2b$

17) $-\frac{3}{5}k^4(15k^2 + 20k - 3) = \left(-\frac{3}{5}k^4\right)(15k^2) + \left(-\frac{3}{5}k^4\right)(20k) + \left(-\frac{3}{5}k^4\right)(-3) = -9k^6 - 12k^5 + \frac{9}{5}k^4$

19) $(c+4)(6c^2 - 13c + 7) = (c)(6c^2) + (c)(-13c) + (c)(7) + (4)(6c^2) + (4)(-13c) + (4)(7)$
$= 6c^3 - 13c^2 + 7c + 24c^2 - 52c + 28 = 6c^3 + 11c^2 - 45c + 28$

21) $(f-5)(3f^2 + 2f - 4) = (f)(3f^2) + (f)(2f) + (f)(-4) + (-5)(3f^2) + (-5)(2f) + (-5)(-4)$
$= 3f^3 + 2f^2 - 4f - 15f^2 - 10f + 20 = 3f^3 - 13f^2 - 14f + 20$

23) $(4x^3 - x^2 + 6x + 2)(2x - 5)$
$= (4x^3)(2x) + (4x^3)(-5) + (-x^2)(2x) + (-x^2)(-5) + (6x)(2x) + (6x)(-5) + (2)(2x) + (2)(-5)$
$= 8x^4 - 20x^3 - 2x^3 + 5x^2 + 12x^2 - 30x + 4x - 10 = 8x^4 - 22x^3 + 17x^2 - 26x - 10$

25) $\left(\frac{1}{3}y^2 + 4\right)(12y^2 + 7y - 9)$
$= \left(\frac{1}{3}y^2\right)(12y^2) + \left(\frac{1}{3}y^2\right)(7y) + \left(\frac{1}{3}y^2\right)(-9) + (4)(12y^2) + (4)(7y) + (4)(-9)$
$= 4y^4 + \frac{7}{3}y^3 - 3y^2 + 48y^2 + 28y - 36 = 4y^4 + \frac{7}{3}y^3 + 45y^2 + 28y - 36$

27) $(s^2-s+2)(s^2+4s-3)$
$=(s^2)(s^2)+(s^2)(4s)+(s^2)(-3)+(-s)(s^2)+(-s)(4s)+(-s)(-3)+(2)(s^2)+(2)(4s)+(2)(-3)$
$=s^4+4s^3-3s^2-s^3-4s^2+3s+2s^2+8s-6=s^4+3s^3-5s^2+11s-6$

29) $(4h^2-h+2)(-6h^3+5h^2-9h)$
$=(4h^2)(-6h^3)+(4h^2)(5h^2)+(4h^2)(-9h)+(-h)(-6h^3)+(-h)(5h^2)+(-h)(-9h)$
$+(2)(-6h^3)+(2)(5h^2)+(2)(-9h)$
$=-24h^5+20h^4-36h^3+6h^4-5h^3+9h^2-12h^3+10h^2-18h$
$=-24h^5+26h^4-53h^3+19h^2-18h$

31) $(3y-2)(5y^2-4y+3)=(3y)(5y^2)+(3y)(-4y)+(3y)(3)+(-2)(5y^2)+(-2)(-4y)+(-2)(3)$
$=15y^3-12y^2+9y-10y^2+8y-6$
$=15y^3-22y^2+17y-6$

$$\begin{array}{r} 5y^2-4y+3 \\ \times \qquad 3y-2 \\ \hline -10y^2+8y-6 \\ +\ 15y^3-12y^2+9y \quad \\ \hline 15y^3-22y^2+17y-6 \end{array}$$

33) First, Outer, Inner, Last

35) $(w+5)(w+7)=w^2+7w+5w+35=w^2+12w+35$

37) $(r-3)(r+9)=r^2+9r-3r-27=r^2+6r-27$

39) $(y-7)(y-1)=y^2-y-7y+7$
$=y^2-8y+7$

41) $(3p+7)(p-2)=3p^2-6p+7p-14$
$=3p^2+p-14$

43) $(7n+4)(3n+1)=21n^2+7n+12n+4$
$=21n^2+19n+4$

45) $(5-4w)(3-w)=15-5w-12w+4w^2$
$=4w^2-17w+15$

47) $(4a-5b)(3a+4b)$
$=12a^2+16ab-15ab-20b^2$
$=12a^2+ab-20b^2$

49) $(6x+7y)(8x+3y)$
$=48x^2+18xy+56xy+21y^2$
$=48x^2+74xy+21y^2$

51) $\left(v+\frac{1}{3}\right)\left(v+\frac{3}{4}\right)$

$=v^2+\frac{3}{4}v+\frac{1}{3}v+\frac{1}{4}$

$=v^2+\frac{9}{12}v+\frac{4}{12}v+\frac{1}{4}$

$=v^2+\frac{13}{12}v+\frac{1}{4}$

53) $\left(\frac{1}{2}a+5b\right)\left(\frac{2}{3}a-b\right)$

$=\frac{2}{6}a^2-\frac{1}{2}ab+\frac{10}{3}ab-5b^2$

$=\frac{1}{3}a^2-\frac{3}{6}ab+\frac{20}{6}ab-5b^2$

$=\frac{1}{3}a^2+\frac{17}{6}ab-5b^2$

55) a) $P=2(y+5)+2(y-3)$

$=2y+10+2y-6$

$=4y+4$ units

b) $A=(y+5)(y-3)$

$=y^2-3y+5y-15$

$=y^2+2y-15$ sq. units

57) a) $P=2(3m)+2\left(m^2-2m+7\right)$

$=6m+2m^2-4m+14$

$=2m^2+2m+14$ units

b) $A=(3m)\left(m^2-2m+7\right)$

$=3m^3-6m^2+21m$ sq. units

59) $A=\frac{1}{2}(6n-5)n$

$=\frac{1}{2}\left(6n^2-5n\right)$

$=3n^2-\frac{5}{2}n$ sq. units

61) Both are correct.

63) $2(n+3)(4n-5)$

$=(2n+6)(4n-5)$

$=8n^2-10n+24n-30$

$=8n^2+14n-30$

65) $-5z^2(z-8)(z-2)$

$=-5z^2\left(z^2-2z-8z+16\right)$

$=-5z^2\left(z^2-10z+16\right)$

$=-5z^4+50z^3-80z^2$

67) $(c+3)(c+4)(c-1)$

$=\left(c^2+4c+3c+12\right)(c-1)$

$=\left(c^2+7c+12\right)(c-1)$

$=c^3+7c^2+12c-c^2-7c-12$

$=c^3+6c^2+5c-12$

69) $(3x-1)(x-2)(x-6)$

$=(3x-1)\left(x^2-6x-2x+12\right)$

$=(3x-1)\left(x^2-8x+12\right)$

$=3x^3-24x^2+36x-x^2+8x-12$

$=3x^3-25x^2+44x-12$

71) $8p\left(\frac{1}{4}p^2+3\right)(p^2+5)$
$=(2p^3+24p)(p^2+5)$
$=2p^5+10p^3+24p^3+120p$
$=2p^5+34p^3+120p$

73) $(y+5)(y-5)=y^2-5^2=y^2-25$

75) $(a-7)(a+7)=(a+7)(a-7)$
$=a^2-7^2$
$=a^2-49$

77) $(3-p)(3+p)=(3+p)(3-p)$
$=3^2-p^2$
$=9-p^2$

79) $\left(u+\frac{1}{5}\right)\left(u-\frac{1}{5}\right)=u^2-\left(\frac{1}{5}\right)^2=u^2-\frac{1}{25}$

81) $\left(\frac{2}{3}-k\right)\left(\frac{2}{3}+k\right)=\left(\frac{2}{3}+k\right)\left(\frac{2}{3}-k\right)$
$=\left(\frac{2}{3}\right)^2-k^2$
$=\frac{4}{9}-k^2$

83) $(2r+7)(2r-7)=(2r)^2-7^2$
$=4r^2-49$

85) $-(8j-k)(8j+k)$
$=-(8j+k)(8j-k)$
$=-\left[(8j)^2-k^2\right]$
$=-(64j^2-k^2)$
$=k^2-64j^2$

87) $(d+4)^2=d^2+2(d)(4)+4^2$
$=d^2+8d+16$

89) $(n-13)^2=n^2-2(n)(13)+(-13)^2$
$=n^2-26n+169$

91) $(h-0.6)^2=h^2-2(h)(0.6)+(-0.6)^2$
$=h^2-1.2h+0.36$

93) $(3u+1)^2=(3u)^2+2(3u)(1)+1^2=9u^2+6u+1$

95) $(2d-5)^2=(2d)^2-2(2d)(5)+(-5)^2$
$=4d^2-20d+25$

97) $(3c+2d)^2=(3c)^2+2(3c)(2d)+(2d)^2$
$=9c^2+12cd+4d^2$

99) $\left(\frac{3}{2}k+8m\right)^2=\left(\frac{3}{2}k\right)^2+2\left(\frac{3}{2}k\right)(8m)+(8m)^2$
$=\frac{9}{4}k^2+24km+64m^2$

101) $\left[(2a+b)+3\right]^2=(2a+b)^2+2(2a+b)(3)+(3)^2=(2a)^2+2(2a)(b)+b^2+12a+6b+9$
$=4a^2+4ab+b^2+12a+6b+9$

103) $\left[(f-3g)+4\right]\left[(f-3g)-4\right]=(f-3g)^2-4^2=f^2-2(f)(3g)+(-3g)^2-16$
$=f^2-6fg+9g^2-16$

105) No. The order of operations tell usto perform exponents, $(r+2)^2$, before multiplying by 3.

107) $7(y+2)^2=7\left[y^2+2(y)(2)+2^2\right]=7\left[y^2+4y+4\right]=7y^2+28y+28$

109) $4c(c+3)^2=4c\left[c^2+2(c)(3)+3^2\right]=4c\left[c^2+6c+9\right]=4c^3+24c^2+36c$

111) $(r+5)^3=(r+5)^2(r+5)=(r^2+10r+25)(r+5)$

$=r^3+10r^2+5r^2+25r+50r+125=r^3+15r^2+75r+125$

113) $(g-4)^3=(g-4)^2(g-4)=(g^2-8g+16)(g-4)$

$=g^3-4g^2-8g^2+32g+16g-64=g^3-12g^2+48g-64$

115) $(2a-1)^3=(2a-1)^2(2a-1)=\left[(2a)^2-2(2a)(1)+(-1)^2\right](2a-1)=(4a^2-4a+1)(2a-1)$

$=8a^3-4a^2-8a^2+4a+2a-1$

$=8a^3-12a^2+6a-1$

117) $(h+3)^4=(h+3)^2(h+3)^2=(h^2+6h+9)(h^2+6h+9)$

$=h^4+6h^3+9h^2+6h^3+36h^2+54h+9h^2+54h+81$

$=h^4+12h^3+54h^2+108h+81$

119) $(5t-2)^4=(5t-2)^2(5t-2)^2=(25t^2-20t+4)(25t^2-20t+4)$

$=625t^4-500t^3+100t^2-500t^3+400t^2-80t+100t^2-80t+16$

$=625t^4-1000t^3+600t^2-160t+16$

121) No; $(x+2)^2=x^2+4x+4$

123) $(c-12)(c+7)=c^2+7c-12c-84=c^2-5c-84$

125) $4(6-5a)(2a-1)=(24-20a)(2a-1)=48a-24-40a^2+20a=-40a^2+68a-24$

127) $(2k-9)(5k^2+4k-1)=(2k)(5k^2)+(2k)(4k)+(2k)(-1)+(-9)(5k^2)+(-9)(4k)+(-9)(-1)$

$=10k^3+8k^2-2k-45k^2-36k+9=10k^3-37k^2-38k+9$

129) $\left(\frac{1}{6}-h\right)\left(\frac{1}{6}+h\right)=\left(\frac{1}{6}+h\right)\left(\frac{1}{6}-h\right)=\left(\frac{1}{6}\right)^2-h^2=\frac{1}{36}-h^2$

131) $(3c+1)^3=(3c+1)^2(3c+1)=\left((3c)^2+2(3c)(1)+(1)^2\right)(3c+1)$
$$=\left(9c^2+6c+1\right)(3c+1)=27c^3+9c^2+18c^2+6c+3c+1$$
$$=27c^3+27c^2+9c+1$$

133) $\left(\frac{3}{8}p^7\right)\left(\frac{3}{4}p^4\right)=\frac{9}{32}p^{11}$

135) $(a+7b)^2=(a)^2+2(a)(7b)+(7b)^2=a^2+14ab+49b^2$

137) $-5z(z-3)^2=-5z\left[z^2-2(z)(3)+(-3)^2\right]=-5z\left[z^2-6z+9\right]=-5z^3+30z^2-45z$

139) $V=(a+4)^3=(a+4)^2(a+4)=\left(a^2+8a+16\right)(a+4)$
$$=a^3+4a^2+8a^2+32a+16a+64=a^3+12a^2+48a+64 \text{ cubic units}$$

141) $A=\pi(k+5)^2=\pi\left(k^2+10k+25\right)=\pi k^2+10\pi k+25\pi$ sq. units

143) $A=(3c-2)(3c-2)-\frac{1}{2}(6)(c)=(3c)^2-2(3c)(2)+(-2)^2-3c$
$$=9c^2-12c+4-3c=9c^2-15c+4 \text{ sq. units}$$

Section 6.4: Exercises

1) dividend: $6c^3+15c^2-9c$; divisor: $3c$; quotient: $2c^2+5c-3$

3) Answers may vary.Divide each term in the polynomial by the monomial and simplify.

5) $\frac{49p^4+21p^3+28p^2}{7}=\frac{49p^4}{7}+\frac{21p^3}{7}+\frac{28p^2}{7}=7p^4+3p^3+4p^2$

7) $\frac{12w^3-40w^2-36w}{4w}=\frac{12w^3}{4w}-\frac{40w^2}{4w}-\frac{36w}{4w}=3w^2-10w-9$

9) $\frac{22z^6+14z^5-38z^3+2z}{2z}=\frac{22z^6}{2z}+\frac{14z^5}{2z}-\frac{38z^3}{2z}+\frac{2z}{2z}=11z^5+7z^4-19z^2+1$

11) $\frac{9h^8+54h^6-108h^3}{9h^2}=\frac{9h^8}{9h^2}+\frac{54h^6}{9h^2}-\frac{108h^3}{9h^2}=h^6+6h^4-12h$

13) $\frac{36r^7-12r^4+6}{12r}=\frac{36r^7}{12r}-\frac{12r^4}{12r}+\frac{6}{12r}=3r^6-r^3+\frac{1}{2r}$

15) $\dfrac{8d^6-12d^5+18d^4}{2d^4}=\dfrac{8d^6}{2d^4}-\dfrac{12d^5}{2d^4}+\dfrac{18d^4}{2d^4}=4d^2-6d+9$

17) $\dfrac{28k^7+8k^5-44k^4-36k^2}{4k^2}=\dfrac{28k^7}{4k^2}+\dfrac{8k^5}{4k^2}-\dfrac{44k^4}{4k^2}-\dfrac{36k^2}{4k^2}=7k^5+2k^3-11k^2-9$

19) $\left(35d^5-7d^2\right)\div\left(-7d^2\right)=\dfrac{35d^5-7d^2}{-7d^2}=\dfrac{35d^5}{-7d^2}-\dfrac{7d^2}{-7d^2}=-5d^3+1$

21) $\dfrac{10w^5+12w^3-6w^2+2w}{6w^2}=\dfrac{10w^5}{6w^2}+\dfrac{12w^3}{6w^2}-\dfrac{6w^2}{6w^2}+\dfrac{2w}{6w^2}=\dfrac{5}{3}w^3+2w-1+\dfrac{1}{3w}$

23) $\left(12k^8-4k^6-15k^5-3k^4+1\right)\div\left(2k^5\right)=\dfrac{12k^8-4k^6-15k^5-3k^4+1}{2k^5}$

$=\dfrac{12k^8}{2k^5}-\dfrac{4k^6}{2k^5}-\dfrac{15k^5}{2k^5}-\dfrac{3k^4}{2k^5}+\dfrac{1}{2k^5}=6k^3-2k-\dfrac{15}{2}-\dfrac{3}{2k}+\dfrac{1}{2k^5}$

25) $\dfrac{48p^5q^3+60p^4q^2-54p^3q+18p^2q}{6p^2q}=\dfrac{48p^5q^3}{6p^2q}+\dfrac{60p^4q^2}{6p^2q}-\dfrac{54p^3q}{6p^2q}+\dfrac{18p^2q}{6p^2q}$

$=8p^3q^2+10p^2q-9p+3$

27) $\dfrac{14s^6t^6-28s^5t^4-s^3t^3+21st}{7s^2t}=\dfrac{14s^6t^6}{7s^2t}-\dfrac{28s^5t^4}{7s^2t}-\dfrac{s^3t^3}{7s^2t}+\dfrac{21st}{7s^2t}=2s^4t^5-4s^3t^3-\dfrac{1}{7}st^2+\dfrac{3}{s}$

29) The answer is incorrect. When you divide $5p$ by $5p$, you get 1. The quotient should be $8p^2-2p+1$.

31) dividend: $12w^3-2w^2-23w-7$; divisor: $3w+1$; quotient: $4w^2-2w-7$

33) 2

35)

$$\begin{array}{r} 158 \\ 6\overline{)949} \\ \underline{6} \\ 34 \\ \underline{-30} \\ 49 \\ \underline{-48} \\ 1 \end{array}$$

$Answer: 158\dfrac{1}{6}$

37)

$$\begin{array}{r} 437 \\ 9\overline{)3937} \\ \underline{-36} \\ 33 \\ \underline{-27} \\ 67 \\ \underline{-63} \\ 4 \end{array}$$

$Answer: 437\dfrac{4}{9}$

39)
$$\begin{array}{r} g+4 \\ g+5\overline{)g^2+9g+20} \\ \underline{-(g^2+5g)} \\ 4g+20 \\ \underline{-(4g+20)} \\ 0 \end{array}$$

41)
$$\begin{array}{r} a+6 \\ a+7\overline{)a^2+13a+42} \\ \underline{-(a^2+7a)} \\ 6a+42 \\ \underline{-(6a+42)} \\ 0 \end{array}$$

43)
$$\begin{array}{r} k-6 \\ k+5\overline{)k^2-k-30} \\ \underline{-(k^2+5k)} \\ -6k-30 \\ \underline{-(-6k-30)} \\ 0 \end{array}$$

45)
$$\begin{array}{r} x+8 \\ x-5\overline{)x^2+3x-40} \\ \underline{-(x^2-5x)} \\ 8x-40 \\ \underline{-(8x-40)} \\ 0 \end{array}$$

47)
$$\begin{array}{r} 2h^2+5h+3 \\ 3h-4\overline{)6h^3+7h^2-17h-4} \\ \underline{-(6h^3-8h^2)} \\ 15h^2-17h \\ \underline{-(15h^2-20h)} \\ 3h-4 \\ \underline{-(3h-12)} \\ 8 \end{array}$$

49)
$$\begin{array}{r} 3p^2+5p-1 \\ 4p+1\overline{)12p^3+23p^2+p-1} \\ \underline{-(12p^3+3p^2)} \\ 20p^2+p \\ \underline{-(20p^2+5p)} \\ -4p-1 \\ \underline{-(-4p-1)} \\ 0 \end{array}$$

51)
$$\begin{array}{r} 7m+12 \\ m-4\overline{)7m^2-16m-41} \\ \underline{-(7m^2-28m)} \\ 12m-41 \\ \underline{-(12m-48)} \\ 7 \end{array}$$

$$\left(7m^2-16m-41\right)\div\left(m-4\right)=7m+12+\frac{7}{m-4}$$

53)
$$\begin{array}{r} 4a^2-7a\ \ +2 \\ 5a-2\overline{\big)\ 20a^3-43a^2+24a-12} \\ \underline{-(20a^3-8a^2)} \\ -35a^2+24a \\ \underline{-\ (-35a^2+14a)} \\ 10a-12 \\ \underline{-\ (10a-4)} \\ -8 \end{array}$$

$$\left(20a^3-43a^2+24a-12\right)\div\left(5a-2\right)$$
$$=4a^2-7a+2-\frac{8}{5a-2}$$

55)
$$\begin{array}{r} n^2-3n+9 \\ n+3\overline{\big)\ n^3+0n^2+0n+27} \\ \underline{-(n^3+3n^2)} \\ -3n^2+0n \\ \underline{-(-3n^2-9n)} \\ 9n+27 \\ \underline{-(9n+27)} \\ 0 \end{array}$$

57)
$$\begin{array}{r} 2r^2+4r+5 \\ 4r-5\overline{\big)\ 8r^3+6r^2+0r-25} \\ \underline{-(8r^3-10r^2)} \\ 16r^2+0r \\ \underline{-\ (16r^2-20r)} \\ 20r-25 \\ \underline{-(20r-25)} \\ 0 \end{array}$$

59)
$$\begin{array}{r} 6x^2-9x+5 \\ 2x+3\overline{\big)\ 12x^3+0x^2-17x+4} \\ \underline{-(12x^3+18x^2)} \\ -18x^2-17x \\ \underline{-\ (-18x^2-27x)} \\ 10x+4 \\ \underline{-(10x+15)} \\ -11 \end{array}$$

$$\left(12x^3-17x+4\right)\div\left(2x+3\right)$$
$$=6x^2-9x+5-\frac{11}{2x+3}$$

61)
$$\begin{array}{r} k^2+k+5 \\ k^2+0k+4\overline{\big)\ k^4+k^3+\ 9k^2+4k+20} \\ \underline{-(k^4+0k^3+4k^2)} \\ k^3\quad +5k^2+4k \\ \underline{-\ (k^3\quad +0k^2+4k)} \\ 5k^2\qquad +20 \\ \underline{-(5k^2\qquad +20)} \\ 0 \end{array}$$

63)
$$\begin{array}{r} 3t^2-8t-6 \\ 5t^2-1\overline{\big)\ 15t^4-40t^3-33t^2+10t+2} \\ \underline{-(15t^4\qquad -3t^2)} \\ -40t^3-30t^2\ +10t \\ \underline{-\ (-40t^3\qquad +8t)} \\ -30t^2\ \ +2t\ +2 \\ \underline{-(-30t^2\qquad +6)} \\ 2t\ -4 \end{array}$$

$$\frac{15t^4-40t^3-33t^2+10t+2}{5t^2-1}=3t^2-8t-6+\frac{2t-4}{5t^2-1}$$

65) No. For example, $\dfrac{12x+8}{3x}=4x+\dfrac{8}{3x}$.
The quotient is not a polynomial because one term has a variable denominator.

67) $\dfrac{50a^4b^4+30a^4b^3-a^2b^2+2ab}{10a^2b^2}$

$=\dfrac{50a^4b^4}{10a^2b^2}+\dfrac{30a^4b^3}{10a^2b^2}-\dfrac{a^2b^2}{10a^2b^2}+\dfrac{2ab}{10a^2b^2}$

$=5a^2b^2+3a^2b-\dfrac{1}{10}+\dfrac{1}{5ab}$

69)

$$\begin{array}{r|l} & -3f^3+6f^2-2f+9 \\ 5f-2 & -15f^4+36f^3-22f^2+49f+5 \\ & -(-15f^4+6f^3) \\ & 30f^3-22f^2 \\ & -(30f^3-12f^2) \\ & -10f^2+49f \\ & -(-10f^2+4f) \\ & 45f+5 \\ & -(45f-18) \\ & 23 \end{array}$$

$\left(-15f^4+36f^3-22f^2+49f+5\right)\div(5f-2)$

$=-3f^3+6f^2-2f+9+\dfrac{23}{5f-2}$

71)

$$\begin{array}{r|l} & 8t+5 \\ t-3 & 8t^2-19t-4 \\ & -(8t^2-24t) \\ & 5t-4 \\ & -(5t-15) \\ & 11 \end{array}$$

$\dfrac{8t^2-19t-4}{t-3}=8t+5+\dfrac{11}{t-3}$

73)

$$\begin{array}{r|l} & 16p^2+12p+9 \\ 4p-3 & 64p^3+0p^2+0p-27 \\ & -(64p^3-48p^2) \\ & 48p^2+0p \\ & -(48p^2-36p) \\ & 36p-27 \\ & -(36p-27) \\ & 0 \end{array}$$

75)

$$\begin{array}{r|l} & 6x^2+x-7 \\ x^2+0x+3 & 6x^4+x^3+11x^2+3x-21 \\ & -(6x^4+0x^3+18x^2) \\ & x^3-7x^2+3x \\ & -(x^3+0x^2+3x) \\ & -7x^2-21 \\ & -(-7x^2-21) \\ & 0 \end{array}$$

77) $\dfrac{-20v^3+35v^2-15v}{-5v}$

$=\dfrac{-20v^3}{-5v}+\dfrac{35v^2}{-5v}-\dfrac{15v}{-5v}$

$=4v^2-7v+3$

79)

$$\begin{array}{r} 5h^2-3h-2 \\ 2h^2-9 \overline{\big) \, 10h^4-6h^3-49h^2+27h+19} \\ \underline{-(10h^4 \quad -45h^2)} \\ -6h^3-4h^2+27h \\ \underline{-(-6h^3 \quad +27h)} \\ -4h^2 \quad +19 \\ \underline{-(-4h^2 \quad +18)} \\ 1 \end{array}$$

$$\frac{10h^4-6h^3-49h^2+27h+19}{2h^2-9}$$

$$=5h^2-3h-2+\frac{1}{2h^2-9}$$

81)

$$\begin{array}{r} m^2 \quad -4 \\ m^2+4 \overline{\big) \, m^4+0m^3+0m^2+0m-16} \\ \underline{-(m^4 \quad +4m^2)} \\ -4m^2+0m-16 \\ \underline{-(-4m^2 \quad -16)} \\ 0 \end{array}$$

83) $\dfrac{45t^4-81t^2-27t^3+8t^6}{-9t^3}$

$$\frac{8t^6+45t^4-27t^3-81t^2}{-9t^3}$$

$$=\frac{8t^6}{-9t^3}+\frac{45t^4}{-9t^3}-\frac{27t^3}{-9t^3}-\frac{81t^2}{-9t^3}$$

$$=-\frac{8}{9}t^3-5t+3+\frac{9}{t}$$

85)

$$\begin{array}{r} \frac{1}{2}x+5 \\ 2x-3 \overline{\big) \, x^2+\frac{17}{2}x-15} \\ \underline{-(x^2-\frac{3}{2}x)} \\ 10x-15 \\ \underline{-(10x-15)} \\ 0 \end{array}$$

87)

$$\begin{array}{r} 3p^2-5p+1 \\ 7p^2+2p-4 \overline{\big) \, 21p^4-29p^3-15p^2+28p+16} \\ \underline{-(21p^4+6p^3-12p^2)} \\ -35p^3-3p^2+28p \\ \underline{-(-35p^3-10p^2+20p)} \\ 7p^2+8p+16 \\ \underline{-(7p^2+2p-4)} \\ 6p+20 \end{array}$$

$$\frac{21p^4-29p^3-15p^2+28p+16}{7p^2+2p-4}$$

$$=3p^2-5p+1+\frac{6p+20}{7p^2+2p-4}$$

89) $w=$ width of the rectangle.

$$6x^2+23x+21=(2x+3)w$$

$$\frac{6x^2+23x+21}{2x+3}=w$$

$$\begin{array}{r} 3x+7 \\ 2x+3 \overline{\big) \, 6x^2+23x+21} \\ \underline{-(6x^2+9x)} \\ 14x+21 \\ \underline{-\ (14x+21)} \\ 0 \end{array}$$

width $=3x+7$

91) $b =$ base of the triangle

$$15n^3 - 18n^2 + 6n = \frac{1}{2}(\text{base})(\text{height})$$
$$15n^3 - 18n^2 + 6n = \frac{1}{2}(b)(n)$$
$$2(15n^3 - 18n^2 + 6n) = (b)(n)$$
$$\frac{30n^3 - 36n^2 + 12n}{n} = b$$
$$\frac{30n^3}{n} + \frac{-36n^2}{n} + \frac{12n}{n} = b$$
$$30n^2 - 36n + 12 = b$$
$$base = 30n^2 - 36n + 12$$

93) distance $= (\text{rate})(\text{time})$

Let r be the rate.

$$3x^3 + 5x^2 - 26x + 8 = (r)(x+4)$$
$$\frac{3x^3 + 5x^2 - 26x + 8}{x+4} = r$$

$$\begin{array}{r} 3x^2 - 7x + 2 \\ x+4\overline{)3x^3 + 5x^2 - 26x + 8} \\ \underline{-(3x^3 + 12x^2)} \\ -7x^2 - 26x \\ \underline{-(-7x^2 - 28x)} \\ 2x + 8 \\ \underline{-(2x+8)} \\ 0 \end{array}$$

rate $= 3x^2 - 7x + 2$ mph

Chapter 6 Review

1) $\frac{2^{11}}{2^6} = 2^{11-6} = 2^5 = 32$

3) $\left(\frac{2}{5}\right)^{-3} = \left(\frac{5}{2}\right)^3 = \frac{125}{8}$

5) $(p^7)^4 = p^{7\cdot 4} = p^{28}$

7) $\frac{60t^9}{12t^3} = 5t^{9-3} = 5t^6$

9) $(-7c)(6c^8) = -42c^{1+8} = -42c^9$

11) $\frac{k^7}{k^{12}} = k^{7-12} = k^{-5} = \frac{1}{k^5}$

13) $(-2r^2s)^3(6r^{-9}s)$

$$= (-8r^6s^3)(6r^{-9}s)$$
$$= -48r^{-3}s^{3+1}$$
$$= -\frac{48s^4}{r^3}$$

15) $\left(\frac{2xy^{-8}}{3x^{-2}y^6}\right)^{-2} = \left(\frac{3x^{-2}y^6}{2xy^{-8}}\right)^2$

$$= \left(\frac{3y^{14}}{2x^3}\right)^2$$
$$= \frac{9y^{28}}{4x^6}$$

17) $\frac{m^{-1}n^8}{mn^{14}} = m^{-1-1}n^{8-14} = m^{-2}n^{-6} = \frac{1}{m^2n^6}$

19) $A = (3f)(4f)$

$$= 12f^2 \text{ sq. units}$$

$$P = 2(3f) + 2(4f)$$
$$= 6f + 8f$$
$$= 14f \text{ units}$$

21) $y^{4a} \cdot y^{3a} = y^{4a+3a} = y^{7a}$

23) $\frac{r^{11x}}{r^{2x}} = r^{11x-2x} = r^{9x}$

25) Degree of polynomial = 3

Term	Coeff	Degree
$7s^3$	7	3
$-9s^2$	-9	2
s	1	1
6	6	0

27) $2r^2 - 8r - 11$

$$2(-3)^2 - 8(-3) - 11$$
$$= 2(9) + 24 - 11$$
$$= 31$$

29) a) $h(x) = 5x^2 - 3x - 6$

$$h(-2) = 5(-2)^2 - 3(-2) - 6$$
$$h(-2) = 5(4) + 6 - 6$$
$$h(-2) = 20 - 6 + 6 = 20$$

b) $h(x) = 5x^2 - 3x - 6$

$$h(0) = 5(0)^2 - 3(0) - 6$$
$$h(0) = -6$$

31) $(6c^2 + 2c - 8) - (8c^2 + c - 13)$

$$= (6c^2 + 2c - 8) + (-8c^2 - c + 13)$$
$$= -2c^2 + c + 5$$

33)
$$\begin{array}{r} 6.7j^3 - 1.4j^2 + j - 5.3 \\ +\ \ 3.1j^3 + 5.7j^2 + 2.4j + 4.8 \\ \hline 9.8j^3 + 4.3j^2 + 3.4j - 0.5 \end{array}$$

35) $\left(\frac{3}{5}k^2 + \frac{1}{2}k + 4\right) - \left(\frac{1}{10}k^2 + \frac{3}{2}k - 2\right) = \left(\frac{3}{5}k^2 + \frac{1}{2}k + 4\right) + \left(-\frac{1}{10}k^2 - \frac{3}{2}k + 2\right)$

$$= \frac{5}{10}k^2 - k + 6 = \frac{1}{2}k^2 - k + 6$$

37) $(x^2y^2 + 2x^2y - 4xy + 11) - (4x^2y^2 - 7x^2y + xy + 5)$

$$= (x^2y^2 + 2x^2y - 4xy + 11) + (-4x^2y^2 + 7x^2y - xy - 5) = -3x^2y^2 + 9x^2y - 5xy + 6$$

39) $(6m + 2n - 17) + (-3m + 2n + 14) = 3m + 4n - 3$

41) $[(7x - 16) + (8x^2 - 15x + 6)] - (2x^2 + 3x + 18) = [8x^2 - 8x - 10] + (-2x^2 - 3x - 18)$

$$= 6x^2 - 11x - 28$$

43) $P = 2(d^2 + 6d + 2) + 2(d^2 - 3d + 1) = 2d^2 + 12d + 4 + 2d^2 - 6d + 2 = 4d^2 + 6d + 6$ units

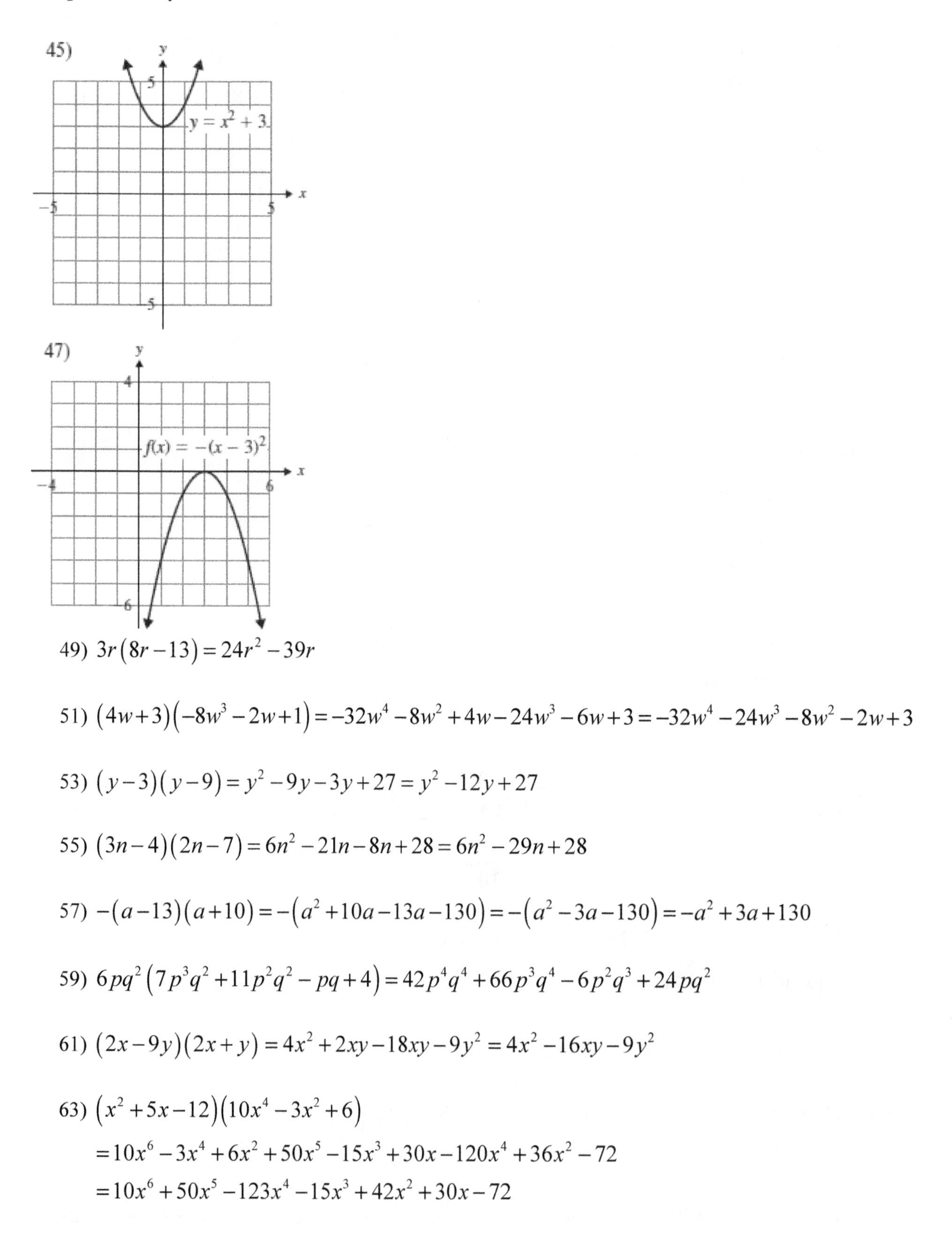

49) $3r(8r-13)=24r^2-39r$

51) $(4w+3)(-8w^3-2w+1)=-32w^4-8w^2+4w-24w^3-6w+3=-32w^4-24w^3-8w^2-2w+3$

53) $(y-3)(y-9)=y^2-9y-3y+27=y^2-12y+27$

55) $(3n-4)(2n-7)=6n^2-21n-8n+28=6n^2-29n+28$

57) $-(a-13)(a+10)=-(a^2+10a-13a-130)=-(a^2-3a-130)=-a^2+3a+130$

59) $6pq^2(7p^3q^2+11p^2q^2-pq+4)=42p^4q^4+66p^3q^4-6p^2q^3+24pq^2$

61) $(2x-9y)(2x+y)=4x^2+2xy-18xy-9y^2=4x^2-16xy-9y^2$

63) $(x^2+5x-12)(10x^4-3x^2+6)$
$=10x^6-3x^4+6x^2+50x^5-15x^3+30x-120x^4+36x^2-72$
$=10x^6+50x^5-123x^4-15x^3+42x^2+30x-72$

65) $4f^2(2f-7)(f-6)=4f^2\left(2f^2-12f-7f+42\right)=4f^2\left(2f^2-19f+42\right)$
$=8f^4-76f^3+168f^2$

67) $(z+3)(z+1)(z+4)=(z+3)\left(z^2+4z+z+4\right)=(z+3)\left(z^2+5z+4\right)$
$=z^3+5z^2+4z+3z^2+15z+12=z^3+8z^2+19z+12$

69) $\left(\frac{2}{7}d+3\right)\left(\frac{1}{2}d-8\right)=\frac{1}{7}d^2-\frac{16}{7}d+\frac{3}{2}d-24=\frac{1}{7}d^2-\frac{32}{14}m+\frac{21}{14}d-24=\frac{1}{7}d^2-\frac{11}{14}d-24$

71) $(c+4)^2=c^2+8c+16$

73) $(4p-3)^2=16p^2-24p+9$

75) $(x-3)^3=(x-3)^2(x-3)=\left(x^2-6x+9\right)(x-3)=x^3-3x^2-6x^2+1+9x-27=x^3-9x^2+27x-27$

77) $\left[(m-3)+n\right]^2=(m-3)^2+2(m-3)(n)+(n)^2=(m)^2-2(m)(3)+(-3)^2+2mn-6n+n^2$
$=m^2-6m+9+2mn-6n+n^2$

79) $(p-13)(p+13)=(p+13)(p-13)$
$=(p)^2-(13)^2$
$=p^2-169$

81) $\left(\frac{9}{2}+\frac{5}{6}x\right)\left(\frac{9}{2}-\frac{5}{6}x\right)=\frac{81}{4}-\frac{25}{36}x^2$

83) $\left(3a-\frac{1}{2}b\right)\left(3a+\frac{1}{2}b\right)$
$=\left(3a+\frac{1}{2}b\right)\left(3a-\frac{1}{2}b\right)$
$=9a^2-\frac{1}{4}b^2$

85) $3u(u+4)^2=3u\left(u^2+8u+16\right)$
$=3u^3+24u^2+48u$

87) a) $A=(2n+11)(n-2)$
$=2n^2-4n+11n-22$
$=2n^2+7n-22$ sq. units

b) $P=2(2n+11)+2(n-2)$
$=4n+22+2n-4$
$=6n+18$ units

89) $\frac{12t^6-30t^5-15t^4}{3t^4}=\frac{12t^6}{3t^4}-\frac{30t^5}{3t^4}-\frac{15t^4}{3t^4}$
$=4t^2-10t-5$

91)
$$\begin{array}{r|l} & w+5 \\ \hline w+4 & w^2+9w+20 \\ & -(w^2+4w) \\ \hline & 5w+20 \\ & -(5w+20) \\ \hline & 0 \end{array}$$

93)
$$\begin{array}{r} 4r^2+r-3 \\ 2r+5\overline{\big)\ 8r^3+22r^2-r-15} \\ \underline{-(8r^3+20r^2)} \\ 2r^2-r \\ \underline{-(2r^2+5r)} \\ -6r-15 \\ \underline{-(-6r-15)} \\ 0 \end{array}$$

95)
$$\frac{14t^4+28t^3-21t^2+20t}{14t^3}$$
$$=\frac{14t^4}{14t^3}+\frac{28t^3}{14t^3}-\frac{21t^2}{14t^3}+\frac{20t}{14t^3}$$
$$=t+2-\frac{3}{2t}+\frac{10}{7t^2}$$

97)
$$\begin{array}{r} 2v\ \ -1 \\ 4v+9\overline{\big)\ 8v^2+14v-3} \\ \underline{-(8v^2+18v)} \\ -4v-3 \\ \underline{-(-4v-9)} \\ 6 \end{array}$$
$$\left(8v^2+14v-3\right)\div\left(4v+9\right)$$
$$=2v-1+\frac{6}{4v+9}$$

99)
$$\begin{array}{r} 3v^2-7v+8 \\ 2v^2+0v^3+3\overline{\big)\ 6v^4-14v^3+25v^2-21v+24} \\ \underline{-(6v^4+0v^3\ \ +9v^2)} \\ -14v^3+16v^2-21v \\ \underline{-(-14v^3+0v^2-21v)} \\ 16v^2\qquad +24 \\ \underline{-(16v^2\qquad +24)} \\ 0 \end{array}$$

101)
$$\begin{array}{r} c^2+2c+4 \\ c-2\overline{\big)\ c^3+0c^2+0c-8} \\ \underline{-(c^3-2c^2)} \\ 2c^2+0c \\ \underline{-(2c^2-4c)} \\ 4c-8 \\ \underline{-(4c-8)} \\ 0 \end{array}$$

103)
$$\begin{array}{r} 6k^2-4k+7 \\ 3k+2\overline{\big)\ 18k^3+0k^2+13k-4} \\ \underline{-(18k^3+12k^2)} \\ -12k^2+13k \\ \underline{-(-12k^2-8k)} \\ 21k-\ 4 \\ \underline{-(-21k+14)} \\ -18 \end{array}$$

$$\frac{18k^3+13k-4}{3k+2}$$
$$=6k^2-4k+7-\frac{18}{3k+2}$$

105)
$$\left(20x^4y^4-48x^2y^4-12xy^2+15x\right)\div\left(-12xy^2\right)$$
$$=\frac{20x^4y^4-48x^2y^4-12xy^2+15x}{-12xy^2}$$
$$=\frac{20x^4y^4}{-12xy^2}-\frac{48x^2y^4}{-12xy^2}-\frac{12xy^2}{-12xy^2}+\frac{15x}{-12xy^2}$$
$$=-\frac{5}{3}x^3y^2+4xy^2+1-\frac{5}{4y^2}$$

107) Let b = the base.
$$12a^2+3a=\frac{1}{2}b(3a)$$
$$24a^2+6a=b(3a)$$
$$\frac{24a^2+6a}{3a}=b$$
$$\frac{24a^2}{3a}+\frac{6a}{3a}=b$$
$$8a+2=b$$

109)
$$\begin{array}{rr} & 18c^3+7c^2-11c+2 \\ + & 2c^3-19c^2\qquad\;\; -1 \\ \hline & 20c^3-12c^2-11c+1 \end{array}$$

111) $(12-7w)(12+7w)$
$$=(12+7w)(12-7w)$$
$$=(12)^2-(7w)^2$$
$$=144-49w^2$$

113) $5\left(-2r^7t^9\right)^3$
$$=5(-2)^3\left(r^7\right)^3\left(t^9\right)^3$$
$$=5(-8)r^{7\cdot3}t^{9\cdot3}$$
$$=-40r^{21}t^{27}$$

115) $\left(39a^6b^6+21a^4b^5-5a^3b^4+a^2b\right)\div\left(3a^3b^3\right)$
$$=\frac{39a^6b^6+21a^4b^5-5a^3b^4+a^2b}{3a^3b^3}$$
$$=\frac{39a^6b^6}{3a^3b^3}+\frac{21a^4b^5}{3a^3b^3}-\frac{5a^3b^4}{3a^3b^3}+\frac{a^2b}{3a^3b^3}$$
$$=13a^3b^3+7ab^2-\frac{5}{3}b+\frac{1}{3ab^2}$$

117) $(h-5)^3=(h-5)^2(h-5)$
$$=\left(h^2-10h+25\right)(h-5)$$
$$=h^3-5h^2-10h^2+50h+25h-125$$
$$=h^3-15h^2+75h-125$$

119)
$$\begin{array}{r} 2c^2-8c+9 \\ c+4\overline{\big)\;2c^3+0c^2-23c+41} \\ \underline{-(2c^3+8c^2)} \\ -8c^2-23c \\ \underline{-(-8c^2-32c)} \\ 9c+41 \\ \underline{-\;(9c+36)} \\ 5 \end{array}$$

$$\left(2c^3-23c+41\right)\div(c+4)$$
$$=2c^2-8c+9+\frac{5}{c+4}$$

121) $\left(\frac{5}{y^4}\right)^{-3}=\left(\frac{y^4}{5}\right)^3$
$$=\frac{y^{12}}{125}$$

123)

$$\begin{array}{r} 2p^2+3p-5 \\ 3p^2+p-4\overline{\big)\ 6p^4+11p^3-20p^2-17p+20} \\ \underline{-(6p^4+2p^3-8p^2)} \\ 9p^3-12p^2-17p \\ \underline{-(9p^3+3p^2-12p)} \\ -15p^2-5p+20 \\ \underline{-(-15p^2-5p+20)} \\ 0 \end{array}$$

Chapter 6 Test

1) $\left(\frac{3}{4}\right)^{-3}=\left(\frac{4}{3}\right)^3=\frac{64}{27}$

3) $(8p^3)(-4p^6)=-32p^9$

5) $\frac{g^{11}h^{-4}}{g^7h^6}=g^{11-7}h^{-4-6}=g^4h^{-10}=\frac{g^4}{h^{10}}$

7) a) -1 b) 3

9) $-2r^2+7s,\ r=-4;\ s=5$

$-2(-4)^2+7(5)=-32+35=3$

11) $(7a^3b^2+9a^2b^2-4ab+8)$

$+(5a^3b^2-12a^2b^2+ab+1)$

$=12a^3b^2-3a^2b^2-3ab+9$

13) $3(-c^3+3c-6)-4(2c^3+3c^2+7c-1)$

$=-3c^3+9c-18-8c^3-12c^2-28c+4$

$=-11c^3-12c^2-19c-14$

15) $(4g+3)(2g+1)=8g^2+4g+6g+3$

$=8g^2+10g+3$

17) $(3x-7y)(2x+y)=6x^2+3xy-14xy-7y^2$

$=6x^2-11xy-7y^2$

19) $2y(y+6)^2=2y(y^2+12y+36)$

$=2y^3+24y^2+72y$

21) $\left(\frac{4}{3}x+y\right)^2=\frac{16}{9}x^2+\frac{8}{3}xy+y^2$

23)

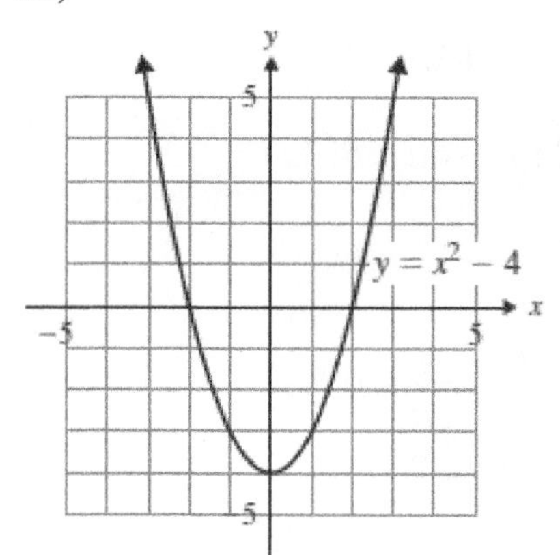

25) $\frac{24m^6-40m^5+8m^4-6m^3}{8m^4}$

$=\frac{24m^6}{8m^4}-\frac{40m^5}{8m^4}+\frac{8m^4}{8m^4}-\frac{6m^3}{8m^4}$

$=3m^2-5m+1-\frac{3}{4m}$

27)

$$\begin{array}{r} y^2+3y+9 \\ y-3\overline{\big)\ y^3+0y^2+0y-27} \\ \underline{-(y^3-3y^2)} \\ 3y^2+0y \\ \underline{-(3y^2-9y)} \\ 9y-27 \\ \underline{-(9y-27)} \\ 0 \end{array}$$

29) a) $A = (3d+1)(d-5)$
$= 3d^2 - 15d + d - 5$
$= 3d^2 - 14d - 5$ sq. units

b) $P = 2(3d+1) + 2(d-5)$
$= 6d + 2 + 2d - 10$
$= 8d - 8$ units

Cumulative Review: Chapters 1-6

1) a) $\{41, 0\}$ b) $\{-15, 41, 0\}$
c) $\left\{\frac{3}{8}, -15, 2.1, 41, 0.\overline{52}, 0\right\}$

3) $3\frac{1}{8} \div 1\frac{7}{24} = \frac{25}{8} \div \frac{31}{24}$
$= \frac{25}{8} \cdot \frac{24}{31}$
$= \frac{25}{1} \cdot \frac{3}{31}$
$= \frac{75}{31}$ or $2\frac{13}{31}$

5) $c^{10} \cdot c^7 = c^{10+7} = c^{17}$

7) $-\frac{18}{7}m - 9 = 21$

$-\frac{18}{7}m = 30$

$m = -\frac{30 \cdot 7}{18} = -\frac{35}{3}$; $\left\{-\frac{35}{3}\right\}$

9) $5y - 16 \geq 8y - 1$
$-3y - 16 \geq -1$
$-3y \geq 15$
$y \leq -5$ $(-\infty, -5]$

11) $3x - 8y = 24$
x-int: Let $y = 0$, and solve for x.
$3x - 8(0) = 24$
$3x = 24$
$x = 8$ $(8, 0)$
y-int: Let $x = 0$, and solve for y.

$3(0) - 8y = 24$
$-8y = 24$
$y = -3$ $(0, -3)$

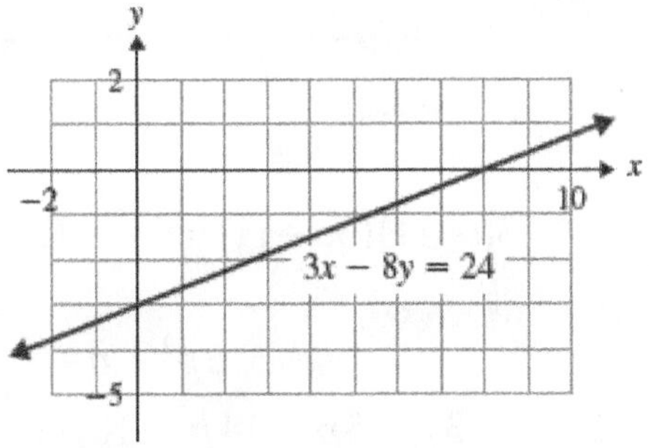

13) $m = \frac{-11-7}{2-(-4)} = \frac{-18}{6} = -3,$
$(x_1, y_1) = (-4, 7)$
$(y - y_1) = m(x - x_1)$
$(y - 7) = -3(x - (-4))$
$y - 7 = -3(x + 4)$
$y - 7 = -3x - 12$
$3x + y = -5$

15) $x + 2y = -4$
$2(x + 2y) = 2(-4)$
$2x + 4y = -8$
Add the equations.
$3x - 4y = -17$
$+ \quad 2x + 4y = -8$
$5x = -25$
$x = -5$

$$\begin{aligned} x+2y&=-4 \\ (-5)+2y&=-4 \\ 2y&=1 \\ y&=\frac{1}{2} \end{aligned}$$

$$\left(-5, \frac{1}{2}\right)$$

17) $(6q^2+7q-1)-4(2q^2-5q+8)$
$+3(-9q-4)$
$=6q^2+7q-1-8q^2+20q-32-27q-12$
$=-2q^2-45$

19) $(3a-11)(3a+11)=9a^2-121$

21)
$$\begin{array}{r} 5p^2+p-7 \\ p-3\overline{)\ 5p^3-14p^2-10p+5} \\ \underline{-(5p^3-15p^2)} \\ p^2-10p \\ \underline{-(p^2-3p)} \\ -7p+5 \\ \underline{-(-7p+21)} \\ -16 \end{array}$$

$(5p^3-14p^2-10p+5)\div(p-3)$
$=5p^2+p-7-\dfrac{16}{p-3}$

23) $5c(c-4)^2=5c(c^2-8c+16)$
$=5c^3-40c^2+80c$

25) $g(x)=-3x^2+2x+9,\ x=-2$

$g(-2)=-3(-2)^2+2(-2)+9=-12-4+9=-7$

Section 7.1 Exercises

1) $28 = 7 \cdot 2 \cdot 2$
$21c = 7 \cdot 3 \cdot c$
GCF of 28 and $21c$ is 7

3) $18p^3 = 2 \cdot 3 \cdot 3 \cdot p \cdot p \cdot p,$
$12p^2 = 2 \cdot 2 \cdot 3 \cdot p \cdot p$
GCF of $18p^3$ and $12p^2$
is $2 \cdot 3 \cdot p \cdot p = 6p^2$

5) $4n^6$ 7) $5a^2b$

9) $21r^3s^2$ 11) ab

13) $(k-9)$

15) Answers may vary

17) *Yes*

19) *No*

21) *Yes*

23) $2(w+5)$

25) $9(2z^2-1)$

27) $10m(10m^2-3)$

29) $r^2(r^7+1)$

31) $\frac{1}{5}y(y+4)$

33) Does not factor

35) GCF is $5n^3$
$10n^5 - 5n^4 + 40n^3$
$= (5n^3)(2n^2) - (5n^3)(n) + (5n^3)(8)$
$= 5n^3(2n^2 - n + 8)$

37) GCF is $8p^3$
$40p^6 + 40p^5 - 8p^4 + 8p^3$
$= (8p^3)(5p^3) + (8p^3)(5p^2)$
$- (8p^3)(p) + (8p^3)(1)$
$= 8p^3(5p^3 + 5p^2 - p + 1)$

39) GCF is $9a^2b$
$63a^3b^3 - 36a^3b^2 + 9a^2b$
$= (9a^2b)(7ab^2) - (9a^2b)(4ab) + (9a^2b)(1)$
$= 9a^2b(7ab^2 - 4ab + 1)$

41) GCF is -6
$-30m - 42 = -6(5m+7)$

43) GCF is $-4w^3$
$-12w^5 - 16w^3$
$= -4w^3(3w^2+4)$

45) GCF is -1
$-k + 3 = -1(k-3)$

47) $u(t-5) + 6(t-5)$
$= (t-5)(u+6)$

49) $y(6x+1) - z(6x+1)$
$= (6x+1)(y-z)$

51) $p(q+12) + (q+12)$
$= (q+12)(p+1)$

53) $5h^2(9k+8) - (9k+8)$
$= (9k+8)(5h^2-1)$

55) Factor out a from the first two terms
Factor out 7 from the last two terms
$ab+2a+7b+14$
$=a(b+2)+7(b+2)=(a+7)(b+2)$

57) Factor out r from the first two terms
Factor out -9 from the last two terms
$3rt+4r-27t-36$
$=r(3t+4)-9(3t+4)=(r-9)(3t+4)$

59) Factor out $4b$ from the first two terms
Factor out c^2 from the last two terms
$8b^2+20bc+2bc^2+5c^3$
$=4b(2b+5c)+c^2(2b+5c)$
$=(4b+c^2)(2b+c)$

61) $fg-7f+4g-28$
$=f(g-7)+4(g-7)$
$=(f+4)(g-7)$

63) $st-10s-6t+60$
$=s(t-10)-6(t-10)$
$=(t-10)(t-6)$

65) $5tu+6t-5u-6$
$=t(5u+6)-1(5u+6)$
$=(5u+6)(t-1)$

67) $36g^4+3gh-96g^3h-8h^2$
$=3g(12g^3+h)-8h(12g^3+h)$
$=(3g-8h)(12g^3+h)$

69) Answers may vary.

71) $4xy+12x+20y+60=4(xy+3x+5y+15)$
Group the terms and factor out the GCF from each group.
$4(x+5)(y+3)$

73) $3cd+6c+21d+42$
$=3(cd+2c+7d+14)$
$=3[c(d+2)+7(d+2)]$
$=3(c+7)(d+2)$

75) $2p^2q-10p^2-8pq+40p$
$=2p(pq-5p-4q+20)$
$=2p[p(q-5)-4(q-5)]$
$=2p(q-5)(p-4)$

77) $10st+5s-12t-6$
$=5s(2t+1)-6(2t+1)$
$=(5s-6)(2t+1)$

79) $3a^3-21a^2b-2ab+14b^2$
$=3a^2(a-7b)-2b(a-7b)$
$=(a-7b)(3a^2-2)$

81) $8u^2v^2+16u^2v+10uv^2+20uv$
$=2uv(4uv+8u+5v+10)$
$=2uv[4u(v+2)+5(v+2)]$
$=2uv(4u+5)(v+2)$

83) $3mn+21m+10n+70$
$=3m(n+7)+10(n+7)$
$=(3m+10)(n+7)$

85) $16b-24=8(2b-3)$

87) $cd+6c-4d-24$
$=c(d+6)-4(d+6)$
$=(c-4)(d+6)$

89)
$$6a^4b+12a^4-8a^3b-16a^3$$
$$=2a^3(3ab+6a-4b-8)$$
$$=2a^3[3a(b+2)-4(b+2)]$$
$$=2a^3(3a-4)(b+2)$$

91) $7cd+12+28c+3d$
$$7cd+28c+3d+12$$
$$=7c(d+4)+3(d+4)$$
$$=(7c+3)(d+4)$$

93) $dg-d+g-1$
$$=d(g-1)+1(g-1)$$
$$=(d+1)(g-1)$$

95) $x^4y^2+12x^3y^3$
$$=x^3y^2(x+12y)$$

97) $4nm+8m+12n+24$
$$=4[nm+2m+3n+6]$$
$$=4[m(n+2)+3(n+2)]$$
$$=4(m+3)(n+2)$$

99) $-6p^2-20p+2=-2(3p^2+10p-1)$

Section 7.2 Exercises

1) a) $5,2$ b) $-8,7$

c) $5,-1$ d) $-9,-4$

3) They are negative.

5) Can I factor out a GCF?

7) Can I factor again? If so, factor again.

9) $n^2+7n+10=(n+5)(n+2)$

11) $c^2-16c+60=(c-6)(c-10)$

13) $x^2+x-12=(x-3)(x+4)$

15) $g^2+8g+12=(g+6)(g+2)$

17) $y^2+10y+16=(y+8)(y+2)$

19) $w^2-17w+72=(w-9)(w-8)$

21) $b^2-3b-4=(b-4)(b+1)$

23) $z^2+6z-11$ prime

25) $c^2-13c+36=(c-9)(c-4)$

27) $m^2+4m-60=(m+10)(m-6)$

29) $r^2-4r-96=(r+8)(r-12)$

31) $q^2+12q+42=$ prime

33) $x^2+16x+64=(x+8)(x+8)$ or $(x+8)^2$

35) $n^2-2n+1=(n-1)(n-1)$ or $(n-1)^2$

37) $24+14d+d^2=d^2+14d+24$
$$=(d+12)(d+2)$$

39) $-56+12a+a^2=a^2+12a-56$
$$=\text{prime}$$

41) $2k^2-22k+48=2(k^2-11k+24)$
$$=2(k-8)(k-3)$$

43) $50h+35h^2+5h^3=5h(10+7h+h^2)$
$$=5h(h^2+7h+10)$$
$$=5h(h+2)(h+5)$$

45) $r^4 + r^3 - 132r^2 = r^2(r^2 + r - 132)$
$= r^2(r+12)(r-11)$

47) $7q^3 - 49q^2 - 42q = 7q(q^2 - 7q - 6)$

49) $3z^4 + 24z^3 + 48z^2 = 3z^2(z^2 + 8z + 16)$
$= 3z^2(z+4)(z+4)$
or $3z^2(z+4)^2$

51) $xy^3 - 2xy^2 - 63xy = xy(y^2 - 2y - 63)$
$xy(y-9)(y+7)$

53) $-m^2 - 12m - 35 = -(m^2 + 12m + 35)$
$= -(m+5)(m+7)$

55) $-c^2 - 3c + 28 = -(c^2 + 3c - 28)$
$= -(c+7)(c-4)$

57) $-z^2 + 13z - 30 = -(z^2 - 13 + 30)$
$= -(z-10)(z-3)$

59) $-p^2 + p + 56 = -(p^2 - p - 56)$
$= -(p-8)(p+7)$

61) $x^2 + 7xy + 12y^2 = (x+4y)(x+3y)$

63) $c^2 - 7cd - 8d^2 = (c-8d)(c+d)$

65) $u^2 - 14uv + 45v^2 = (u-9v)(u-5v)$

67) $m^2 + 4mn - 21n^2 = (m+7n)(m-3n)$

69) $a^2 + 24ab + 144b^2 = (a+12b)(a+12b)$ or
$= (a+12b)^2$

71) No;
$3x^2 + 21x + 30 = 3(x^2 + 7x + 10)$
$= 3(x+5)(x+2)$

73) yes

75) $2x^2 + 16x + 30 = 2(x^2 + 8x + 15)$
$= 2(x+3)(x+5)$

77) $n^2 - 6n + 8 = (n-4)(n-2)$

79) $m^2 + 7mn - 44n^2 = (m+11n)(m-4n)$

81) $h^2 - 10h + 32 =$ prime

83) $4q^3 - 28q^2 + 48q = 4q(q^2 - 7q + 12)$
$= 4q(q-4)(q-3)$

85) $-k^2 - 18k - 81 = -1(k^2 + 18k + 81)$
$= -1(k+9)(k+9)$
or $-(k+9)^2$

87) $4h^5 + 32h^4 + 28h^3 = 4h^3(h^2 + 8h + 7)$
$= 4h^3(h+7)(h+1)$

89) $k^2 + 21k + 108 = (k+12)(k+9)$

91) $p^3q - 17p^2q^2 + 70pq^3 = pq(p^2 - 17pq + 70q^2)$
$= pq(p-10q)(p-7q)$

93) $a^2 + 9ab + 24b^2 =$ prime

95) $x^2 - 13xy + 12y^2 = (x-12y)(x-y)$

97) $5v^5 + 55v^4 - 45v^3 = 5v^3(v^2 + 11v - 9)$

99) $6x^3y^2 - 48x^2y^2 - 54xy^2$
$= 6xy^2(x^2 - 8x - 9)$
$= 6xy^2(x-9)(x+1)$

101) $36-13z+z^2=z^2-13z+36$
$=(z-9)(z-4)$

103) $a^2b^2+13ab+42=(ab+7)(ab+6)$

105) $(x+y)z^2+7(x+y)z-30(x+y)$
$=(x+y)(z^2+7z-30)$
$=(x+y)(z+10)(z-3)$

107) $(a-b)c^2-11(a-b)c+28(a-b)$
$=(a-b)(c^2-11c+28)$
$=(a-b)(c-7)(c-4)$

109) $(p+q)r^2+24(p+q)r+144(p+q)$
$=(p+q)(r^2+24r+144)$
$=(p+q)(r+12)(r+12)$ or
$=(p+q)(r+12)^2$

111) $n^2+2n-24=(n+6)(n-4)$

Section 7.3 Exercises

1) a) $10, -5$ b) $-27, -1$

c) $6, 2$ d) $-12, 6$

3) $3c^2+12c+8c+32$
$=3c(c+4)+8(c+4)$
$=(3c+8)(c+4)$

5) $6k^2-6k-7k+7$
$=6k(k-1)-7(k-1)$
$=(6k-7)(k-1)$

7) $6x^2-27xy+8xy-36y^2$
$=3x(2x-9y)+4y(2x-9y)$
$=(3x+4y)(2x-9y)$

9) Can I factor out a GCF?

11) $(4k+9)(k+2)$
$=4k^2+8k+9k+18$
$=4k^2+17k+18$

13) $5t^2+13t+6=(5t+3)(\quad)$
$=5t^2+10t+3t+6$
$=5t(t+2)+3(t+2)$
$=(5t+3)(t+2)$

15) $6a^2-11a-10=(2a-5)(3a+2)$

17) $12x^2-25xy+7y^2$
$=(4x-7y)(3x-y)$

19) $2h^2+13h+15=2h^2+10h+3h+15$
$=2h(h+5)+3(h+5)$
$=(2h+3)(h+5)$

21) $7y^2-11y+4=7y^2-7y-4y+4$
$=7y(y-1)-4(y-1)$
$=(7y-4)(y-1)$

23) $5b^2+9b-18=5b^2+15b-6b-18$
$=5b(b+3)-6(b+3)$
$=(5b-6)(b+3)$

25) $6p^2+p-2=6p^2-3p+4p-2$
$=3p(2p-1)+2(2p-1)$
$=(3p+2)(2p-1)$

27) $4t^2+16t+15=4t^2+10t+6t+15$
$=2t(2t+5)+3(2t+5)$
$=(2t+5)(2t+3)$

29) $9x^2-13xy+4y^2=9x^2-9xy-4xy+4y^2$
$=9x(x-y)-4y(x-y)$
$=(9x-4y)(x-y)$

31) because 2 can be factored out of $(2x-4)$, but 2 cannot be factored out of $(2x^2+13x-24)$

33) $2r^2+9r+10=(2r+5)(r+2)$

35) $3u^2-23u+30=(3u-5)(u-6)$

37) $7a^2+31a-20=(7a-4)(a+5)$

39) $6y^2+23y+10=(3y+10)(2y+1)$

41) $9w^2+20w-21=(9w-7)(w+3)$

43) $8c^2-42c+27=(4c-3)(2c-9)$

45) $4k^2+40k+99=(2k+11)(2k+9)$

47) $20b^2-32b-45=(10b+9)(2b-5)$

49) $2r^2+13rt-24t^2=(2r-3t)(r+8t)$

51) $6a^2-25ab+4b^2=(6a-b)(a-4b)$

53) $(4z-3)(z+2)$; the answer is the same.

55) $3p^2-16p-12=(3p+2)(p-6)$

57) $4k^2+15k+9=(4k+3)(k+3)$

59) $30w^3+76w^2+14w=2w(15w^2+38w+7)$
$=2w(5w+1)(3w+7)$

61) $21r^2-90r+24=3(7r^2-30r+8)$
$=3(7r-2)(r-4)$

63) $6y^2-10y+3=$ prime

65) $42b^2+11b-3=(7b+3)(6b-1)$

67) $7x^2-17xy+6y^2=(7x-3y)(x-2y)$

69) $2d^2+2d-40=2(d^2+d-20)$
$=2(d+5)(d-4)$

71) $30r^4t^2+23r^3t^2+3r^2t^2$
$=r^2t^2(30r^2+23r+3)$
$=r^2t^2(6r+1)(5r+3)$

73) $9k^2-42k+49=(3k-7)^2$

75) $2m^2(n+9)-5m(n+9)-7(n+9)$
$=(n+9)(2m^2-5m-7)$
$=(n+9)(2m-7)(m+1)$

77) $6v^2(u+4)^2+23v(u+4)^2+20(u+4)^2$
$=(u+4)^2(6v^2+23v+20)$
$=(u+4)^2(2v+5)(3v+4)$

79) $15b^2(2a-1)^4-28b(2a-1)^4+12(2a-1)^4$
$=(2a-1)^4(15b^2-28b+12)$
$=(2a-1)^4(5b-6)(3b-2)$

81) $-n^2-8n+48=-(n^2+8n-48)$
$=-(n+12)(n-4)$

83) $-7a^2+4a+3=-1(7a^2-4a-3)$
$=-1(7a-3)(a+1)$

85) $-10z^2+19z-6=-(10z^2-19z+6)$
$=-(5z-2)(2z-3)$

87) $-20m^3-120m^2-135m$
$=-5m(4m^2+24m+27)$
$=-5m(2m+9)(2m+3)$

89) $-6a^3b+11a^2b^2+2ab^3$
$=-ab(6a^2-11ab-2b^2)$
$=-ab(6a+b)(a-2b)$

Section 7.4 Exercises

1) a) $7^2=49$ b) $9^2=81$

c) $6^2=36$ d) $10^2=100$

e) $5^2=25$ f) $4^2=16$

g) $11^2=121$ h) $\left(\frac{1}{3}\right)^2=\frac{1}{9}$

i) $\left(\frac{3}{8}\right)^2=\frac{9}{64}$

3) a) $(c^2)^2=c^4$ b) $(3r)^2=9r^2$

c) $(7k)^2=49k^2$ d) $(6m^2)^2=36m^4$

e) $\left(\frac{1}{2}\right)^2=\frac{1}{4}$ f) $\left(\frac{12}{5}\right)^2=\frac{144}{5}$

5) $y^2+12y+36$

7) The middle term does not equal $2(2a)(-3)$. It would have to equal $-12a$ to be a perfect square trinomial.

9) $h^2+10h+25=(h)^2+(2\cdot h\cdot 5)+(5)^2$
$=(h+5)^2$

11) $b^2-14b+49=(b)^2-(2\cdot b\cdot 7)+(7)^2$
$=(b-7)^2$

13) $4w^2+4w+1$
$=(2w)^2+(2\cdot 2w\cdot 1)+(1)^2$
$=(2w+1)^2$

15) $9k^2-24k+16$
$=(3k)^2-(2\cdot 3k\cdot 4)+(4)^2$
$=(3k-4)^2$

17) $c^2+c+\frac{1}{4}$
$=(c)^2+\left(2\cdot c\cdot\frac{1}{2}\right)+\left(\frac{1}{2}\right)^2$
$=\left(c+\frac{1}{2}\right)^2$

19) $k^2-\frac{14}{5}k+\frac{49}{25}$
$=(k)^2-\left(2\cdot k\cdot\frac{7}{5}\right)+\left(\frac{7}{5}\right)^2$
$=\left(k-\frac{7}{5}\right)^2$

21) $a^2+8ab+16b^2$
$=(a)^2+(2\cdot a\cdot 4b)+(4b)^2$
$=(a+4b)^2$

23) $25m^2-30mn+9n^2$
$=(5m)^2-(2\cdot 5m\cdot 3n)+(3n)^2$
$=(5m-3n)^2$

25) $4f^2+24f+36$
$=4(f^2+6f+9)$
$=4\left[(f)^2+(2\cdot f\cdot 3)+(3)^2\right]$
$=4(f+3)^2$

27) $5a^4-30a^3+45a^2$
$=5a^2(a^2-6a+9)$
$=5a^2\left[(a)^2-(2\cdot p\cdot 3)+(3)^2\right]$
$=5a^2(a-3)^2$

29) $-16y^2-80y-100$
$=-4(4y^2+20y+25)$
$=-4\left[(2y)^2+(2\cdot 2y\cdot 5)+(5)^2\right]$
$=-4(2y+5)^2$

31) $75h^3-6h^2+12h=3h(25h^2-2h+4)$

33) a) x^2-81 b) $81-x^2$

35) $w^2-64=(w)^2-(8)^2=(w+8)(w-8)$

37) $121-p^2=(11)^2-(p)^2$
$=(11+p)(11-p)$

39) $64c^2-25b^2=(8c)^2-(5b)^2=(8c+5b)(8c-5b)$

41) $k^2-4=(k+2)(k-2)$

43) $c^2-25=(c+5)(c-5)$

45) $w^2+9=$ prime

47) $x^2-\frac{1}{9}=(x+\frac{1}{3})(x-\frac{1}{3})$

49) $a^2-\frac{4}{49}=(a+\frac{2}{7})(a-\frac{2}{7})$

51) $144-v^2=(12+v)(12-v)$

53) $1-h^2=(1+h)(1-h)$

55) $\frac{36}{25}-b^2=(\frac{6}{5}+b)(\frac{6}{5}-b)$

57) $100m^2-49=(10m+7)(10m-7)$

59) $169k^2-1=(13k+1)(13k-1)$

61) $4y^2+49=$ prime

63) $\frac{1}{9}t^2-\frac{25}{4}=(\frac{1}{3}t+\frac{5}{2})(\frac{1}{3}t-\frac{5}{2})$

65) $w^4-100=(w^2+10)(w^2-10)$

67) $36c^2-d^4=(6c+d^2)(6c-d^2)$

69) $r^4-1=(r^2+1)(r^2-1)$
$=(r^2+1)(r+1)(r-1)$

71) $r^4-81t^4=(r^2+9t^2)(r^2-9t^2)$
$=(r^2+9t^2)(r+3t)(r-3t)$

73) $5u^2-45=5(u^2-9)$
$=5(u+3)(u-3)$

75) $2n^2-288=2(n^2-144)$
$=2(n+12)(n-12)$

77) $12z^4 - 75z^2 = 3z^2(4z^2 - 25)$
$= 3z^2(2z+5)(2z-5)$

79) a) $4^3 = 64$ b) $1^3 = 1$

c) $10^3 = 1000$ d) $3^3 = 27$

e) $5^3 = 125$ f) $2^3 = 8$

81) a) $(m)^3 = m^3$ b) $(3t)^3 = 27t^3$

c) $(2b)^3 = 8b^3$ d) $(h^2)^3 = h^6$

83) $y^3 + 8 = (y)^3 + (2)^3$
$= (y+2)(y^2 - 2y + 4)$

85) $t^3 + 64 = (t)^3 + (4)^3$
$= (t+4)(t^2 - 4t + 16)$

87) $z^3 - 1 = (z)^3 - (1)^3$
$= (z-1)(z^2 + z + 1)$

89) $27m^3 - 125$
$= (3m)^3 - (5)^3$
$= (3m-5)(9m^2 + 15m + 25)$

91) $125y^3 - 8 = (5y)^3 - (2)^3$
$= (5y-2)(25y^2 + 10y + 4)$

93) $1000c^3 - d^3$
$= (10c)^3 - (d)^3$
$= (10c - d)(100c^2 + 10cd + d^2)$

95) $8j^3 + 27k^3$
$= (2j)^3 + (3k)^3$
$= (2j + 3k)(4j^2 - 6jk + 9k^2)$

97) $64x^3 + 125y^3$
$= (4x)^3 + (5y)^3$
$= (4x + 5y)(16x^2 - 20xy + 25y^2)$

99) $6c^3 + 48 = 6(c^3 + 8)$
$= 6[(c)^3 + (2)^3]$
$= 6(c+2)(c^2 - 2c + 4)$

101) $7v^3 - 7000w^3$
$= 7(v^3 - 1000w^3)$
$= 7[(v)^3 - (10w)^3]$
$= 7(v - 10w)(v^2 + 10vw + 100w^2)$

103) $h^6 - 64 = (h^3)^2 - (8)^2 = (h^3 + 8)(h^3 - 8) = [(h)^3 + (2)^3][(h)^3 - (2)^3]$
$= (h+2)(h^2 - 2h + 4)(h-2)(h^2 + 2h + 4)$

105) $(d+4)^2-(d-3)^2=\left[(d+4)+(d-3)\right]\left[(d+4)-(d-3)\right]$
$=(2d+1)(d+4-d+3)$
$=7(2d+1)$

107) $(3k+1)^2-(k+5)^2=\left[(3k+1)+(k+5)\right]\left[(3k+1)-(k+5)\right]$
$=(4k+6)(3k+1-k-5)$
$=(4k+6)(2k-4)=4(2k+3)(k-2)$

109) $(r-2)^3+27=(r-2)^3+(3)^3$
$=(r-2+3)[(r-2)^2-(r-2)(3)+9]$
$=(r+1)(r^2-4r+4-3r+6+9)$
$=(r+1)(r^2-7r+19)$

111) $(c+4)^3-125=(c+4)^3-(5)^3$
$=(c+4-5)[(c+4)^2+5(c+4)+25]$
$=(c-1)(c^2+8c+16+5c+20+25)$
$=(c-1)(c^2+13c+61)$

Putting It All Together

1) $c^2+15c+56=(c+7)(c+8)$

3) $uv+6u+9v+54$
$=u(v+6)+9(v+6)$
$=(u+9)(v+6)$

5) $2p^2-13p+21=(2p-7)(p-3)$

7) $9v^5+90v^4-54v^3=9v^3(v^2+10v-6)$

9) $24q^3+52q^2-32q=4q\left(6q^3+13q^2-8q\right)$
$=4q(3q+8)(2q-1)$

11) $g^3+125=(g+5)(g^2-5g+25)$

13) $144-w^2=(12+w)(12-w)$

15) $9r^2 + 12rt + 4t^2 = (3r + 2t)^2$

17) $7n^4 - 63n^3 - 70n^2 = 7n^2\left(n^2 - 9n - 10\right)$
$= 7n^2(n - 10)(n + 1)$

19) $9h^2 + 25 =$ prime

21) $40x^3 - 135 = 5(8x^3 - 27)$
$= 5(2x - 3)(4x^2 + 6x + 9)$

23) $m^2 - \dfrac{1}{100} = (m + \dfrac{1}{10})(m - \dfrac{1}{10})$

25) $20x^2y + 6 - 24x^2 - 5y$
$= 20x^2y - 24x^2 - 5y + 6$
$= 4x^2(5y - 6) - 1(5y - 6)$
$= (4x^2 - 1)(5y - 6)$
$= (2x + 1)(2x - 1)(5y - 6)$

27) $p^2 + 17pq + 30q^2 = (p + 15q)(p + 2q)$

29) $t^2 - 2t - 16 =$ prime

31) $50n^2 - 40n + 8 = 2\left(25n^2 - 20n + 4\right)$
$= 2(5n - 2)^2$

33) $36r^2 + 57rs + 21s^2 = 3(12r^2 + 19rs + 7s^2)$
$= 3(12r + 7s)(r + s)$

35) $81x^4 - y^4 = (9x^2 + y^2)(9x^2 - y^2)$
$= (9x^2 + y^2)(3x + y)(3x - y)$

37) $2a^2 - 10a - 72 = 2\left(a^2 - 5a - 36\right)$
$= 2\left(a - 9\right)\left(a + 4\right)$

39) $h^2 - \dfrac{2}{5}h + \dfrac{1}{25} = (h - \dfrac{1}{5})^2$

41) $16uv + 24u - 10v - 15$
$= 8u(2v + 3) - 5(2v + 3)$
$= (8u - 5)(2v + 3)$

43) $8b^2 - 14b - 15 = \left(4b + 3\right)\left(2b - 5\right)$

45) $8y^4z^3 - 28y^3z^3 - 40y^3z^2 + 4y^2z^2$
$= 4y^2z^2(2y^2z - 7yz - 10y + 1)$

47) $2a^2 - 7a + 8 =$ prime

49) $16u^2 + 40uv + 25v^2 = (4u + 5v)^2$

51) $24k^2 + 31k - 15 = (8k - 3)(3k + 5)$

53) $5s^3 - 320t^3 = 5\left(s^3 - 64t^3\right)$
$= 5(s - 4t)(s^2 + 4st + 16t^2)$

55) $ab - a - b + 1 = a(b - 1) - 1(b - 1)$
$= (a - 1)(b - 1)$

57) $7h^2 - 7 = 7\left(h^2 - 1\right)$
$= 7\left(h + 1\right)\left(h - 1\right)$

59) $6m^2 - 60m + 150 = 6\left(m^2 - 10m + 25\right)$
$= 6(m - 5)^2$

61) $121z^2 - 169 = (11z + 13)(11z - 13)$

63) $-12r^2 - 75r - 18 = -3\left(4r^2 + 25r + 6\right)$
$= -3(4r + 1)(r + 6)$

65) $n^3 + 1 = (n + 1)(n^2 - n + 1)$

67) $81u^4 - v^4$
$= \left(9u^2 + v^2\right)\left(9u^2 - v^2\right)$
$= \left(9u^2 + v^2\right)\left(3u + v\right)\left(3u - v\right)$

69) $13h^2+15h+2=(13h+2)(h+1)$

71) $5t^7-8t^4=t^4(5t^3-8)$

73) $d^2-7d-30=(d-10)(d+3)$

75) $z^2+144=$ prime

77) $r^2+2r+1=(r+1)^2$

79) $49n^2-100=(7n+10)(7n-10)$

81) $(2z+1)y^2+6(2z+1)y-55(2z+1)$
$=(2z+1)(y^2+6y-55)$
$=(2z+1)(y+11)(y-5)$

83) $(t-3)^2+3(t-3)-4$
$=[(t-3)+4][(t-3)-1]$
$=(t+1)(t-4)$

85) $(z+7)^2-11(z+7)+28$
$=[(z+7)-7][(z+7)-4]$
$=z(z+3)$

87) $(a+b)^2-(a-b)^2$
$=[(a+b)+(a-b)][(a+b)-(a-b)]$
$=[2a][2b]$
$=4ab$

89) $(5p-2q)^2-(2p+q)^2$
$=[(5p-2q)+(2p+q)][(5p-2q)-(2p+q)]$
$=(7p-q)(3p-3q)$
$=3(7p-q)(p-q)$

91) $(r+2)^3+27=(r+2+3)[(r+2)^2-3(r+2)+9]$
$=(r+5)(r^2+4r+4-3r-6+9)$
$=(r+5)(r^2+r+7)$

93) $(k-7)^3-1=(k-7-1)[(k-7)^2+(k-7)+1]$
$=(k-8)(k^2-14k+49+k-7+1)$
$=(k-8)(k^2-13k+43)$

95) $a^2-8a+16-b^2=(a-4)^2-b^2$
$=(a-4+b)(a-4-b)$

97) $s^2+18s+81-t^2=(s+9)^2-t^2$
$=(s+t+9)(s-t+9)$

Section 7.5 Exercises

1) $ax^2+bx+c=0$

3) *a*) $5x^2+3x-7=0$ quadratic
b) $6(p+1)=0$ linear
c) $(n+4)(n-9)=8$ quadratic
d) $2w+3(w-5)=4w+9$ linear

5) It says that if the product of two quantities equals to 0, then one or both of the quantities must be zero.

7) $(z+11)(z-4)=0$
$z+11=0$ or $z-4=0$
$z=-11$ or $z=4$ $\{-11,4\}$

9) $(2r-3)(r-10)=0$
$2r-3=0$ or $r-10=0$
$r=\frac{3}{2}$ or $r=10$ $\left\{\frac{3}{2},10\right\}$

11) $d(d-12)=0$

$d=0$ or $d-12=0$

$d=0$ or $d=12$ $\{0,12\}$

13) $(3x+5)^2=0$

$3x+5=0$

$x=-\frac{5}{3}$ $\left\{\frac{5}{3}\right\}$

15) $(9h+2)(2h+1)=0$

$9h+2=0$ or $2h+1=0$

$h=-\frac{2}{9}$ or $h=-\frac{1}{2}$

$\left\{-\frac{2}{9},-\frac{1}{2}\right\}$

17) $(m+\frac{1}{4})(m-\frac{2}{5})=0$

$m+\frac{1}{4}=0$ or $m-\frac{2}{5}=0$

$m=-\frac{1}{4}$ or $m=\frac{2}{5}$

$\left\{-\frac{1}{4},\frac{2}{5}\right\}$

19) $n(n-4.6)=0$

$n=0$ or $n-4.6=0$

$n=0$ or $n=4.6$ $\{0,4.6\}$

21) No, the product of the factors must equal zero.

23) $p^2+8p+12=0$

$(p+6)(p+2)=0$

$p=-6$ or $p=-2$

$\{-6,-2\}$

25) $t^2-t-110=0$

$(t-11)(t+10)=0$

$t=11$ or $t=-10$ $\{11,-10\}$

27) $3a^2-10a+8=0\rightarrow$

$(3a-4)(a-2)=0$

$a=\frac{4}{3}$ or $a=2$ $\left\{\frac{4}{3},2\right\}$

29) $12z^2+z-6=0$

$(4z+3)(3z-2)=0$

$z=-\frac{3}{4}$ or $z=\frac{2}{3}$

$\left\{-\frac{3}{4},\frac{2}{3}\right\}$

31) $r^2=60-7r$

$r^2+7r-60=0$

$(r+12)(r-5)=0$

$r=-12$ or $r=5$

$\{-12,5\}$

33) $d^2-15d=-54$

$d^2-15d+54=0$

$(d-9)(d-6)=0$

$d=9$ or $d=6$

$\{9,6\}$

35) $x^2-64=0$

$(x+8)(x-8)=0$

$x=-8$ or $x=8$

$\{-8,8\}$

37) $49 = 100u^2$

$100u^2 - 49 = 0$

$(10u+7)(10u-7) = 0$

$u = -\frac{7}{10}$ or $u = \frac{7}{10}$ $\left\{-\frac{7}{10}, \frac{7}{10}\right\}$

39) $22k = -10k^2 - 12$

$10k^2 + 22k + 12 = 0$ Divide by 2

$5k^2 + 11k + 6 = 0$

$(5k+6)(k+1) = 0$

$k = -\frac{6}{5}$ or $k = -1$

$\left\{-\frac{5}{6}, -1\right\}$

41) $v^2 = 4v$

$v^2 - 4v = 0$

$v(v-4) = 0$

$v = 0$ or $v = 4$ $\{0, 4\}$

43) $(z+3)(z+1) = 15$

$z^2 + z + 3z + 3 = 15$

$z^2 + 4z - 12 = 0$

$(z+6)(z-2) = 0$

$z = -6$ or $z = 2$

$\{-6, 2\}$

45) $t(19-t) = 84$

$19t - t^2 = 84$

$t^2 - 19t + 84 = 0$

$(t-12)(t-7) = 0$

$t = 12$

or $t = 7$ $\{12, 7\}$

47) $6k(k+4) + 3 = 5(k^2 - 12) + 8k$

$6k^2 + 24k + 3 = 5k^2 - 60 + 8k$

$k^2 + 16k + 63 = 0$

$(k+9)(k+7) = 0$

$k = -9$ or $k = -7$ $\{-9, -7\}$

49) $3(n^2 - 15) + 4n = 4n(n-3) + 19$

$3n^2 - 45 + 4n = 4n^2 - 12n + 19$

$n^2 - 16n + 64 = 0$

$(n-8)^2 = 0$

$n = 8$ $\{8\}$

51) $\frac{1}{2}(m+1)^2 = -\frac{3}{4}m(m+5) - \frac{5}{2}$

Multiply by 4

$2(m+1)^2 = -3m(m+5) - 10$

$2m^2 + 4m + 2 = -3m^2 - 15m - 10$

$5m^2 + 19m + 12 = 0$

$(5m+4)(m+3) = 0$

$m = -\frac{4}{5}$ or $m = -3$

$\left\{-\frac{4}{5}, -3\right\}$

53) No. You cannot divide an equation by a variable because you may eliminate a solution and may be dividing by zero.

55) $7w(8w-9)(w+6) = 0$

$7w = 0$ or $8w - 9 = 0$

or $w + 6 = 0$

$w = 0$ or $w = \frac{9}{8}$ or $w = -6$

$\left\{0, \frac{9}{8}, -6\right\}$

57) $(6m+7)(m^2-5m+6)=0$
$(6m+7)(m-2)(m-3)=0$
$6m+7=0$ or $m-2=0$ or $m-3=0$
$m=-\frac{7}{6}$ or $m=2$ or $m=3$
$\left\{-\frac{7}{6},2,3\right\}$

59) $49h=h^3$
$h^3-49h=0$
$h(h^2-49)=0$
$h(h+7)(h-7)=0$
$h=0$ or $h=-7$ or $h=7$
$\{0,-7,7\}$

61) $5w^2+36w=w^3$
$w^3-5w^2-36w=0$
$w(w^2-5w-36)=0$
$w(w-9)(w+4)=0$
$w=0$ or $w=9$ or $w=-4$
$\{0,9,-4\}$

63) $60a=44a^2-8a^3$
$8a^3-44a^2+60a=0$
$4a(2a^2-11a+15)=0$
$4a(2a-5)(a-3)=0$
$a=0$ or $a=\frac{5}{2}$ or $a=3$
$\left\{0,\frac{5}{2},3\right\}$

65) $162b^3-8b=0$
$2b(81b^2-4)=0$
$2b(9b+2)(9b-2)=0$
$b=0$ or $b=-\frac{2}{9}$ or $b=\frac{2}{9}$
$\left\{0,-\frac{2}{9},\frac{2}{9}\right\}$

67) $-63=4y(y-8)$
$4y^2-32y+63=0$
$(2y-7)(2y-9)=0$
$y=\frac{7}{2}$ or $y=\frac{9}{2}$
$\left\{\frac{7}{2},\frac{9}{2}\right\}$

69) $\frac{1}{2}d(2-d)-\frac{3}{2}=\frac{2}{5}d(d+1)-\frac{7}{5}$
Multiply by 10
$5d(2-d)-15=4d(d+1)-14$
$10d-5d^2-15=4d^2+4d-14$
$9d^2-6d+1=0$
$(3d-1)^2=0$
$d=\frac{1}{3}$ $\left\{\frac{1}{3}\right\}$

71) $a^2-a=30$
$a^2-a-30=0$
$(a-6)(a+5)=0$
$a=6$
or $a=-5$ $\{6,-5\}$

73) $48t = 3t^3$

$3t^3 - 48t = 0$

$3t(t^2 - 16) = 0$

$3t(t+4)(t-4) = 0$

$t = 0$ or $t = -4$

or $t = 4$ $\{0, -4, 4\}$

75) $104r + 36 = 12r^2$

$12r^2 - 104r - 36 = 0$ Divide by 4

$3r^2 - 26r - 9 = 0$

$(3r+1)(r-9) = 0$

$r = -\frac{1}{3}$

or $r = 9$ $\left\{-\frac{1}{3}, 9\right\}$

77) $w^2 - 121 = 0$

$(w+11)(w-11) = 0$

$w = -11$ or $w = 11$

$\{-11, 11\}$

79) $(2n-5)(n^2 - 6n + 9) = 0$

$(2n-5)(n-3)^2 = 0$

$n = \frac{5}{2}$

or $n = 3$

$\left\{\frac{5}{2}, 3\right\}$

81) $(2d-5)^2 - (d+6)^2 = 0$

$(2d - 5 + d + 6)(2d - 5 - d - 6) = 0$

$(3d+1)(d-11) = 0$

$d = -\frac{1}{3}$

or $d = 11$

$\left\{-\frac{1}{3}, 11\right\}$

83) $(11z-4)^2 - (2z+5)^2 = 0$

$(11z - 4 + 2z + 5)(11z - 4 - 2z - 5) = 0$

$(13z+1)(9z-9) = 0$

$z = -\frac{1}{13}$ or $z = 1$ $\{-\frac{1}{13}, 1\}$

85) $2p^2(p-4) + 9p(p-4) + 9(p-4) = 0$

$(p-4)(2p^2 + 9p + 9) = 0$

$(p-4)(p+3)(2p+3) = 0$

$p = 4$

or $p = -3$

or $p = -\frac{3}{2}$

$\left\{4, -3, -\frac{3}{2}\right\}$

87) $10c^2(2c-7)+7(2c-7)=37c(2c-7)$

$10c^2(2c-7)-37c(2c-7)+7(2c-7)=0$

$(2c-7)(10c^2-37c+7)=0$

$(2c-7)(5c-1)(2c-7)=0$

$(2c-7)^2(5c-1)=0$

$c=\frac{7}{2}$ or $c=\frac{1}{5}$

$\left\{\frac{7}{2},\frac{1}{5}\right\}$

89) $h^3+8h^2-h-8=0$

$h^2(h+8)-1(h+8)=0$

$(h+8)(h^2-1)=0$

$(h+8)(h+1)(h-1)=0$

$h=-8$ or $h=-1$ or $h=1$ $\quad\{-8,-1,1\}$

91) $f(x)=x^2+10x+16;\ f(x)=0$

$x^2+10x+16=0$

$(x+8)(x+2)=0$

$x=-8$ or $x=-2$

93) $g(a)=2a^2-13a+24;\quad g(a)=4$

$2a^2-13a+24=4$

$2a^2-13a+20=0$

$(2a-5)(a-4)=0$

$a=\frac{5}{2}$ or $a=4$

95) $P(a)=a^2-12;\quad p(a)=13$

$a^2-12=13$

$a^2-25=0$

$(a+5)(a-5)=0$

$a=5$ or $a=-5$

97) $h(t)=3t^3-21t^2+18t;\quad h(t)=0$

$3t^3-21t^2+18t=0$

$3t(t^2-7t+6)=0$

$3t(t-6)(t-1)=0$

$t=0$ or $t=6$ or $t=1$

Section 7.6 Exercises

1) $x+3=$ length of rectangle

$x=$ width of rectangle

$\text{Area}=(\text{length})(\text{width})$

$28=x(x+3)$

$28=x^2+3x$

$0=x^2+3x-28$

$0=(x+7)(x-4)$

$x-4=0$ or $x+7=0$

$x=4 \qquad x=-7$

length $=4+3=7$ in; width $=4$ in

3) $2x+1=$ base of triangle

$x+3=$ height of triangle

$\text{Area}=\frac{1}{2}(\text{base})(\text{height})$

$44=\frac{1}{2}(2x+1)(x+3)$

$88=2x^2+7x+3$

$0=2x^2+7x-85$

$0=(2x+17)(x-5)$

$2x+17=0$ or $x-5=0$

$2x=-17 \qquad x=5$

$x=-\frac{17}{2}$

base $=2(5)+1=11$ cm;

height $=5+3=8$ cm

5) $x-1=$ base of parallelogram

$\frac{1}{2}x-1=$ height of parallelogram

$$\text{Area}=(\text{base})(\text{height})$$
$$36=(x-1)\left(\frac{1}{2}x-1\right)$$
$$72=(x-1)(x-2)$$
$$72=x^2-3x+2$$
$$0=x^2-3x-70$$
$$0=(x-10)(x+7)$$
$$x-10=0 \text{ or } x+7=0$$
$$x=10 \qquad x=-7$$

base $=10-1=9$ in;

width $=\frac{1}{2}(10)-1=4$ in

7) $x+1=$ width of box

$x-2=$ height of box

$$\text{Volume}=(\text{length})(\text{width})(\text{height})$$
$$648=12(x+1)(x-2)$$
$$648=12(x^2-x-2)$$
$$54=x^2-x-2$$
$$0=x^2-x-56$$
$$0=(x-8)(x+7)$$
$$x-8=0 \text{ or } x+7=0$$
$$x=8 \qquad x=-7$$

width $=8+1=9$ in;

height $=8-2=6$ in

9) $w=$ the width of the sign

$2w=$ the length of the sign

$$\text{Area}=(\text{length})(\text{width})$$
$$8=2w\cdot w$$
$$8=2w^2$$
$$0=2w^2-8$$
$$0=w^2-4$$
$$0=(w+2)(w-2)$$
$$w+2=0 \quad \text{or} \quad w-2=0$$
$$w=-2 \qquad w=2$$

width $=2$ ft; length $=2\cdot 2=4$ ft

11) $w=$ the width of the granite

$w+3.5=$ the length of the granite

$$\text{Area}=(\text{Length})(\text{Width})$$
$$15=w\cdot(w+3.5)$$
$$15=w^2+3.5w \quad \text{Multiply by 2}$$
$$30=2w^2+7w$$
$$0=2w^2+7w-30$$
$$0=(2w-5)(w+6)$$
$$2w-5=0 \quad \text{or} \quad w+6=0$$
$$w=2.5 \qquad w=-6$$

length $=2.5+3.5=6.0$ ft; width $=2.5$ ft

13) $l=$ length of case

$l-1=$ width of case

$$\text{Volume}=(\text{length})(\text{width})(\text{height})$$
$$90=3l(l-1)$$
$$90=3l^2-3l$$
$$30=l^2-l$$
$$0=l^2-l-30$$
$$0=(l+5)(l-6)$$
$$l+5=0 \text{ or } l-6=0$$
$$l=-5 \qquad l=6$$

length $=6$ in; width $=6-1=5$ in

15) $b = \text{base of triangle}$

$b+1 = \text{height of triangle}$

$\text{Area} = \frac{1}{2}(\text{base})(\text{height})$

$21 = \frac{1}{2}(b)(b+1)$

$42 = b^2 + b$

$0 = b^2 + b - 42$

$0 = (b+7)(b-6)$

$b+7=0$ or $b-6=0$

$b=-7 \qquad b=6$

base $= 6$ cm; height $= 6+1 = 7$ cm

17) $x = \text{the first integer}$

$x+1 = \text{the second integer}$

$x(x+1) = 5(x+x+1)-13$

$x^2 + x = 5(2x+1)-13$

$x^2 + x = 10x+5-13$

$x^2 - 9x + 8 = 0$

$(x-8)(x-1) = 0$

$x-8=0$ or $x-1=0$

$x=8$ or $x=1$

$x=8$, then $8+1=9$

$x=1$, then $1+1=2$

8 and 9; or 1 and 2

19) $x = \text{the first even integer}$

$x+2 = \text{the second even integer}$

$x+4 = \text{the third even integer}$

$x(x+2) = 2(x+x+2+x+4)$

$x^2 + 2x = 2(3x+6)$

$x^2 + 2x = 6x + 12$

$x^2 - 4x - 12 = 0$

$(x-6)(x+2) = 0$

$x=6$ or $x=-2$

$x=-2$, then $-2+2=0$ and $-2+4=2$

$x=6$, then $6+2=8$ and $6+4=10$

$-2,0,2$ or $6,8,10$

21) $x = \text{the first odd integer}$

$x+2 = \text{the second odd integer}$

$x+4 = \text{the third odd integer}$

$(x+2)(x+4) = 3(x+x+2+x+4)+18$

$x^2 + 6x + 8 = 3(3x+6)+18$

$x^2 + 6x + 8 = 9x + 36$

$x^2 - 3x - 28 = 0$

$(x-7)(x+4) = 0$

$x = 7$

$x=7$, then $7+2=9$ and $7+4=11$

7,9,11

23) Answers may vary.

25) $a^2 + b^2 = c^2$

$a^2 + (12)^2 = (15)^2$

$a^2 + 144 = 225$

$a^2 - 81 = 0$

$(a+9)(a-9) = 0$

$a+9=0$ or $a-9=0$

$a=-9 \qquad a=9$

The length of the missing side is 9.

27) $a^2 + b^2 = c^2$

$a^2 + (12)^2 = (13)^2$

$a^2 + 144 = 169$

$a^2 - 25 = 0$

$(a+5)(a-5) = 0$

$a+5=0$ or $a-5=0$

$a=-5 \qquad a=5$

The length of the missing side is 5.

29)
$$a^2+b^2=c^2$$
$$(16)^2+(12)^2=c^2$$
$$256+144=c^2$$
$$400=c^2$$
$$0=c^2-400$$
$$0=(c+20)(c-20)$$
$$c+20=0 \quad \text{or} \quad c-20=0$$
$$c=-20 \qquad c=20$$
The length of the missing side is 20.

31)
$$a^2+b^2=c^2$$
$$x^2+(x-1)^2=(x+1)^2$$
$$x^2+x^2-2x+1=x^2+2x+1$$
$$x^2-4x=0$$
$$x(x-4)=0$$
$$x=4$$
$$x=4; x-1=3; x+1=5$$
$3,4,5$

33)
$$a^2+b^2=c^2$$
$$(\frac{1}{2}x)^2+(x+2)^2=(x+3)^2$$
$$\frac{1}{4}x^2+x^2+4x+4=x^2+6x+9$$
Multiply by 4
$$x^2+4x^2+16x+16=4x^2+24x+36$$
$$x^2-8x-20=0$$
$$(x-10)(x+2)=0$$
$$x=10$$
$$\frac{1}{2}x=5; x+2=12; x+3=13$$
$5,12,13$

35) $x =$ Shortest leg of the triangle

$x+2 =$ Longer leg

$x+4 =$ Hypotenuse
$$a^2+b^2=c^2$$
$$(x+2)^2+x^2=(x+4)^2$$
$$x^2+4x+4+x^2=x^2+8x+16$$
$$x^2-4x-12=0$$
$$(x-6)(x+2)=0$$
$$x=6$$
$$x+4=10$$
The Hypotenuse is 10 in.

37) $x =$ distance from bottom of the ladder to the wall

$x+7 =$ distance from top of the ladder to bottom

$13 =$ The length of the ladder - hypotenuse

$$a^2+b^2=c^2$$
$$x^2+(x+7)^2=13^2$$
$$x^2+x^2+14x+49=169$$
$$2x^2+14x-120=0$$
$$x^2+7x-60=0$$
$$(x+12)(x-5)=0$$
$$x=-12 \text{ or } x=5$$

The distance from the bottom of the ladder to the wall = 5 ft.

39)

$x =$ distance traveled by Lance

$x+2 =$ distance between Lance and Alberto

$4 =$ distance traveled by Alberto

$$a^2+b^2=c^2$$
$$x^2+4^2=(x+2)^2$$
$$x^2+16=x^2+4x+4$$
$$12=4x$$
$$x=3$$

The distance between Alberto and Lance is 5 Miles.

41) a) Let $t=0$ and solve for h.

$$h=-16(0)^2+144=144$$

The initial height of the rock is 144 ft.

b) Let $h=80$ and solve for t.

$$80=-16t^2+144$$
$$-64=-16t^2$$
$$4=t^2$$
$$0=t^2-4$$
$$0=(t+2)(t-2)$$
$$t+2=0 \quad \text{or} \quad t-2=0$$
$$t=-2 \qquad t=2$$

The rock will be 80 ft above the water after 2 seconds.

c) Let $h=0$ and solve for t.

$$0=-16t^2+144$$
$$0=t^2-9$$
$$0=(t+3)(t-3)$$
$$t+3=0 \quad \text{or} \quad t-3=0$$
$$t=-3 \qquad t=3$$

The rock will hit the water after 3 seconds.

43) a) Let $t = 3$

$y = -16(3)^2 + 144(3)$

$y = -16(9) + 432$

$y = 288$ ft

b) Let $t = 3$

$x = 39(3)$

$x = 117$ ft

c) Let $t = 4.5$

$y = -16(4.5)^2 + 144(4.5)$

$y = -16(20.25) + 648$

$y = 324$ ft

d) Let $t = 4.5$

$x = 39(4.5)$

$x = 175.5 \approx 176$ ft

45) a) Let $t = 0$

$h = -16t^2 + 96t$

$h = -16(0)^2 + 96(0)$

$h = 0$ ft

b) Let $h = 128$

$128 = -16t^2 + 96t$

Divide by 16

$8 = -t^2 + 6t$

$t^2 - 6t + 8 = 0$

$(t-4)(t-2) = 0$

$t = 4$ sec or $t = 2$ sec

c) Let $t = 3$

$h = -16(3)^2 + 96(3)$

$h = -144 + 288$

$h = 144$ ft

d) Let $h = 0$

$0 = -16t^2 + 96t$

$16t^2 - 96t = 0$

$16t(t-6) = 0$

$t = 6$ sec

47) a) Let $p = 15$

$R(15) = -9p^2 + 324p$

$R = -9(15)^2 + 324(15)$

$R = -9(225) + 4860$

$R = -2025 + 4860 =$ \$2835

b) Let $p = 20$

$R(20) = -9(20)^2 + 324(20)$

$R = -9(400) + 6480$

$R = -3600 + 6480 =$ \$2880

c) Let $R = 2916$

$2916 = -9p^2 + 324p$

Divde by 9

$324 = -p^2 + 36p$

$p^2 - 36p + 324 = 0$

$(p-18)^2 = 0$

$p =$ \$18

Chapter 7 Review

1) GCF of 40 and 56 is 8.

3) GCF of $15h^2$, $45h^5$ and $20h^3$ is $5h^3$

5) GCF $= 9$

$63t + 45 = 9(7t) + 9(5)$

$= 9(7t + 5)$

7) $\text{GCF} = 2p^4$
$2p^6 - 20p^5 + 2p^4$
$= 2p^4(p^2) - 2p^4(10p) + 2p^4(1)$
$= 2p^4\left(p^2 - 10p + 1\right)$

9) $\text{GCF} = (n-5)$
$n(m+8) - 5(m+8)$
$(m+8)(n-5)$

11) $-15r^3 - 40r^2 + 5r$
$= -5r(3r^2 + 8r - 1)$

13) $ab + 2a + 9b + 18$
$= a(b+2) + 9(b+2)$
$= (a+9)(b+2)$

15) $4xy - 28y - 3x + 21$
$= 4y(x-7) - 3(x-7)$
$= (4y-3)(x-7)$

17) $q^2 + 10q + 24 = (q+4)(q+6)$

19) $z^2 - 6z - 72 = (z-12)(z+6)$

21) $m^2 - 13mn + 30n^2 = (m-10n)(m-3n)$

23) $4v^2 - 24v - 64 = 4(v-8)(v+2)$

25) $9w^4 + 9w^3 - 18w^2 = 9w^2\left(w^2 + w - 2\right)$
$= 9w^2(w+2)(w-1)$

27) $3r^2 - 23r + 14 = (3r-2)(r-7)$

29) $4p^2 - 8p - 5 = (2p-5)(2p+1)$

31) $12c^2 + 38c + 20 = 2\left(6c^2 + 19c + 10\right)$
$= 2(3c+2)(2c+5)$

33) $10x^2 + 39xy - 27y^2 = (5x-3y)(2x+9y)$

35) $w^2 - 49 = (w+7)(w-7)$

37) $64t^2 - 25u^2 = (8t+5u)(8t-5u)$

39) $4b^2 + 9 =$ prime

41) $64x^3 - 4x = 4x\left(16x^2 - 1\right)$
$= 4x\left(4x-1\right)\left(4x+1\right)$

43) $r^2 + 12r + 36 = (r+6)^2$

45) $20k^2 - 60k + 45 = 5\left(4k^2 - 12k + 9\right)$
$= 5(2k-3)^2$

47) $v^3 - 27 = (v-3)(v^2 + 3v + 9)$

49) $125x^3 + 64y^3 = (5x+4y)(25x^2 - 20xy + 16y^2)$

51) $10z^2 - 7z - 12 = (5z+4)(2z-3)$

53) $9k^4 - 16k^2 = k^2\left(9k^2 - 16\right)$
$= k^2(3k+4)(3k-4)$

55) $d^2 - 17d + 60 = (d-12)(d-5)$

57) $3a^2b + a^2 - 12b - 4$
$= a^2(3b+1) - 4(3b+1)$
$= (a^2 - 4)(3b+1)$
$= (a+2)(a-2)(3b+1)$

59) $48p^3 - 6q^3 = 6\left(8p^3 - q^3\right)$
$= 6(2p-q)(4p^2 + 2pq + q^2)$

61) $(x+4)^2-(y-5)^2$
$=(x+4+y-5)(x+4-y+5)$
$=(x+y-1)(x-y+9)$

63) $25c^2-20c+4=(5c-2)^2$

65) $y(3y+7)=0$
$y=0$ or $3y+7=0$
$y=0$ or $y=-\frac{7}{3}$ $\left\{0,-\frac{7}{3}\right\}$

67) $2k^2+18=13k$
$2k^2-13k+18=0$
$(2k-9)(k-2)=0$
$k=\frac{9}{2}$ or $k=2$ $\left\{\frac{9}{2},2\right\}$

69) $h^2+17h+72=0$
$(h+9)(h+8)=0$
$h=-9$ or $h=-8$ $\{-9,-8\}$

71) $121=81r^2$
$0=81r^2-121$
$(9r+11)(9r-11)=0$
$r=-\frac{11}{9}$ or $r=\frac{11}{9}$
$\left\{-\frac{11}{9},\frac{11}{9}\right\}$

73) $3m^2-120=18m$
$3m^2-18m-120=0$
$3(m+4)(m-10)=0$
$m=-4$ or $m=10$ $\{-4,10\}$

75) $(w+3)(w+8)=-6$
$w^2+8w+3w+24=-6$
$w^2+11w+30=0$
$(w+6)(w+5)=0$
$w=-6$ or $w=-5$ $\{-6,-5\}$

77) $(5z+4)(3z^2-7z+4)=0$
$(5z+4)(z-1)(3z-4)=0$
$z=-\frac{4}{5}$ or $z=1$ or $z=\frac{4}{3}$
$\left\{-\frac{4}{5},1,\frac{4}{3}\right\}$

79) $3v+(v-3)^2=5(v^2-4v+1)+8$
$3v+v^2-6v+9=5v^2-20v+5+8$
$0=4v^2-17v+4$
$(4v-1)(v-4)=0$
$v=\frac{1}{4}$ or $v=4$ $\left\{\frac{1}{4},4\right\}$
$\left\{0,\frac{5}{2},3\right\}$

81) $45p^3-20p=0$
$5p(9p^2-4)=0$
$5p(3p+2)(3p-2)=0$
$p=0$ or $p=-\frac{2}{3}$
or $p=\frac{2}{3}$
$\left\{0,-\frac{2}{3},\frac{2}{3}\right\}$

83) $4x+1=$ Base of the triangle
$x+2=$ Height of the triangle
$18=$ Area of the triangle
$\frac{1}{2}(4x+1)(x+2)=18$
$4x^2+8x+x+2=36$
$4x^2+9x-34=0$
$(x-2)(4x+17)=0$
$x=2$ or $x=-\frac{17}{4}$
Base of Triangle: 9 cm; Height: 4 cm

85) $x=$ Height
$x+4=$ Base
$12=$ Area
$x(x+4)=12$
$x^2+4x-12=0$
$(x+6)(x-2)=0$
$x=-6$ or $x=2$
Base: 6 ft; Height: 2 ft

87) $a^2+b^2=c^2$
$a^2+(8)^2=(17)^2$
$a^2+64=289$
$a^2-225=0$
$(a+15)(a-15)=0$
$a=-15$ or $a=15$
The missing side is 15.

89) $x=$ width
$x+1.5=$ length
$10=$ area
$x(x+1.5)=10$
$x^2+1.5x=10$
$2x^2+3x-20=0$
$(2x-5)(x+4)=0$
$x=2.5$ or $x=-4$
Length: 4 ft; Width: 2.5 ft

91) $x=$ First Integer
$x+1=$ Second Integer
$x+2=$ Third Integer
$x+(x+1)+(x+2)=x^2-1$
$0=x^2-3x-4$
$(x-4)(x+1)=0$
$x=4$ or $x=-1$
The integers are; $-1,0,1$ or $4,5,6$

93) $x=$ Distance traveled by Desmond
$x+1=$ Distance traveled by Marcus
$x+2=$ Distance between Marcus and Desmond
$x+(x+1)^2=(x+2)^2$
$x^2+x^2+2x+1=x^2+4x+4$
$x^2-2x-3=0$
$(x-3)(x+1)=0$
$x=3$ or $x=-1$
Distance traveled by Desmond: 3 Miles

Chapter 7 Test

1) See if you can factor out a GCF.

3) $36-v^2=(6+v)(6-v)$

5) $20a^3b^4+36a^2b^3+4ab^2$
$=4ab^2(5a^2b^2+9ab+1)$

7) $64t^3-27u^3=(4t-3u)(16t^2+12tu+9u^2)$

9) $36r^2-60r+25=(6r-5)^2$

11) $x^2-3xy-18y^2=(x-6y)(x+3y)$

13) $p^2(q-4)^2+17p(q-4)^2+30(q-4)^2$
$=(q-4)^2(p^2+17p+30)$
$=(q-4)^2(p+15)(p+2)$

15) $k^8+k^5=k^5(k^3+1)$
$=k^5(k+1)(k^2-k+1)$

17) $144r=r^3$
$r(r^2-144)=0$
$r(R+12)(r-12)=0$
$r=-12$ or $r=0$ or $r=12$
$\{-12,0,12,\}$

19) $(y-7)(y-5)=3$
$y^2-5y-7y+35=3$
$y^2-12y+32=0$
$(y-8)(y-4)=0$
$y=8$ or $y=4$ $\{8,4\}$

21) $20k^2-52k=24$
$20k^2-52k-24=0$
$4(5k+2)(k-3)=0$
$k=-\frac{2}{5}$ or $k=3$ $\left\{-\frac{2}{5},3\right\}$

23) $x=$ the first odd integer
$x+2=$ the second odd integer
$x+4=$ the third odd integer
$x+x+2+x+4=(x+4)(x+2)-110$
$3x+6=(x^2+6x+8)-110$
$3x+6=x^2+6x-102$
$0=x^2+3x-108$
$0=(x+12)(x-9)$
$x+12=0$ or $x-9=0$
$x=-12$ or $x=9$
The odd integers are 9,11,13

25) $x=$ Width
$7x=$ Length
$252=$ Area
$7x^2=252$
$7(x^2-36)=0$
$7(x+6)(x-6)=0$
$x=-6$ or $x=6$
Width: 6 ft; Length: 42 ft.

27) a) Let $t=1$
$h=-16t^2+200t$
$h=-16(1)^2+200(1)$
$h=184$ ft

b) Let $t=4$
$h=-16(4)^2+200(4)$
$h=-256+800$
$h=544$ ft

c) Let h = 400

$400 = -16(t)^2 + 200(t)$

$16t^2 - 200t + 400 = 0$

$8(2t^2 - 25t + 50) = 0$

$8(2t-5)(t-10) = 0$

$t = 2.5$ or $t = 10$

When $t = 2.5$ sec and $t = 10.0$ sec

d) Let $h = 0$ and solve for t.

$0 = -16t^2 + 200t$

$16t^2 - 200t = 0$

$8t(2t-25) = 0$

$t = 12.5$

When $t = 12.5$ sec

9)

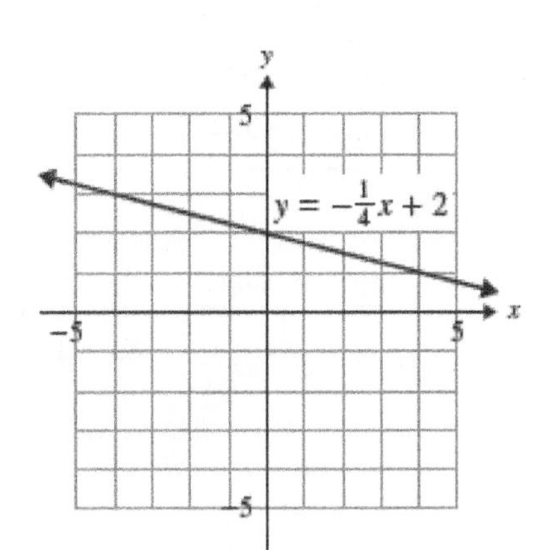

Cumulative Review: Chapters 1-7

1) $\frac{2}{9} - \frac{5}{6} + \frac{1}{3} = \frac{4}{18} - \frac{15}{18} + \frac{6}{18}$

$= \frac{4-15+6}{18}$

$= -\frac{5}{18}$

3) $2(-3p^4q)^2 = 2(9p^8q^2) = 18p^8q^2$

5) $0.0000839 = 8.39\text{x}10^{-5}$

7) $A = \frac{1}{2}h(b_1 + b_2)$

$\frac{2A}{h} = b_1 + b_2$

$b_2 = \frac{2A}{h} - b_1$

$b_2 = \frac{2A - b_1 h}{h}$

11) $4 + 2(1-3x) + 7y = 3x - 5y$

$4 + 2 - 6x + 7y = 3x - 5y$

$\boxed{\text{I}}\quad 9x - 12y = -6$

$10y + 9 = 3(2x+5) + 2y$

$10y + 9 = 6x + 15 + 2y$

$\boxed{\text{II}}\quad 6x - 8y = -6$

Add Equations $(\text{I})\cdot 2$ and $(\text{II})\cdot(-3)$

$$\begin{array}{r} 18x - 24y = -12 \\ -18x + 24y = \ \ 18 \\ \hline 0 \neq \ \ 6 \end{array}$$

No Solution $\varnothing$

13) $(4w-7)(2w+3)$

$= 8w^2 + 12w - 14w - 21$

$= 8w^2 - 2w - 21$

$(6z-5)(2z^2+7z-3) =$

$$\begin{array}{r} 12z^3 + 42z^2 - 18z \qquad \\ -10z^2 - 35z + 15 \\ \hline 12z^3 + 32z^2 - 53z + 15 \end{array}$$

15)

17)
$$\begin{array}{r} 4x^2+7x-2 \\ 4x-7\overline{\smash{)}\,16x^3+\ 0x^2\ -57x+14} \\ -(16x^3-28x^2) \\ \hline 28x^2-57x \\ -\ \ (\ 28x^2-49x) \\ \hline -8x+14 \\ -\ \ (-8x+14) \\ \hline 0 \end{array}$$

19) $6c^2+15c-54=3\left(2c^2+5c-18\right)$
$=3(2c+9)(c-2)$

21) xy^2+4y^2-x-4
$=y^2(x+4)-1(x+4)$
$=(y^2-1)(x+4)$
$=(y+1)(y-1)(x+4)$

23) $h^3+125=(h+5)\left(h^2-5h+25\right)$

25) $24n^3=54n$
$24n^3-54n=0$
$6n\left(4n^2-9\right)=0$
$6n(2n+3)(2n-3)=0$
$n=0$ or $n=-\dfrac{3}{2}$ or $n=\dfrac{3}{2}$ $\quad\left\{0,-\dfrac{3}{2},\dfrac{3}{2}\right\}$

Chapter 8: Rational Expressions

Section 8.1: Exercises

1) when its denominator equals zero

3) a) $\frac{2(3)-1}{5(3)+2}=\frac{6-1}{15+2}=\frac{5}{17}$

b) $\frac{2(-2)-1}{5(-2)+2}=\frac{-4-1}{-10+2}=\frac{-5}{-8}=\frac{5}{8}$

5) a) $\frac{[4(1)]^2}{(1)^2-(1)-12}=\frac{(4)^2}{1-1-12}$

$=-\frac{16}{12}=-\frac{4}{3}$

b)

$\frac{[4(-3)]^2}{(-3)^2-(-3)-12}=\frac{(-12)^2}{9+3-12}$

$=\frac{144}{0}=\text{undefined}$

7) a) $\frac{15+5(1)}{16-(1)^2}=\frac{15+5}{16-1}=\frac{20}{15}=\frac{4}{3}$

b) $\frac{15+5(-3)}{16-(-3)^2}=\frac{15-15}{16-9}=\frac{0}{7}=0$

9) Set the denominator equal to zero and solve for the variable. That value cannot be substituted into the expression because it will make the denominator equal to zero.

11) a) $m+4=0$
$m=-4$

b) $3m=0$
$m=0$

13) a) $2w-7=0$
$2w=7$
$w=\frac{7}{2}$

b) $4w+1=0$
$4w=-1$
$w=-\frac{1}{4}$

15) a) $11v-v^2=0$
$v(11-v)=0$
$v=0$ or $11-v=0$
$v=11$

b) $5v-9=0$
$5v=9$
$v=\frac{9}{5}$

17) a) It never equals zero.

b) $p=0$

19) a) $7k=0$
$k=0$

b) $k^2+9k+20=0$
$(k+4)(k+5)=0$
$k+4=0$ or $k+5=0$
$k=-4 \qquad k=-5$

21) a) $c+20=0$
$c=-20$

b) $2c^2+3c-9=0$
$(2c-3)(c+3)=0$
$2c-3=0$ or $c+3=0$
$2c=3 \qquad c=-3$
$c=\frac{3}{2}$

23) a) $g^2+9g+18=0$

$(g+6)(g+3)=0$

$g+6=0$ or $g+3=0$

$g=-6 \qquad g=-3$

b) $9g=0$

$g=0$

25) a) $4y=0$

$y=0$

b)

y^2+9 never equals zero.
Never undefined. Any real number can be substituted for y.

27) $$\frac{7x\cancel{(x-11)}}{3\cancel{(x-11)}}=\frac{7x}{3}$$

29) $$\frac{24g^2}{56g^4}=\frac{3}{7g^2}$$

31) $$\frac{4d-20}{5d-25}=\frac{4\cancel{(d-5)}}{5\cancel{(d-5)}}=\frac{4}{5}$$

33) $$\frac{-14h-56}{6h+24}=\frac{-14\cancel{(h+4)}}{6\cancel{(h+4)}}=-\frac{14}{6}=-\frac{7}{3}$$

35) $$\frac{39u^2+26}{30u^2+20}=\frac{13\cancel{(3u^2+2)}}{10\cancel{(3u^2+2)}}=\frac{13}{10}$$

37) $$\frac{g^2-g-56}{g+7}=\frac{(g-8)\cancel{(g+7)}}{\cancel{g+7}}=g-8$$

39) $$\frac{t-5}{t^2-25}=\frac{\cancel{t-5}}{\cancel{(t-5)}(t+5)}=\frac{1}{t+5}$$

41) $$\frac{3c^2+28c+32}{c^2+10c+16}=\frac{(3c+4)\cancel{(c+8)}}{(c+2)\cancel{(c+8)}}=\frac{3c+4}{c+2}$$

43) $$\frac{q^2-25}{2q^2-7q-15}=\frac{(q+5)\cancel{(q-5)}}{(2q+3)\cancel{(q-5)}}=\frac{q+5}{2q+3}$$

45) $$\frac{w^3+125}{5w^2-25w+125}=\frac{(w+5)(w^2-5w+25)}{5(w^2-5w+25)}=\frac{w+5}{5}$$

47) $$\frac{9c^2-27c+81}{c^3+27}=\frac{9\cancel{(c^2-3c+9)}}{(c+3)\cancel{(c^2-3c+9)}}=\frac{9}{c+3}$$

49) $$\frac{4u^2-20u+4uv-20v}{13u+13v}=\frac{4(u^2-5u+uv-5v)}{13(u+v)}=\frac{4[u(u-5)+v(u-5)]}{13(u+v)}=\frac{4(u-5)\cancel{(u+v)}}{13\cancel{(u+v)}}=\frac{4(u-5)}{13}$$

51) $\dfrac{m^2-n^2}{m^3-n^3}=\dfrac{(m+n)\cancel{(m-n)}}{\cancel{(m-n)}(m^2+mn+n^2)}$

$=\dfrac{m+n}{m^2+mn+n^2}$

53) -1

55) $\dfrac{8-q}{q-8}=-1$

57) $\dfrac{m^2-121}{11-m}=\dfrac{(m+11)\cancel{(m-11)}}{\cancel{11-m}}$

$=-(m+11)$

$=-m-11$

59) $\dfrac{36-42x}{7x^2+8x-12}=\dfrac{6\cancel{(6-7x)}}{\cancel{(7x-6)}(x+2)}$

$=-\dfrac{6}{x+2}$

61) $\dfrac{16-4b^2}{b-2}=\dfrac{4(4-b^2)}{b-2}$

$=\dfrac{4(2+b)(2-b)}{b-2}$

$=-\dfrac{4(b+2)\cancel{(b-2)}}{\cancel{b-2}}$

$=-4(b+2)$

63) $\dfrac{y^3-3y^2+2y-6}{21-7y}$

$=\dfrac{y^2(y-3)+2(y-3)}{7(3-y)}=-\dfrac{y^2+2}{7}$

65) $\dfrac{18c+45}{12c^2+18c-30}=\dfrac{9\cancel{(2c+5)}}{6(c-1)\cancel{(2c+5)}}$

$=\dfrac{3}{2(c-1)}$

67) $\dfrac{r^3-t^3}{t^2-r^2}=-\dfrac{\cancel{(r-t)}(r^2+rt+t^2)}{(r+t)\cancel{(r-t)}}$

$=-\dfrac{r^2+rt+t^2}{r+t}$

69) $\dfrac{b^2+6b-72}{4b^2+52b+48}=\dfrac{(b+12)(b-6)}{4(b^2+13b+12)}$

$=\dfrac{\cancel{(b+12)}(b-6)}{4\cancel{(b+12)}(b+1)}$

$=\dfrac{b-6}{4(b+1)}$

71) $\dfrac{28h^4-56h^3+7h}{7h}=\dfrac{\cancel{7h}(4h^3-8h^2+1)}{\cancel{7h}}$

$=4h^3-8h^2+1$

73) $\dfrac{14-6w}{12w^3-28w^2}=\dfrac{2(7-3w)}{4w^2(3w-7)}$

$=-\dfrac{2\cancel{(3w-7)}}{4w^2\cancel{(3w-7)}}$

$=-\dfrac{1}{2w^2}$

75) $\dfrac{-5v-10}{v^3-v^2-4v+4}=\dfrac{-5(v+2)}{v^2(v-1)-4(v-1)}$

$=\dfrac{-5(v+2)}{(v^2-4)(v-1)}$

$=\dfrac{-5\cancel{(v+2)}}{\cancel{(v+2)}(v-2)(v-1)}$

$=-\dfrac{5}{(v-2)(v-1)}$

77) Possible answers:

$\dfrac{-u-7}{u-2}, \dfrac{-(u+7)}{u-2}, \dfrac{u+7}{2-u},$

$\dfrac{u+7}{-(u-2)}, \dfrac{u+7}{-u+2}$

79) Possible answers:

$\dfrac{-9+5t}{2t-3}, \dfrac{5t-9}{2t-3}, \dfrac{-(9-5t)}{2t-3},$

$\dfrac{9-5t}{-2t+3}, \dfrac{9-5t}{3-2t}, \dfrac{9-5t}{-(2t-3)}$

81) Possible answers:

$-\dfrac{12m}{m^2-3}, \dfrac{12m}{-(m^2-3)}, \dfrac{12m}{-m^2+3}, \dfrac{12m}{3-m^2}$

83) a)

$$\begin{array}{r} 4y-3 \\ y-2\overline{\big)4y^2-11y+6} \\ -\underline{(4y^2-8y)} \\ -3y+6 \\ -\underline{(-3y+6)} \\ 0 \end{array}$$

b) $\dfrac{4y^2-11y+6}{y-2}=\dfrac{(4y-3)(y-2)}{y-2}$

$=4y-3$

85) a)

$$\begin{array}{r} 4a^2-10a+25 \\ 2a+5\overline{\big)8a^3+0a^2+0a+125} \\ -\underline{(8a^3+20a^2)} \\ -20a^2+0a \\ -\underline{(-20a^2-50a)} \\ 50a+125 \\ -\underline{(50a+125)} \\ 0 \end{array}$$

b) $\dfrac{8a^3+125}{2a+5}$

$=\dfrac{(2a+5)(4a^2-10a+25)}{2a+5}$

$=4a^2-10a+25$

87) $l=\dfrac{3x^2+8x+2}{x+2}$

$=\dfrac{(3x+2)(x+2)}{x+2}=3x+2$

89) $$w=\frac{2c^3+4c^2+8c+16}{c^2+4}$$
$$=\frac{c^2(2c+4)+4(2c+4)}{c^2+4}$$
$$=\frac{(2c+4)(c^2+4)}{c^2+4}=2c+4$$

91) $$h=\frac{2(3k^2+13k+4)}{2k+8}$$
$$=\frac{2(k+4)(3k+1)}{2(k+4)}=3k+1$$

Section 8.2: Exercises

1) $$\frac{5}{6}\cdot\frac{7}{9}=\frac{35}{54}$$

3) $$\frac{6}{15}\cdot\frac{25}{42}=\frac{\overset{1}{\cancel{6}}}{\underset{3}{\cancel{15}}}\cdot\frac{\overset{5}{\cancel{25}}}{\underset{7}{\cancel{42}}}=\frac{5}{21}$$

5) $$\frac{16b^5}{3}\cdot\frac{4}{36b}=\frac{\overset{4b^4}{\cancel{16b^5}}}{3}\cdot\frac{4}{\underset{9}{\cancel{36b}}}=\frac{16b^4}{27}$$

7) $$\frac{21s^4}{15t^2}\cdot\frac{5t^4}{42s^{10}}=\frac{\overset{1}{\cancel{21s^4}}}{\underset{3}{\cancel{15t^2}}}\cdot\frac{\overset{t^2}{\cancel{5t^4}}}{\underset{2s^6}{\cancel{42s^{10}}}}=\frac{t^2}{6s^6}$$

9) $$\frac{9c^4}{42c}\cdot\frac{35}{3c^3}=\frac{\overset{3c}{\cancel{9c^4}}}{\underset{6c}{\cancel{42c}}}\cdot\frac{\overset{5}{\cancel{35}}}{\underset{1}{\cancel{3c^3}}}=\frac{\overset{5}{\cancel{15c}}}{\underset{2}{\cancel{6c}}}=\frac{5}{2}$$

11) $$\frac{5t^2}{(3t-2)^2}\cdot\frac{3t-2}{10t^3}=\frac{\cancel{5t^2}}{(3t-2)^{\cancel{2}}}\cdot\frac{\cancel{3t-2}}{\underset{2t}{\cancel{10t^3}}}$$
$$=\frac{1}{2t(3t-2)}$$

13) $$\frac{4u-5}{9u^3}\cdot\frac{6u^5}{(4u-5)^3}=\frac{\cancel{4u-5}}{\underset{3}{\cancel{9u^3}}}\cdot\frac{\overset{2u^2}{\cancel{6u^5}}}{(4u-5)^{\cancel{3}}}$$
$$=\frac{2u^2}{3(4u-5)^2}$$

15) $$\frac{6}{n+5}\cdot\frac{n^2+8n+15}{n+3}$$
$$=\frac{6}{\cancel{n+5}}\cdot\frac{\cancel{(n+5)}\cancel{(n+3)}}{\cancel{n+3}}=6$$

17) $$\frac{18y-12}{4y^2}\cdot\frac{y^2-4y-5}{3y^2+y-2}$$
$$=\frac{\overset{3}{\cancel{6}}\cancel{(3y-2)}}{\underset{2}{\cancel{4}}y^2}\cdot\frac{(y-5)\cancel{(y+1)}}{\cancel{(3y-2)}\cancel{(y+1)}}$$
$$=\frac{3(y-5)}{2y^2}$$

19) $$(c-6)\cdot\frac{5}{c^2-6c}$$
$$=\cancel{(c-6)}\cdot\frac{5}{c\cancel{(c-6)}}$$
$$=\frac{5}{c}$$

21) $\frac{7x}{11-x}\cdot(x^2-121)$

$=-\frac{7x}{(x-11)}\cdot(x+11)(x-11)$

$=-7x(x+11)$

23) $\frac{20}{9}\div\frac{10}{27}=\frac{20}{9}\cdot\frac{27}{10}=6$

25) $42\div\frac{9}{2}=42\cdot\frac{2}{9}=\frac{28}{3}$

27) $\frac{12}{5m^5}\div\frac{21}{8m^{12}}=\frac{12}{5m^5}\cdot\frac{8m^{12}}{21}=\frac{32m^7}{35}$

29) $-\frac{50g}{7h^3}\div\frac{15g^4}{14h}=-\frac{50g}{7h^3}\cdot\frac{14h}{15g^4}$

$=-\frac{20}{3g^3h^2}$

31) $\frac{2(k-2)}{21k^6}\div\frac{(k-2)^2}{28}$

$=\frac{2(k-2)}{21k^6}\cdot\frac{28}{(k-2)^2}=\frac{8}{3k^6(k-2)}$

33) $\frac{16q^5}{p+7}\div\frac{2q^4}{(p+7)^2}=\frac{16q^5}{p+7}\cdot\frac{(p+7)^2}{2q^4}$

$=8q(p+7)$

35) $\frac{q+8}{q}\div\frac{q^2+q-56}{5}$

$=\frac{q+8}{q}\cdot\frac{5}{q^2+q-56}$

$=\frac{q+8}{q}\cdot\frac{5}{(q-7)(q+8)}$

$=\frac{5}{q(q-7)}$

37) $\frac{z^2+18z+80}{2z+1}\div(z+8)^2$

$=\frac{z^2+18z+80}{2z+1}\cdot\frac{1}{(z+8)^2}$

$=\frac{(z+8)(z+10)}{2z+1}\cdot\frac{1}{(z+8)(z+8)}$

$=\frac{z+10}{(2z+1)(z+8)}$

39) $\frac{36a-12}{16}\div(9a^2-1)$

$=\frac{36a-12}{16}\cdot\frac{1}{(9a^2-1)}$

$=\frac{12(3a-1)}{16}\cdot\frac{1}{(3a+1)(3a-1)}$

$=\frac{12}{16(3a+1)}=\frac{3}{4(3a+1)}$

41) $\dfrac{7n^2-14n}{8n}\div\dfrac{n^2+4n-12}{4n+24}$

$=\dfrac{7n^2-14n}{8n}\cdot\dfrac{4n+24}{n^2+4n-12}$

$=\dfrac{7\cancel{n}\cancel{(n-2)}}{\underset{2}{\cancel{8}}\cancel{n}}\cdot\dfrac{\cancel{4}\cancel{(n+6)}}{\cancel{(n+6)}\cancel{(n-2)}}$

$=\dfrac{7}{2}$

43) $\dfrac{4c-9}{2c^2-8c}\div\dfrac{12c-27}{c^2-3c-4}$

$=\dfrac{4c-9}{2c^2-8c}\cdot\dfrac{c^2-3c-4}{12c-27}$

$=\dfrac{\cancel{4c-9}}{2c\cancel{(c-4)}}\cdot\dfrac{\cancel{(c-4)}(c+1)}{3\cancel{(4c-9)}}=\dfrac{c+1}{6c}$

45) Answers may vary.

47) Let ? be the missing polynomial. Then

$\dfrac{9h+45}{\cancel{h^4}^{1}}\cdot\dfrac{\cancel{h^3}}{?}=\dfrac{8}{h(h-2)}$

$\dfrac{9h+45}{h(?)}=\dfrac{9}{h(h-2)}$

$\dfrac{9(h+5)}{h(?)}=\dfrac{9}{h(h-2)}$

$\dfrac{\cancel{9}(h+5)\cancel{h}(h-2)}{\cancel{9}\cancel{h}}=?$

$?=(h+5)(h-2)$

$?=h^2-2h+5h-10$

$?=h^2+3h-10$

49) Let ? be the missing binomial. Then

$\dfrac{4z^2-49}{z^2-3z-40}\div\dfrac{?}{z+5}=\dfrac{2z+7}{8-z}$

$\dfrac{4z^2-49}{z^2-3z-40}\cdot\dfrac{z+5}{?}=\dfrac{2z+7}{8-z}$

$\dfrac{\cancel{(z+5)}(2z+7)(2z-7)}{(z-8)\cancel{(z+5)}?}=\dfrac{2z+7}{8-z}$

$\dfrac{(2z+7)(2z-7)}{(z-8)?}=\dfrac{2z+7}{8-z}$

$(2z+7)(2z-7)(8-z)=(2z+7)(z-8)?$

$\dfrac{\cancel{(2z+7)}(2z-7)(8-z)}{\cancel{(2z+7)}(z-8)}=?$

$-\dfrac{(2z-7)\cancel{(z-8)}}{\cancel{(z-8)}}=?$

$?=7-2z$

51) $\dfrac{\frac{25}{42}}{\frac{8}{21}}=\dfrac{25}{42}\div\dfrac{8}{21}=\dfrac{25}{\underset{2}{\cancel{42}}}\cdot\dfrac{\overset{1}{\cancel{21}}}{8}=\dfrac{25}{16}$

53) $\dfrac{\frac{5}{24}}{\frac{15}{4}}=\dfrac{5}{24}\div\dfrac{15}{4}=\dfrac{\overset{1}{\cancel{5}}}{\underset{6}{\cancel{24}}}\cdot\dfrac{\overset{1}{\cancel{4}}}{\underset{3}{\cancel{15}}}=\dfrac{1}{18}$

55) $\dfrac{\frac{3d+7}{24}}{\frac{3d+7}{6}}=\dfrac{3d+7}{24}\div\dfrac{3d+7}{6}$

$=\dfrac{\cancel{3d+7}}{\underset{4}{\cancel{24}}}\cdot\dfrac{\cancel{6}}{\cancel{3d+7}}=\dfrac{1}{4}$

57) $$\frac{\frac{16r+24}{r^3}}{\frac{12r+18}{r}} = \frac{16r+24}{r^3} \div \frac{12r+18}{r}$$
$$= \frac{16r+24}{r^3} \cdot \frac{r}{12r+18}$$
$$= \frac{\overset{4}{\cancel{8}}\cancel{(2r+3)}}{\underset{r^2}{\cancel{r^3}}} \cdot \frac{\cancel{r}}{\underset{3}{\cancel{6}}\cancel{(2r+3)}}$$
$$= \frac{4}{3r^2}$$

59) $$\frac{\frac{a^2-25}{3a^{11}}}{\frac{4a+20}{a^3}}$$
$$= \frac{a^2-25}{3a^{11}} \div \frac{4a+20}{a^3}$$
$$= \frac{a^2-25}{3a^{11}} \cdot \frac{a^3}{4a+20}$$
$$= \frac{(a-5)\cancel{(a+5)}}{3\underset{a^8}{\cancel{a^{11}}}} \cdot \frac{\cancel{a^3}}{4\cancel{(a+5)}} = \frac{a-5}{12a^8}$$

61) $$\frac{\frac{16x^2-25}{x^7}}{\frac{36x-45}{6x^3}}$$
$$= \frac{16x^2-25}{x^7} \div \frac{36x-45}{6x^3}$$
$$= \frac{16x^2-25}{x^7} \cdot \frac{6x^3}{36x-45}$$
$$= \frac{(4x+5)\cancel{(4x-5)}}{\underset{x^4}{\cancel{x^7}}} \cdot \frac{\overset{2}{\cancel{6}}\cancel{x^3}}{\underset{3}{\cancel{9}}\cancel{(4x-5)}} = \frac{2(4x+5)}{3x^4}$$

63) $$\frac{c^2+c-30}{9c+9} \cdot \frac{c^2+2c+1}{c^2-25}$$
$$= \frac{\cancel{(c-5)}(c+6)}{9\cancel{(c+1)}} \cdot \frac{(c+1)^{\cancel{2}}}{(c+5)\cancel{(c-5)}}$$
$$= \frac{(c+6)(c+1)}{9(c+5)}$$

65) $$\frac{3x+2}{9x^2-4} \div \frac{4x}{15x^2-7x-2}$$
$$= \frac{3x+2}{9x^2-4} \cdot \frac{15x^2-7x-2}{4x}$$
$$= \frac{\cancel{3x+2}}{\cancel{(3x+2)}\cancel{(3x-2)}} \cdot \frac{(5x+1)\cancel{(3x-2)}}{4x}$$
$$= \frac{5x+1}{4x}$$

67) $$\frac{3k^2-12k}{12k^2-30k-72} \cdot (2k+3)^2$$
$$= \frac{3k(k-4)}{6(2k^2-5k-12)} \cdot (2k+3)^2$$
$$= \frac{\cancel{3}k\cancel{(k-4)}}{\underset{2}{\cancel{6}}\cancel{(2k+3)}\cancel{(k-4)}} \cdot (2k+3)^{\cancel{2}}$$
$$= \frac{k(2k+3)}{2}$$

69) $\dfrac{7t^6}{t^2-4} \div \dfrac{14t^2}{3t^2-7t+2}$

$= \dfrac{7t^6}{t^2-4} \cdot \dfrac{3t^2-7t+2}{14t^2}$

$= \dfrac{\cancel{7t^6}^{\,t^4}}{(t+2)\cancel{(t-2)}} \cdot \dfrac{(3t-1)\cancel{(t-2)}}{\cancel{14t^2}_{\,2}}$

$= \dfrac{t^4(3t-1)}{2(t+2)}$

71) $\dfrac{4h^3}{h^2-64} \cdot \dfrac{8h-h^2}{12}$

$= \dfrac{\cancel{4}^{\,1}h^3}{(h+8)\cancel{(h-8)}} \cdot \dfrac{h\cancel{(8-h)}^{\,-1}}{\cancel{12}_{\,3}}$

$= -\dfrac{h^4}{3(h+8)}$

73) $\dfrac{54x^8}{22x^3y^2} \div \dfrac{36xy^5}{11x^2y} = \dfrac{\cancel{54x^8}^{\,3x^7}}{\cancel{22x^3y^2}_{\,2xy^2}} \cdot \dfrac{\cancel{11x^2}^{\,1}y}{\cancel{36xy^5}_{\,2y^5}}$

$= \dfrac{3x^7y}{4xy^7}$

$= \dfrac{3x^6}{4y^6}$

75) $\dfrac{r^3+8}{r+2} \cdot \dfrac{7}{3r^2-6r+12}$

$= \dfrac{\cancel{(r+2)}\cancel{(r^2-2r+4)}}{\cancel{r+2}} \cdot \dfrac{7}{3\cancel{(r^2-2r+4)}}$

$= \dfrac{7}{3}$

77) $\dfrac{a^2-4a}{6a+54} \cdot \dfrac{a^2+13a+36}{16-a^2}$

$= \dfrac{a\cancel{(a-4)}^{\,-1}}{6\cancel{(a+9)}} \cdot \dfrac{\cancel{(a+9)}\cancel{(a+4)}}{\cancel{(4+a)}\cancel{(4-a)}} = -\dfrac{a}{6}$

79) $\dfrac{2a^2}{a^2+a-20} \cdot \dfrac{a^3+5a^2+4a+20}{2a^2+8}$

$= \dfrac{\cancel{2}a^2}{(a-4)\cancel{(a+5)}} \cdot \dfrac{\cancel{(a+5)}\cancel{(a^2+4)}}{\cancel{2}\cancel{(a^2+4)}}$

$= \dfrac{a^2}{a-4}$

81) $\dfrac{30}{4y^2-4x^2} \div \dfrac{10x^2+10xy+10y^2}{x^3-y^3}$

$= \dfrac{30}{4(y^2-x^2)} \cdot \dfrac{x^3-y^3}{10x^2+10xy+10y^2}$

$= \dfrac{\cancel{30}^{\,3}}{4(y+x)\cancel{(y-x)}} \cdot \dfrac{\cancel{(x-y)}^{\,-1}\cancel{(x^2+xy+y^2)}}{\cancel{10}\cancel{(x^2+xy+y^2)}}$

$= -\dfrac{3}{4(x+y)}$

83) $\dfrac{3m^2+8m+4}{4} \div (12m+8)$

$= \dfrac{3m^2+8m+4}{4} \cdot \dfrac{1}{12m+8}$

$= \dfrac{\cancel{(3m+2)}(m+2)}{4} \cdot \dfrac{1}{4\cancel{(3m+2)}}$

$= \dfrac{m+2}{16}$

85) $$\frac{4j^2-21j+5}{j^3}\div\left(\frac{3j+2}{j^3-j^2}\cdot\frac{j^2-6j+5}{j}\right)=\frac{4j^2-21j+5}{j^3}\div\left(\frac{3j+2}{j^2\cancel{(j-1)}}\cdot\frac{(j-5)\cancel{(j-1)}}{j}\right)$$
$$=\frac{(4j-1)(j-5)}{j^3}\div\frac{(3j+2)(j-5)}{j^3}$$
$$=\frac{(4j-1)\cancel{(j-5)}}{\cancel{j^3}}\cdot\frac{\cancel{j^3}}{(3j+2)\cancel{(j-5)}}=\frac{4j-1}{3j+2}$$

87) $$\frac{x}{3x^2-15x+75}\div\left(\frac{4x+20}{x+9}\cdot\frac{x^2-81}{x^3+125}\right)=\frac{x}{3(x^2-5x+25)}\div\left(\frac{4\cancel{(x+5)}}{\cancel{x+9}}\cdot\frac{\cancel{(x+9)}(x-9)}{\cancel{(x+5)}(x^2-5x+25)}\right)$$
$$=\frac{x}{3(x^2-5x+25)}\div\frac{4(x-9)}{x^2-5x+25}$$
$$=\frac{x}{3\cancel{(x^2-5x+25)}}\cdot\frac{\cancel{x^2-5x+25}}{4(x-9)}=\frac{x}{12(x-9)}$$

89) $$l=\frac{\frac{3x}{2y^2}}{\frac{y}{8x^4}}=\frac{3x}{\cancel{2}y^2}\cdot\frac{\overset{4}{\cancel{8}}x^4}{y}=\frac{12x^5}{y^3}$$

Section 8.3: Exercises

1) $12=2\cdot2\cdot3$
$15=3\cdot5$
$\text{LCD}=2^2\cdot3\cdot5=4\cdot3\cdot5=60$

3) $40=2\cdot2\cdot2\cdot5$
$10=2\cdot5$
$12=2\cdot2\cdot3$
$\text{LCD}=2^3\cdot5\cdot3=8\cdot15=120$

5) $\text{LCD}=n^{11}$

7) $\text{LCD}=28r^7$

9) $\text{LCD}=36z^5$

11) $\text{LCD}=110m^4$

13) $\text{LCD}=24x^3y^2$

15) $\text{LCD}=11(z-3)$

17) $\text{LCD}=w(2w+1)$

19) Factor the denominators.

21) $5c-5=5(c-1)$
$2c-2=2(c-1)$
$\text{LCD}=10(c-1)$

23) $9p^4-6p^3=3p^3(3p-2)$
$3p^6-2p^5=p^5(3p-2)$
$\text{LCD}=3p^5(3p-2)$

25) $m-7$ and $m-3$ are different factors. The LCD will be the product of these factors. $\text{LCD}=(m-7)(m-3)$

27) $z^2+11z+24=(z+3)(z+8)$
$z^2+5z-24=(z-3)(z+8)$
$\text{LCD}=(z+3)(z+8)(z-3)$

29) $t^2-3t-18=(t-6)(t+3)$
$t^2-36=(t+6)(t-6)$
$t^2+9t+18=(t+6)(t+3)$
$\text{LCD}=(t-6)(t+3)(t+6)$

31) $\text{LCD}=a-8$ or $8-a$

33) $\text{LCD}=x-y$ or $y-x$

35) Multiply the numerator and denominator of the fraction by $x-3$.

37) $\frac{7}{12}\cdot\frac{4}{4}=\frac{28}{48}$

39) $\frac{8}{z}\cdot\frac{9}{9}=\frac{72}{9z}$

41) $\frac{3}{8k}\cdot\frac{7k^3}{7k^3}=\frac{21k^3}{56k^4}$

43) $\frac{6}{5t^5u^2}\cdot\frac{2t^2u^3}{2t^2u^3}=\frac{12t^2u^3}{10t^7u^5}$

45) $\frac{7}{3r+4}\cdot\frac{r}{r}=\frac{7r}{r(3r+4)}$

47) $\frac{v}{4(v-3)}\cdot\frac{4v^5}{4v^5}=\frac{4v^6}{16v^5(v-3)}$

49) $\frac{9x}{x+6}\cdot\frac{x-5}{x-5}=\frac{9x(x-5)}{(x+6)(x-5)}$
$=\frac{9x^2-45x}{(x+6)(x-5)}$

51) $\frac{z-3}{2z-5}\cdot\frac{z+8}{z+8}=\frac{(z-3)(z+8)}{(2z-5)(z+8)}$
$=\frac{z^2+5z-24}{(2z-5)(z+8)}$

53) $\frac{5}{3-p}\cdot\frac{-1}{-1}=\frac{-5}{-(3-p)}=-\frac{5}{p-3}$

55) $-\frac{8c}{6c-7}\cdot\frac{-1}{-1}=-\frac{-8c}{-(6c-7)}=\frac{8c}{7-6c}$

57) $\text{LCD}=30$
$\frac{8}{15}\cdot\frac{2}{2}=\frac{16}{30}$
$\frac{1}{6}\cdot\frac{5}{5}=\frac{5}{30}$

59) $\text{LCD}=u^3$
$\frac{4}{u}\cdot\frac{u^2}{u^2}=\frac{4u^2}{u^3}$
$\frac{8}{u^3}$ is already written with the LCD.

61) $\text{LCD}=24n^6$
$\frac{9}{8n^6}\cdot\frac{3}{3}=\frac{27}{24n^6}$
$\frac{2}{3n^2}\cdot\frac{8n^4}{8n^4}=\frac{16n^4}{24n^6}$

63) $\text{LCD} = 4a^4b^5$

$$\frac{6}{4a^3b^5}\cdot\frac{a}{a}=\frac{6a}{4a^4b^5}$$

$$\frac{6}{a^4b}\cdot\frac{4b^4}{4b^4}=\frac{24b^4}{4a^4b^5}$$

65) $\text{LCD} = 5(r-4)$

$$\frac{r}{5}\cdot\frac{r-4}{r-4}=\frac{r^2-4r}{5(r-4)}$$

$$\frac{2}{r-4}\cdot\frac{5}{5}=\frac{10}{5(r-4)}$$

67) $\text{LCD} = d(d-9)$

$$\frac{3}{d}\cdot\frac{d-9}{d-9}=\frac{3d-27}{d(d-9)}$$

$$\frac{7}{d-9}\cdot\frac{d}{d}=\frac{7d}{d(d-9)}$$

69) $\text{LCD} = m(m+7)$

$$\frac{m}{m+7}\cdot\frac{m}{m}=\frac{m^2}{m(m+7)}$$

$$\frac{3}{m}\cdot\frac{m+7}{m+7}=\frac{3m+21}{m(m+7)}$$

71) $\frac{a}{30a-15}=\frac{a}{15(2a-1)}$

$$\frac{1}{12a-6}=\frac{1}{6(2a-1)}$$

$\text{LCD} = 30(2a-1)$

$$\frac{a}{15(2a-1)}\cdot\frac{2}{2}=\frac{2a}{30(2a-1)}$$

$$\frac{1}{6(2a-1)}\cdot\frac{5}{5}=\frac{5}{30(2a-1)}$$

73) $\text{LCD} = (k-9)(k+3)$

$$\frac{9}{k-9}\cdot\frac{k+3}{k+3}=\frac{9k+27}{(k-9)(k+3)}$$

$$\frac{5k}{k+3}\cdot\frac{k-9}{k-9}=\frac{5k^2-45k}{(k-9)(k+3)}$$

75) $\text{LCD} = (a+2)(3a+4)$

$$\frac{3}{a+2}\cdot\frac{3a+4}{3a+4}=\frac{9a+12}{(a+2)(3a+4)}$$

$$\frac{2a}{3a+4}\cdot\frac{a+2}{a+2}=\frac{2a^2+4a}{(a+2)(3a+4)}$$

77) $\frac{9y}{y^2-y-42}=\frac{9y}{(y-7)(y+6)}$

$$\frac{3}{2y^2+12y}=\frac{3}{2y(y+6)}$$

$\text{LCD} = 2y(y+6)(y-7)$

$$\frac{9y}{(y-7)(y+6)}\cdot\frac{2y}{2y}=\frac{18y^2}{2y(y-7)(y+6)}$$

$$\frac{3}{2y(y+6)}\cdot\frac{(y-7)}{(y-7)}=\frac{3y-21}{2y(y-7)(y+6)}$$

79) $\frac{c}{c^2+9c+18}=\frac{c}{(c+6)(c+3)}$

$$\frac{11}{c^2+12c+36}=\frac{11}{(c+6)^2}$$

$\text{LCD} = (c+6)^2(c+3)$

$$\frac{c}{(c+6)(c+3)}\cdot\frac{c+6}{c+6}=\frac{c^2+6c}{(c+6)^2(c+3)}$$

$$\frac{11}{(c+6)^2}\cdot\frac{c+3}{c+3}=\frac{11c+33}{(c+6)^2(c+3)}$$

81) $\dfrac{11}{g-3}$

$\dfrac{4}{9-g^2}=\dfrac{4}{(3+g)(3-g)}$

$=-\dfrac{4}{(g+3)(g-3)}$

$\text{LCD}=(g+3)(g-3)$

$\dfrac{11}{(g-3)}\cdot\dfrac{g+3}{g+3}=\dfrac{11g+33}{(g-3)(g+3)}$

$-\dfrac{4}{(g+3)(g-3)}$ is written with the LCD.

83) $\dfrac{4}{3x-4}$

$\dfrac{7x}{16-9x^2}=\dfrac{7x}{(4+3x)(4-3x)}$

$=-\dfrac{7x}{(3x+4)(3x-4)}$

$\text{LCD}=(3x+4)(3x-4)$

$\dfrac{4}{3x-4}\cdot\dfrac{3x+4}{3x+4}=\dfrac{12x+16}{(3x+4)(3x-4)}$

$-\dfrac{7x}{(3x+4)(3x-4)}$ is written with the LCD.

85) $\dfrac{2}{z^2+3z}=\dfrac{2}{z(z+3)}$

$\dfrac{6}{3z^2+9z}=\dfrac{6}{3z(z+3)}$

$\dfrac{8}{z^2+6z+9}=\dfrac{8}{(z+3)^2}$

$\text{LCD}=3z(z+3)^2$

$\dfrac{2}{z(z+3)}\cdot\dfrac{3(z+3)}{3(z+3)}=\dfrac{6(z+3)}{3z(z+3)^2}$

$=\dfrac{6z+18}{3z(z+3)^2}$

$\dfrac{6}{3z(z+3)}\cdot\dfrac{z+3}{z+3}=\dfrac{6z+18}{3z(z+3)^2}$

$\dfrac{8}{(z+3)^2}\cdot\dfrac{3z}{3z}=\dfrac{24z}{3z(z+3)^2}$

87) $\dfrac{t}{t^2-13t+30}=\dfrac{t}{(t-10)(t-3)}$

$\dfrac{6}{t-10}$

$\dfrac{7}{t^2-9}=\dfrac{7}{(t+3)(t-3)}$

$\text{LCD}=(t+3)(t-3)(t-10)$

$\dfrac{t}{(t-10)(t-3)}\cdot\dfrac{t+3}{t+3}=\dfrac{t^2+3t}{(t+3)(t-3)(t-10)}$

$\dfrac{6}{t-10}\cdot\dfrac{(t+3)(t-3)}{(t+3)(t-3)}=\dfrac{6(t^2-9)}{(t+3)(t-3)(t-10)}$

$=\dfrac{6t^2-54}{(t+3)(t-3)(t-10)}$

$\dfrac{7}{(t+3)(t-3)}\cdot\dfrac{t-10}{t-10}=\dfrac{7t-70}{(t+3)(t-3)(t-10)}$

89) $-\frac{9}{h^3+8}, \frac{2h}{5h^2-10h+20}$

$h^3+8=(h+2)(h^2-2h+4)$

$5h^2-10h+20=5(h^2-2h+4)$

LCD: $5(h+2)(h^2-2h+4)$

$-\frac{9}{h^3+8}=-\frac{9}{(h+2)(h^2-2h+4)}\cdot\frac{5}{5}$

$=-\frac{45}{5(h+2)(h^2-2h+4)}$

$\frac{2h}{5h^2-10h+20}=\frac{2h}{5(h^2-2h+4)}\cdot\frac{h+2}{h+2}$

$\frac{2h^2+4h}{5(h+2)(h^2-2h+4)}$

Section 8.4: Exercises

1) $\frac{5}{16}+\frac{9}{16}=\frac{5+9}{16}=\frac{14}{16}=\frac{7}{8}$

3) $\frac{11}{14}-\frac{3}{14}=\frac{11-3}{14}=\frac{8}{14}=\frac{4}{7}$

5) $\frac{5}{p}-\frac{23}{p}=\frac{5-23}{p}=-\frac{18}{p}$

7) $\frac{7}{3c}+\frac{8}{3c}=\frac{7+8}{3c}=\frac{15}{3c}=\frac{5}{c}$

9) $\frac{6}{z-1}+\frac{z}{z-1}=\frac{z+6}{z-1}$

11) $\frac{8}{x+4}+\frac{2x}{x+4}=\frac{8+2x}{x+4}=\frac{2(4+x)}{x+4}=2$

13) $\frac{25t+17}{t(4t+3)}-\frac{5t+2}{t(4t+3)}=\frac{25t+17-5t-2}{t(4t+3)}$

$=\frac{20t+15}{t(4t+3)}$

$=\frac{5(4t+3)}{t(4t+3)}=\frac{5}{t}$

15) $\frac{d^2+15}{(d+5)(d+2)}+\frac{8d-3}{(d+5)(d+2)}$

$=\frac{d^2+15+8d-3}{(d+5)(d+2)}$

$=\frac{d^2+8d+12}{(d+5)(d+2)}$

$=\frac{(d+6)(d+2)}{(d+5)(d+2)}=\frac{d+6}{d+5}$

17) a) $18b^4$

b) Multiply $\frac{4}{9b^2}$ of both the numerator and denominator by $2b^2$, and multiply the numerator and denominator of $\frac{5}{6b^4}$ by 3.

c) $\frac{4}{9b^2}\cdot\frac{2b^2}{2b^2}=\frac{8b^2}{18b^4}$

$\frac{5}{6b^4}\cdot\frac{3}{3}=\frac{15}{18b^4}$

19) $\frac{3}{8}+\frac{2}{5}=\frac{15}{40}+\frac{16}{40}=\frac{31}{40}$

21) $\frac{4t}{3}+\frac{3}{2}=\frac{8t}{6}+\frac{9}{6}$
$=\frac{8t+9}{6}$

23) $\frac{10}{3h^3}+\frac{2}{5h}=\frac{50}{15h^3}+\frac{6h^2}{15h^3}=\frac{50+6h^2}{15h^3}$

25) $\frac{3}{2f^2}-\frac{7}{f}=\frac{3}{2f^2}-\frac{14f}{2f^2}=\frac{3-14f}{2f^2}$

27) $\frac{13}{y+3}+\frac{3}{y}=\frac{13y}{y(y+3)}+\frac{3(y+3)}{y(y+3)}$
$=\frac{13y+3y+9}{y(y+3)}$
$=\frac{16y+9}{y(y+3)}$

29) $\frac{15}{d-8}-\frac{4}{d}=\frac{15d}{d(d-8)}-\frac{4(d-8)}{d(d-8)}$
$=\frac{15d-4d+32}{d(d-8)}$
$=\frac{11d+32}{d(d-8)}$

31) $\frac{9}{c-4}+\frac{6}{c+8}$
$=\frac{9(c+8)}{(c-4)(c+8)}+\frac{6(c-4)}{(c-4)(c+8)}$
$=\frac{9c+72+6c-24}{(c-4)(c+8)}=\frac{15c+48}{(c-4)(c+8)}$
$=\frac{3(5c+16)}{(c-4)(c+8)}$

33) $\frac{m}{3m+5}-\frac{2}{m-10}$
$=\frac{m(m-10)}{(3m+5)(m-10)}-\frac{2(3m+5)}{(3m+5)(m-10)}$
$=\frac{m^2-10m-6m-10}{(3m+5)(m-10)}=\frac{m^2-16m-10}{(3m+5)(m-10)}$

35) $\frac{8u+2}{u^2-1}+\frac{3u}{u+1}=\frac{8u+2}{(u+1)(u-1)}+\frac{3u}{u+1}$
$=\frac{8u+2}{(u+1)(u-1)}+\frac{3u(u-1)}{(u+1)(u-1)}$
$=\frac{8u+2+3u^2-3u}{(u+1)(u-1)}$
$=\frac{3u^2+5u+2}{(u+1)(u-1)}$
$=\frac{\cancel{(u+1)}(3u+2)}{\cancel{(u+1)}(u-1)}=\frac{3u+2}{u-1}$

37) $\frac{7g}{g^2+10g+16}+\frac{3}{g^2-64}$
$=\frac{7g}{(g+8)(g+2)}+\frac{3}{(g+8)(g-8)}$
$=\frac{7g(g-8)}{(g+2)(g+8)(g-8)}+\frac{3(g+2)}{(g+2)(g+8)(g-8)}$
$=\frac{7g^2-56g+3g+6}{(g+2)(g+8)(g-8)}=\frac{7g^2-53g+6}{(g+2)(g+8)(g-8)}$

39) $\dfrac{5a}{a^2-6a-27}-\dfrac{2a+1}{a^2+2a-3}=\dfrac{5a}{(a-9)(a+3)}-\dfrac{2a+1}{(a-1)(a+3)}$

$$=\frac{5a(a-1)}{(a-9)(a+3)(a-1)}-\frac{(2a+1)(a-9)}{(a-9)(a+3)(a-1)}$$

$$=\frac{5a^2-5a-(2a^2-17a-9)}{(a-9)(a+3)(a-1)}$$

$$=\frac{3a^2+12a+9}{(a-9)(a+3)(a-1)}=\frac{3(a^2+4a+3)}{(a-9)(a+3)(a-1)}$$

$$=\frac{3\cancel{(a+3)}(a+1)}{(a-9)\cancel{(a+3)}(a-1)}=\frac{3(a+1)}{(a-9)(a-1)}$$

41) $\dfrac{2x}{x^2+x-20}-\dfrac{4}{x^2+2x-15}=\dfrac{2x}{(x-4)(x+5)}-\dfrac{4}{(x-3)(x+5)}$

$$=\frac{2x(x-3)}{(x-4)(x+5)(x-3)}-\frac{4(x-4)}{(x-4)(x+5)(x-3)}$$

$$=\frac{2x^2-6x-4x+16}{(x-4)(x+5)(x-3)}$$

$$=\frac{2x^2-10x+16}{(x-4)(x+5)(x-3)}=\frac{2(x^2-5x+8)}{(x-4)(x+5)(x-3)}$$

43) $\dfrac{4b+1}{3b-12}+\dfrac{5b}{b^2-b-12}=\dfrac{4b+1}{3(b-4)}+\dfrac{5b}{(b-4)(b+3)}$

$$=\frac{(4b+1)(b+3)}{3(b-4)(b+3)}+\frac{15b}{3(b-4)(b+3)}$$

$$=\frac{4b^2+13b+3+15b}{3(b-4)(b+3)}=\frac{4b^2+28b+3}{3(b-4)(b+3)}$$

45) No. If the sum is rewritten as $\dfrac{9}{x-6}-\dfrac{4}{x-6}$, then the LCD $=x-6$.

If the sum is rewritten as $\dfrac{-9}{6-x}+\dfrac{4}{6-x}$, then the LCD $=6-x$.

47) $\dfrac{16}{q-4}+\dfrac{10}{4-q}=\dfrac{16}{q-4}-\dfrac{10}{q-4}=\dfrac{6}{q-4}$ or $-\dfrac{6}{4-q}$

49) $\dfrac{11}{f-7}-\dfrac{15}{7-f}=\dfrac{11}{f-7}+\dfrac{15}{f-7}=\dfrac{26}{f-7}$ or $-\dfrac{26}{7-f}$

51) $\dfrac{7}{x-4}+\dfrac{x-1}{4-x}=\dfrac{7}{x-4}-\dfrac{x-1}{x-4}=\dfrac{7-x+1}{x-4}=\dfrac{8-x}{x-4}$ or $\dfrac{x-8}{4-x}$

53) $\dfrac{8}{3-a}+\dfrac{a+5}{a-3}=\dfrac{8}{3-a}-\dfrac{a+5}{3-a}=\dfrac{8-a-5}{3-a}=\dfrac{3-a}{3-a}=1$

55) $\dfrac{3}{2u-3v}-\dfrac{6u}{3v-2u}=\dfrac{3}{2u-3v}+\dfrac{6u}{2u-3v}=\dfrac{3+6u}{2u-3v}=\dfrac{3(1+2u)}{2u-3v}$ or $-\dfrac{3(1+2u)}{3v-2u}$

57) $\dfrac{8}{x^2-9}+\dfrac{2}{3-x}=\dfrac{8}{(x+3)(x-3)}-\dfrac{2}{x-3}=\dfrac{8}{(x+3)(x-3)}-\dfrac{2(x+3)}{(x+3)(x-3)}$

$$=\frac{8-2x-6}{(x+3)(x-3)}=\frac{2-2x}{(x+3)(x-3)}=-\frac{2(x-1)}{(x+3)(x-3)}$$

59) $\dfrac{a}{4a^2-9}-\dfrac{4}{3-2a}=\dfrac{a}{(2a+3)(2a-3)}+\dfrac{4}{(2a-3)}=\dfrac{a}{(2a+3)(2a-3)}+\dfrac{4(2a+3)}{(2a+3)(2a-3)}$

$$=\frac{a+8a+12}{(2a+3)(2a-3)}=\frac{9a+12}{(2a+3)(2a-3)}=\frac{3(3a+4)}{(2a+3)(2a-3)}$$

61) $\dfrac{5}{a^2-2a}+\dfrac{8}{a}-\dfrac{10a}{a-2}=\dfrac{5}{a(a-2)}+\dfrac{8}{a}-\dfrac{10a}{a-2}=\dfrac{5}{a(a-2)}+\dfrac{8(a-2)}{a(a-2)}-\dfrac{a\cdot 10a}{a(a-2)}$

$$=\frac{5+8a-16-10a^2}{a(a-2)}=\frac{-10a^2+8a-11}{a(a-2)}$$

63) $\dfrac{3b-1}{b^2+8b}+\dfrac{b}{3b^2+25b+8}+\dfrac{2}{3b^2+b}=\dfrac{3b-1}{b(b+8)}+\dfrac{b}{(3b+1)(b+8)}+\dfrac{2}{b(3b+1)}$

$$=\frac{(3b-1)(3b+1)}{b(b+8)(3b+1)}+\frac{b\cdot b}{b(3b+1)(b+8)}+\frac{2(b+8)}{b(3b+1)(b+8)}$$

$$=\frac{(9b^2-1)+b^2+(2b+16)}{b(b+8)(3b+1)}=\frac{10b^2+2b+15}{b(b+8)(3b+1)}$$

65) $$\frac{c}{c^2-8c+16}-\frac{5}{c^2-c-12}=\frac{c}{(c-4)^2}-\frac{5}{(c-4)(c+3)}$$
$$=\frac{c(c+3)}{(c-4)^2(c+3)}-\frac{5(c-4)}{(c-4)^2(c+3)}$$
$$=\frac{c^2+3c-5c+20}{(c-4)^2(c+3)}=\frac{c^2-2c+20}{(c-4)^2(c+3)}$$

67) $$\frac{9}{4a+4b}+\frac{8}{a-b}-\frac{6a}{a^2-b^2}=\frac{9}{4(a+b)}+\frac{8}{a-b}-\frac{6a}{(a+b)(a-b)}$$
$$=\frac{9(a-b)}{4(a+b)(a-b)}+\frac{8\cdot 4(a+b)}{4(a+b)(a-b)}-\frac{4\cdot 6a}{4(a+b)(a-b)}$$
$$=\frac{9a-9b+32a+32b-24a}{4(a+b)(a-b)}=\frac{17a+23b}{4(a+b)(a-b)}$$

69) $$\frac{2v+1}{6v^2-29v-5}-\frac{v-2}{3v^2-13v-10}=\frac{2v+1}{(6v+1)(v-5)}-\frac{v-2}{(3v+2)(v-5)}$$
$$=\frac{(2v+1)(3v+2)}{(6v+1)(3v+2)(v-5)}-\frac{(6v+1)(v-2)}{(6v+1)(3v+2)(v-5)}$$
$$=\frac{6v^2+4v+3v+2-\left(6v^2-12v+v-2\right)}{(6v+1)(3v+2)(v-5)}$$
$$=\frac{\cancel{6v^2}+7v+2-\cancel{6v^2}+11v+2}{(6v+1)(3v+2)(v-5)}=\frac{18v+4}{(6v+1)(3v+2)(v-5)}$$
$$=\frac{2(9v+2)}{(6v+1)(3v+2)(v-5)}$$

71) $$\frac{g-5}{5g^2-30g}+\frac{g}{2g^2-17g+30}-\frac{6}{2g^2-5g}$$
$$=\frac{g-5}{5g(g-6)}+\frac{g}{(2g-5)(g-6)}-\frac{6}{g(2g-5)}$$
$$=\frac{(g-5)(2g-5)}{5g(g-6)(2g-5)}+\frac{5g\cdot g}{5g(g-6)(2g-5)}-\frac{6\cdot 5(g-6)}{5g(g-6)(2g-5)}$$
$$=\frac{2g^2-15g+25+5g^2-30g+180}{5g(g-6)(2g-5)}=\frac{7g^2-45g+205}{5g(g-6)(2g-5)}$$

73) a) $A=\left(\frac{k-4}{\cancel{4}}\right)\left(\frac{\overset{2}{\cancel{8}}}{k+1}\right)=\frac{2(k-4)}{k+1}$ sq. units

b) $P=2\left(\frac{k-4}{4}\right)+2\left(\frac{8}{k+1}\right)=\frac{k-4}{2}+\frac{16}{k+1}=\frac{(k-4)(k+1)}{2(k+1)}+\frac{2\cdot16}{2(k+1)}$

$=\frac{(k-4)(k+1)+32}{2(k+1)}=\frac{k^2-3k+28}{2(k+1)}$ units

75) a) $A=\left(\frac{6}{h^2+9h+20}\right)\left(\frac{h}{h+5}\right)=\left(\frac{6}{(h+5)(h+4)}\right)\left(\frac{h}{h+5}\right)=\frac{6h}{(h+5)^2(h+4)}$ sq. units

b) $P=2\left(\frac{6}{h^2+9h+20}\right)+2\left(\frac{h}{h+5}\right)=\frac{12}{(h+5)(h+4)}+\frac{2h}{h+5}$

$=\frac{12}{(h+5)(h+4)}+\frac{2h(h+4)}{(h+5)(h+4)}=\frac{12+2h^2+8h}{(h+5)(h+4)}$

$=\frac{2h^2+8h+12}{(h+5)(h+4)}=\frac{2(h^2+4h+6)}{(h+5)(h+4)}$ units

77) $P=\frac{1}{4x}+\frac{3}{2x^2}+\frac{12}{x}=\frac{1\cdot x}{4x^2}+\frac{3\cdot2}{4x^2}+\frac{12\cdot4x}{4x^2}$

$=\frac{x+6+48x}{4x^2}=\frac{49x+6}{4x^2}$ units

Putting It All Together

1) a) $\frac{-3+3}{3(-3)+4}=\frac{0}{-9+4}=\frac{0}{-5}=0$

b) $\frac{2+3}{3(2)+4}=\frac{5}{6+4}=\frac{5}{10}=\frac{1}{2}$

3) a) $\frac{5(-3)-3}{(-3)^2+10(-3)+21}=\frac{-15-3}{9-30+21}$

$=\frac{-18}{-21+21}=\frac{-18}{0}$

undefined

b) $\frac{5(2)-3}{(2)^2+10(2)+21}=\frac{10-3}{4+20+21}=\frac{7}{45}$

5) a)
$$w^2-36=0$$
$$(w+6)(w-6)=0$$
$$w+6=0 \text{ or } w-6=0$$
$$w=-6 \qquad w=6$$

b) $5w=0$
$$w=0$$

7) a)
$$b^2+2b-8=0$$
$$(b+4)(b-2)=0$$
$$b+4=0 \text{ or } b-2=0$$
$$b=-4 \qquad b=2$$

b) $3-5b=0$
$$-5b=-3$$
$$b=\frac{3}{5}$$

9) a) $5r=0$
$$r=0$$

b) It never equals zero.

11) $\frac{12w^{16}}{3w^5}=4w^{11}$

13)
$$\frac{m^2+6m-27}{2m^2+2m-24}=\frac{(m+9)\cancel{(m-3)}}{2(m+4)\cancel{(m-3)}}$$
$$=\frac{m+9}{2(m+4)}$$

15)
$$\frac{12-15n}{5n^2+6n-8}=\frac{3(4-5n)}{(5n-4)(n+2)}$$
$$=\frac{3\overset{-1}{\cancel{(4-5n)}}}{\cancel{(5n-4)}(n+2)}$$
$$=-\frac{3}{n+2}$$

17)
$$\frac{4c^2+4c-24}{c+3}\div\frac{3c-6}{8}$$
$$=\frac{4(c^2+c-6)}{c+3}\cdot\frac{8}{3c-6}$$
$$=\frac{4\cancel{(c+3)}\cancel{(c-2)}}{\cancel{c+3}}\cdot\frac{8}{3\cancel{(c-2)}}=\frac{32}{3}$$

19)
$$\frac{4j}{j^2-81}+\frac{3}{j^2-3j-54}$$
$$=\frac{4j}{(j+9)(j-9)}+\frac{3}{(j-9)(j+6)}$$
$$=\frac{4j(j+6)}{(j+9)(j-9)(j+6)}+\frac{3(j+9)}{(j+9)(j-9)(j+6)}$$
$$=\frac{4j^2+24j+3j+27}{(j+9)(j-9)(j+6)}$$
$$=\frac{4j^2+27j+27}{(j+9)(j-9)(j+6)}$$

21)
$$\frac{\overset{y}{\cancel{12y^7}}}{\underset{z^2}{\cancel{4z^6}}}\cdot\frac{\overset{2}{\cancel{8z^4}}}{\underset{6}{\cancel{72y^6}}}=\frac{2y}{6z^2}=\frac{y}{3z^2}$$

23) $$\frac{x}{2x^2-7x-4}-\frac{x+3}{4x^2+4x+1}=\frac{x}{(2x+1)(x-4)}-\frac{x+3}{(2x+1)^2}$$
$$=\frac{x(2x+1)}{(2x+1)^2(x-4)}-\frac{(x+3)(x-4)}{(2x+1)^2(x-4)}$$
$$=\frac{2x^2+x-\left(x^2-x-12\right)}{(2x+1)^2(x-4)}=\frac{x^2+2x+12}{(2x+1)^2(x-4)}$$

25) $$\frac{16-m^2}{m+4}\div\frac{8m-32}{m+7}$$
$$=\frac{16-m^2}{m+4}\cdot\frac{m+7}{8m-32}=\frac{(4+m)(4-m)}{m+4}\cdot\frac{m+7}{8(m-4)}=\frac{\cancel{(4+m)}\,\overset{-1}{\cancel{(4-m)}}}{\cancel{m+4}}\cdot\frac{m+7}{8\cancel{(m-4)}}=-\frac{m+7}{8}$$

27) $$\frac{xy-5x+2y-10}{y^2-25}\div\frac{x^3+8}{19x}$$
$$=\frac{xy-5x+2y-10}{y^2-25}\cdot\frac{19x}{x^3+8}=\frac{\cancel{(x+2)}\,\cancel{(y-5)}}{(y+5)\cancel{(y-5)}}\cdot\frac{19x}{\cancel{(x+2)}\left(x^2-2x+4\right)}=\frac{19x}{(y+5)\left(x^2-2x+4\right)}$$

29) $$\frac{9}{d+3}+\frac{8}{d^2}=\frac{9d^2}{d^2(d+3)}+\frac{8(d+3)}{d^2(d+3)}=\frac{9d^2+8d+24}{d^2(d+3)}$$

31) $$\frac{\dfrac{9k^2-1}{14k}}{\dfrac{3k-1}{21k^4}}=\frac{9k^2-1}{14k}\div\frac{3k-1}{21k^4}$$
$$=\frac{9k^2-1}{14k}\cdot\frac{21k^4}{3k-1}=\frac{\cancel{(3k-1)}(3k+1)}{\underset{2}{\cancel{14k}}}\cdot\frac{\overset{3k^3}{\cancel{21k^4}}}{\cancel{3k-1}}=\frac{3k^3(3k+1)}{2}$$

33) $$\frac{2w}{25-w^2}+\frac{w-3}{w^2-12w+35}=\frac{2w}{(5+w)(5-w)}+\frac{w-3}{(w-5)(w-7)}=\frac{-2w}{(w+5)(w-5)}+\frac{w-3}{(w-5)(w-7)}$$
$$=\frac{-2w(w-7)}{(w+5)(w-5)(w-7)}+\frac{(w+5)(w-3)}{(w+5)(w-5)(w-7)}=\frac{-2w^2+14w+w^2+2w-15}{(w+5)(w-5)(w-7)}=\frac{-w^2+16w-15}{(w+5)(w-5)(w-7)}$$
$$=-\frac{w^2-16w+15}{(w+5)(w-5)(w-7)}=-\frac{(w-15)(w-1)}{(w+5)(w-5)(w-7)}$$

35) $\dfrac{10}{x-8}+\dfrac{4}{x+3}$

$$=\frac{10(x+3)}{(x-8)(x+3)}+\frac{4(x-8)}{(x-8)(x+3)}=\frac{10x+30+4x-32}{(x-8)(x+3)}=\frac{14x-2}{(x-8)(x+3)}=\frac{2(7x-1)}{(x-8)(x+3)}$$

37) $\dfrac{2h^2+11h+5}{8}\div(2h+1)^2$

$$=\frac{2h^2+11h+5}{8}\cdot\frac{1}{(2h+1)^2}=\frac{\cancel{(2h+1)}(h+5)}{8}\cdot\frac{1}{(2h+1)\cancel{(2h+1)}}=\frac{h+5}{8(2h+1)}$$

39) $\dfrac{3m}{7m-4n}-\dfrac{20n}{4n-7m}=\dfrac{3m}{7m-4n}+\dfrac{20n}{7m-4n}=\dfrac{3m+20n}{7m-4n}$

41) $\dfrac{2p+3}{p^2+7p}-\dfrac{4p}{p^2-p-56}+\dfrac{5}{p^2-8p}$

$$=\frac{2p+3}{p(p+7)}-\frac{4p}{(p+7)(p-8)}+\frac{5}{p(p-8)}$$
$$=\frac{(2p+3)(p-8)}{p(p+7)(p-8)}-\frac{4p^2}{p(p+7)(p-8)}+\frac{5(p+7)}{p(p+7)(p-8)}$$
$$=\frac{2p^2-13p-24-4p^2+5p+35}{p(p+7)(p-8)}=\frac{-2p^2-8p+11}{p(p+7)(p-8)}$$

43) $\dfrac{6t+6}{3t^2-24t}\cdot(t^2-7t-8)=\dfrac{6t+6}{3t^2-24t}\cdot(t^2-7t-8)=\dfrac{\overset{2}{\cancel{6}}(t+1)}{\cancel{3}t\cancel{(t-8)}}\cdot\cancel{(t-8)}(t+1)=\dfrac{2(t+1)^2}{t}$

45) $\dfrac{\dfrac{3c^3}{8c+40}}{\dfrac{9c}{c+5}}=\dfrac{3c^3}{8c+40}\div\dfrac{9c}{c+5}=\dfrac{3c^3}{8c+40}\cdot\dfrac{c+5}{9c}=\dfrac{\overset{c^2}{\cancel{3c^3}}}{8\cancel{(c+5)}}\cdot\dfrac{\cancel{c+5}}{\underset{3}{\cancel{9c}}}=\dfrac{c^2}{24}$

47) $\dfrac{f-8}{f-4}-\dfrac{4}{4-f}=\dfrac{f-8}{f-4}+\dfrac{4}{f-4}=\dfrac{f-4}{f-4}=1$

49) $$\left(\frac{3m}{3m-1}-\frac{4}{m+4}\right)\cdot\frac{9m^2-1}{21m^2+28}=\left(\frac{3m(m+4)}{(3m-1)(m+4)}-\frac{4(3m-1)}{(3m-1)(m+4)}\right)\cdot\frac{(3m+1)(3m-1)}{7(3m^2+4)}$$
$$=\left(\frac{3m^2+12m-12m+4}{(3m-1)(m+4)}\right)\cdot\frac{(3m+1)(3m-1)}{7(3m^2+4)}=\frac{\cancel{3m^2+4}}{\cancel{(3m-1)}(m+4)}\cdot\frac{(3m+1)\cancel{(3m-1)}}{7\cancel{(3m^2+4)}}=\frac{3m+1}{7(m+4)}$$

51) $$\frac{3}{k^2+3k}-\frac{4}{3k}+\frac{1}{k+3}=\frac{3}{k(k+3)}-\frac{4}{3k}+\frac{1}{k+3}=\frac{9}{3k(k+3)}-\frac{4(k+3)}{3k(k+3)}+\frac{3k}{3k(k+3)}$$
$$=\frac{9-4k-12+3k}{3k(k+3)}=\frac{-k-3}{3k(k+3)}=\frac{-1\cancel{(k+3)}}{3k\cancel{(k+3)}}=-\frac{1}{3k}$$

53) a) $$A=\left(\frac{z}{z+5}\right)\left(\frac{6}{z+2}\right)=\frac{6z}{(z+5)(z+2)}\text{ sq. units}$$

b) $$P=2\left(\frac{z}{z+5}\right)+2\left(\frac{6}{z+2}\right)=\frac{2z}{z+5}+\frac{12}{z+2}$$
$$=\frac{2z(z+2)}{(z+5)(z+2)}+\frac{12(z+5)}{(z+5)(z+2)}=\frac{2z^2+4z+12z+60}{(z+5)(z+2)}=\frac{2z^2+16z+60}{(z+5)(z+2)}=\frac{2(z^2+8z+30)}{(z+5)(z+2)}\text{ units}$$

Section 8.5: Exercises

1) i) Rewrite it as a division problem, then simplify.

$$\frac{2}{9}\div\frac{5}{18}=\frac{2}{\cancel{9}}\cdot\frac{\overset{2}{\cancel{18}}}{5}=\frac{4}{5}$$

ii) Multiply the numerator and denominator by 18, the LCD of $\frac{2}{9}$ and $\frac{5}{18}$. Then, simplify

$$\frac{\overset{2}{\cancel{18}}\left(\frac{2}{\cancel{9}}\right)}{\cancel{18}\left(\frac{5}{\cancel{18}}\right)}=\frac{4}{5}$$

3) $$\frac{\frac{5}{9}}{\frac{7}{4}}=\frac{5}{9}\cdot\frac{4}{7}=\frac{20}{63}$$

5) $$\frac{\frac{u^4}{v^2}}{\frac{u^3}{v}}=\frac{\cancel{u^4}^{u}}{{}_{v}\cancel{v^2}}\cdot\frac{\cancel{v}}{\cancel{u^3}}=\frac{u}{v}$$

7) $$\frac{\frac{x^4}{y}}{\frac{x^2}{y^2}}=\frac{x^4}{y}\div\frac{x^2}{y^2}=\frac{x^4}{y}\cdot\frac{y^2}{x^2}=x^2y$$

9) $$\frac{\frac{14m^5n^4}{9}}{\frac{35mn^6}{3}}=\frac{14m^5n^4}{9}\div\frac{35mn^6}{3}$$

$$=\frac{\overset{2m^4}{\cancel{14m^5}n^4}}{\underset{3}{\cancel{9}}}\cdot\frac{\cancel{3}}{\underset{5n^2}{\cancel{35mn^6}}}=\frac{2m^4}{15n^2}$$

11) $$\frac{\frac{m-7}{m}}{\frac{m-7}{18}}=\frac{m-7}{m}\div\frac{m-7}{18}$$

$$=\frac{m-7}{m}\cdot\frac{18}{m-7}=\frac{18}{m}$$

13) $$\frac{\frac{g^2-36}{20}}{\frac{g+6}{60}}=\frac{g^2-36}{20}\div\frac{g+6}{60}$$

$$=\frac{(g-6)\cancel{(g+6)}}{\cancel{20}}\cdot\frac{\overset{3}{\cancel{60}}}{\cancel{g+6}}$$

$$=3(g-6)$$

15) $$\frac{\frac{d^3}{16d-24}}{\frac{d}{40d-60}}=\frac{d^3}{16d-24}\div\frac{d}{40d-60}$$

$$=\frac{\overset{d^2}{\cancel{d^3}}}{\underset{2}{\cancel{8}}\cancel{(2d-3)}}\cdot\frac{\overset{5}{\cancel{20}}\cancel{(2d-3)}}{\cancel{d}}$$

$$=\frac{5d^2}{2}$$

17) $$\frac{\frac{c^2-7c-8}{11c}}{\frac{c+1}{c}}=\frac{c^2-7c-8}{11c}\div\frac{c+1}{c}$$

$$=\frac{\cancel{(c+1)}(c-8)}{11\cancel{c}}\cdot\frac{\cancel{c}}{\cancel{c+1}}$$

$$=\frac{c-8}{11}$$

19) $$\frac{\frac{7}{9}-\frac{2}{3}}{3+\frac{1}{9}}=\frac{\frac{7}{9}-\frac{6}{9}}{\frac{27}{9}+\frac{1}{9}}=\frac{\frac{1}{9}}{\frac{28}{9}}$$

$$=\frac{1}{9}\div\frac{28}{9}=\frac{1}{9}\cdot\frac{9}{28}=\frac{1}{28}$$

21) $$\frac{\frac{r}{s}-4}{\frac{3}{s}+\frac{1}{r}}=\frac{\frac{r}{s}-\frac{4s}{s}}{\frac{3r}{rs}+\frac{s}{rs}}=\frac{\frac{r-4s}{s}}{\frac{3r+s}{rs}}$$

$$=\frac{r-4s}{s}\div\frac{3r+s}{rs}$$

$$=\frac{r-4s}{\cancel{s}}\cdot\frac{r\cancel{s}}{3r+s}$$

$$=\frac{r(r-4s)}{3r+s}$$

23) $$\frac{\frac{8}{r^2t}}{\frac{3}{r}-\frac{r}{t}}=\frac{\frac{8}{r^2t}}{\frac{3t}{rt}-\frac{r^2}{rt}}=\frac{\frac{8}{r^2t}}{\frac{3t-r^2}{rt}}$$

$$=\frac{8}{r^2t}\div\frac{3t-r^2}{rt}$$

$$=\frac{8}{\underset{r}{\cancel{r^2}}\cancel{t}}\cdot\frac{\cancel{r}\cancel{t}}{3t-r^2}$$

$$=\frac{8}{r(3t-r^2)}$$

25) $\dfrac{\dfrac{5}{w-1}+\dfrac{3}{w+4}}{\dfrac{6}{w+4}+\dfrac{4}{w-1}}$

$$\frac{\dfrac{5(w+4)}{(w+4)(w-1)}+\dfrac{3(w-1)}{(w+4)(w-1)}}{\dfrac{6(w-1)}{(w+4)(w-1)}+\dfrac{4(w+4)}{(w+4)(w-1)}}$$

$$=\frac{\dfrac{5w+20+3w-3}{(w+4)(w-1)}}{\dfrac{6w-6+4w+16}{(w+4)(w-1)}}$$

$$=\frac{\dfrac{8w+17}{(w+4)(w-1)}}{\dfrac{10w+10}{(w+4)(w-1)}}$$

$$=\frac{8w+17}{(w+4)(w-1)}\div\frac{10w+10}{(w+4)(w-1)}$$

$$=\frac{8w+17}{(w+4)(w-1)}\cdot\frac{(w+4)(w-1)}{10(w+1)}$$

$$=\frac{8w+17}{10(w+1)}$$

27) $\dfrac{\dfrac{7}{9}-\dfrac{2}{3}}{3+\dfrac{1}{9}}=\dfrac{9\left(\dfrac{7}{9}-\dfrac{2}{3}\right)}{9\left(3+\dfrac{1}{9}\right)}=\dfrac{7-6}{27+1}=\dfrac{1}{28}$

29) $\dfrac{\dfrac{r}{s}-4}{\dfrac{3}{s}+\dfrac{1}{r}}=\dfrac{rs\left(\dfrac{r}{s}-4\right)}{rs\left(\dfrac{3}{s}+\dfrac{1}{r}\right)}=\dfrac{r^2-4rs}{3r+s}$

$$=\frac{r(r-4s)}{3r+s}$$

31) $\dfrac{\dfrac{8}{r^2t}}{\dfrac{3}{r}-\dfrac{r}{t}}=\dfrac{r^2t\left(\dfrac{8}{r^2t}\right)}{r^2t\left(\dfrac{3}{r}-\dfrac{r}{t}\right)}=\dfrac{8}{3rt-r^3}=\dfrac{8}{r(3t-r^2)}$

33) $\dfrac{\dfrac{5}{w-1}+\dfrac{3}{w+4}}{\dfrac{6}{w+4}+\dfrac{4}{w-1}}$

$$=\frac{(w+4)(w-1)\left(\dfrac{5}{w-1}+\dfrac{3}{w+4}\right)}{(w+4)(w-1)\left(\dfrac{6}{w+4}+\dfrac{4}{w-1}\right)}$$

$$=\frac{5(w+4)+3(w-1)}{6(w-1)+4(w+4)}$$

$$=\frac{5w+20+3w-3}{6w-6+4w+4}=\frac{8w+17}{10w-2}$$

$$=\frac{8w+17}{2(5w-1)}$$

35) Answers may vary.

37) $\dfrac{\dfrac{a-4}{12}}{\dfrac{a-4}{a}}=\dfrac{a-4}{12}\div\dfrac{a-4}{a}$

$$=\frac{a-4}{12}\cdot\frac{a}{a-4}=\frac{a}{12}$$

39) $\dfrac{\dfrac{3}{n}-\dfrac{4}{n-2}}{\dfrac{1}{n-2}+\dfrac{5}{n}}=\dfrac{n(n-2)\left(\dfrac{3}{n}-\dfrac{4}{n-2}\right)}{n(n-2)\left(\dfrac{1}{n-2}+\dfrac{5}{n}\right)}$

$$=\frac{3(n-2)-4n}{n+5(n-2)}$$

$$=\frac{3n-6-4n}{n+5n-10}$$

$$=\frac{-n-6}{6n-10}$$

$$=-\frac{n+6}{2(3n-5)}$$

41) $$\frac{\frac{6}{w}-w}{1+\frac{6}{w}}=\frac{\frac{6-w^2}{w}}{\frac{w+6}{w}}=\frac{6-w^2}{w}\div\frac{w+6}{w}$$
$$=\frac{6-w^2}{w}\cdot\frac{w}{w+6}=\frac{6-w^2}{w+6}$$

43) $$\frac{\frac{6}{5}}{\frac{9}{15}}=\frac{6}{5}\div\frac{9}{15}=\frac{6}{\cancel{5}}\cdot\frac{\overset{3}{\cancel{15}}}{9}=\frac{18}{9}=2$$

45) $$\frac{1-\frac{4}{t+5}}{\frac{4}{t^2-25}+\frac{t}{t-5}}$$
$$=\frac{(t+5)(t-5)\left(1-\frac{4}{t+5}\right)}{(t+5)(t-5)\left(\frac{4}{(t+5)(t-5)}+\frac{t}{t-5}\right)}$$
$$=\frac{(t+5)(t-5)-4(t-5)}{4+t(t+5)}$$
$$=\frac{t^2-25-4t+20}{4+t^2+5t}=\frac{t^2-4t-5}{t^2+5t+4}$$
$$=\frac{(t-5)(t+1)}{(t+4)(t+1)}=\frac{t-5}{t+4}$$

47) $$\frac{\frac{9}{x}-\frac{9}{y}}{\frac{2}{x^2}-\frac{2}{y^2}}=\frac{x^2y^2\left(\frac{9}{x}-\frac{9}{y}\right)}{x^2y^2\left(\frac{2}{x^2}-\frac{2}{y^2}\right)}$$
$$=\frac{9xy^2-9x^2y}{2y^2-2x^2}$$
$$=\frac{9xy(y-x)}{2(y+x)(y-x)}$$
$$=\frac{9xy}{2(x+y)}$$

49) $$\frac{\frac{24c-60}{5}}{\frac{8c-20}{c^2}}=\frac{24c-60}{5}\div\frac{8c-20}{c^2}$$
$$=\frac{\overset{3}{\cancel{12}}\cancel{(2t-5)}}{5}\cdot\frac{c^2}{\cancel{4}\cancel{(2t-5)}}=\frac{3c^2}{5}$$

51) $$\frac{45\left(\frac{4}{9}+\frac{2}{5}\right)}{45\left(\frac{1}{5}-\frac{2}{3}\right)}=\frac{20+18}{9-30}=-\frac{38}{21}$$

53) $$\frac{\frac{1}{10}}{\frac{7}{8}}=\frac{1}{10}\div\frac{7}{8}=\frac{1}{\underset{5}{\cancel{10}}}\cdot\frac{\overset{4}{\cancel{8}}}{7}=\frac{4}{35}$$

55) $$\frac{\frac{2}{uv^2}}{\frac{6}{v}-\frac{4v}{u}}=\frac{uv^2\left(\frac{2}{uv^2}\right)}{uv^2\left(\frac{6}{v}-\frac{4v}{u}\right)}=\frac{2}{(6uv-4v^3)}$$
$$=\frac{2}{2v(3u-2v^2)}=\frac{1}{v(3u-2v^2)}$$

57) $$\frac{1+\frac{b}{a-b}}{\frac{b}{a^2-b^2}+\frac{1}{a+b}}$$

$$=\frac{(a+b)(a-b)\left(1+\frac{b}{a-b}\right)}{(a+b)(a-b)\left(\frac{b}{(a+b)(a-b)}+\frac{1}{a+b}\right)}$$

$$=\frac{(a+b)(a-b)+b(a+b)}{b+a-b}$$

$$=\frac{a^2-b^2+ab+b^2}{a}=\frac{a^2+ab}{a}$$

$$=\frac{a(a+b)}{a}=a+b$$

59) $$\frac{\frac{x^2-x-42}{2x-14}}{\frac{x^2-36}{8x+16}}=\frac{x^2-x-42}{2x-14}\div\frac{x^2-36}{8x+16}$$

$$=\frac{\cancel{(x-7)}\cancel{(x+6)}}{\cancel{2}\cancel{(x-7)}}\cdot\frac{\overset{4}{\cancel{8}}(x+2)}{\cancel{(x+6)}(x-6)}$$

$$=\frac{4(x+2)}{x-6}$$

61) $$\frac{\frac{y^4}{z^3}}{\frac{y^6}{z^4}}=\frac{y^4}{z^3}\div\frac{y^6}{z^4}=\frac{\cancel{y^4}}{\cancel{z^3}}\cdot\frac{\cancel{z^4}}{\underset{y^2}{\cancel{y^6}}}=\frac{z}{y^2}$$

63) $$\frac{7-\frac{8}{m}}{\frac{7m-8}{11}}=\frac{\frac{7m-8}{m}}{\frac{7m-8}{11}}$$

$$=\frac{7m-8}{m}\div\frac{7m-8}{11}$$

$$=\frac{7m-8}{m}\cdot\frac{11}{7m-8}=\frac{11}{m}$$

65) $$\frac{\frac{1}{h^2-4}+\frac{2}{h+2}}{h-\frac{3}{2}}$$

$$=\frac{\frac{1}{(h+2)(h-2)}+\frac{2}{h+2}}{h-\frac{3}{2}}$$

$$=\frac{2(h+2)(h-2)\left(\frac{1}{(h+2)(h-2)}+\frac{2}{h+2}\right)}{2(h+2)(h-2)\left(h-\frac{3}{2}\right)}$$

$$=\frac{2+4(h-2)}{2h(h+2)(h-2)-3(h+2)(h-2)}$$

$$=\frac{2+4h-8}{(2h-3)(h+2)(h-2)}=\frac{4h-6}{(2h-3)(h+2)(h-2)}$$

$$=\frac{2\cancel{(2h-3)}}{\cancel{(2h-3)}(h+2)(h-2)}=\frac{2}{(h+2)(h-2)}$$

67) $$\frac{\frac{6}{v+3}-\frac{4}{v-1}}{\frac{2}{v-1}+\frac{1}{v+2}}$$

$$=\frac{(v+3)(v+2)(v-1)\left(\frac{6}{v+3}-\frac{4}{v-1}\right)}{(v+3)(v+2)(v-1)\left(\frac{2}{v-1}+\frac{1}{v+2}\right)}$$

$$=\frac{6(v+2)(v-1)-4(v+3)(v+2)}{2(v+3)(v+2)+1(v+3)(v-1)}$$

$$=\frac{(v+2)(6v-6-4v-12)}{(v+3)(2v+4+v-1)}$$

$$=\frac{(v+2)(2v-18)}{(v+3)(3v+3)}=\frac{2(v+2)(v-9)}{3(v+3)(v+1)}$$

Section 8.6: Exercises

1) Eliminate the denominators.

3) difference; $\dfrac{3r+5}{2}-\dfrac{r}{6}$

$$=\frac{3(3r+5)}{6}-\frac{r}{6}$$

$$=\frac{9r+15-r}{6}=\frac{8r+15}{6}$$

5) equation; $\dfrac{3h}{2}+\dfrac{4}{3}=\dfrac{2h+3}{3}$

$$6\left(\frac{3h}{2}+\frac{4}{3}\right)=6\left(\frac{2h+3}{3}\right)$$

$$9h+8=2(2h+3)$$

$$9h+8=4h+6$$

$$5h=-2$$

$$h=-\frac{2}{5} \qquad \left\{-\frac{2}{5}\right\}$$

7) sum; $\dfrac{3}{a^2}+\dfrac{1}{a+11}$

$$=\frac{3(a+11)}{a^2(a+11)}+\frac{a^2}{a^2(a+11)}$$

$$=\frac{3a+33+a^2}{a^2(a+11)}=\frac{a^2+3a+33}{a^2(a+11)}$$

9) equation; $\dfrac{8}{b-11}-5=\dfrac{3}{b-11}$

$$(b-11)\left(\frac{8}{b-11}-5\right)=(b-11)\cdot\frac{3}{b-11}$$

$$8-5(b-11)=3$$

$$8-5b+55=3$$

$$-5b+63=3$$

$$-5b=-60$$

$$b=12 \qquad \{12\}$$

11) $k-2=0 \qquad k=0$

$$k=2$$

13) $p+3=0 \qquad p=0$

$$p=-3$$

$$p^2-9=0$$

$$(p+3)(p-3)=0$$

$$p+3=0 \text{ or } p-3=0$$

$$p=-3 \qquad p=3$$

15) $h^2-5h-36=0$

$$(h-9)(h+4)=0$$

$$h-9=0 \text{ or } h+4=0$$

$$h=9 \qquad h=-4$$

$$3h-27=0$$

$$3h=27$$

$$h=9$$

17) $\dfrac{a}{3}+\dfrac{7}{12}=\dfrac{1}{4}$

$$12\left(\frac{a}{3}+\frac{7}{12}\right)=12\cdot\frac{1}{4}$$

$$4a+7=3$$

$$4a=-4$$

$$a=-1 \qquad \{-1\}$$

19) $\dfrac{1}{4}j-j=-4$

$$4\left(\frac{1}{4}j-j\right)=4(-4)$$

$$j-4j=-16$$

$$-3j=-16$$

$$j=\frac{16}{3} \qquad \left\{\frac{16}{3}\right\}$$

21) $\dfrac{8m-5}{24}=\dfrac{m}{6}-\dfrac{7}{8}$

$24\left(\dfrac{8m-5}{24}\right)=24\left(\dfrac{m}{6}-\dfrac{7}{8}\right)$

$8m-5=4m-21$

$4m=-16$

$m=-4 \qquad \{-4\}$

23) $\dfrac{8}{3x+1}=\dfrac{2}{x+3}$

$8(x+3)=2(3x+1)$

$8x+24=6x+2$

$2x=-22$

$x=-11 \qquad \{-11\}$

25) $\dfrac{r+1}{2}=\dfrac{4r+1}{5}$

$5(r+1)=2(4r+1)$

$5r+5=8r+2$

$-3r=-3$

$r=1 \qquad \{1\}$

27) $\dfrac{23}{z}+8=-\dfrac{25}{z}$

$z\left(\dfrac{23}{z}+8\right)=z\left(-\dfrac{25}{z}\right)$

$23+8z=-25$

$8z=-48$

$z=-6 \qquad \{-6\}$

29) $\dfrac{5q}{q+1}-2=\dfrac{5}{q+1}$

$(q+1)\left(\dfrac{5q}{q+1}-2\right)=(q+1)\left(\dfrac{5}{q+1}\right)$

$5q-2(q+1)=5$

$5q-2q-2=5$

$3q=7$

$q=\dfrac{7}{3} \qquad \left\{\dfrac{7}{3}\right\}$

31) $\dfrac{2}{s+6}+4=\dfrac{2}{s+6}$

$(s+6)\left(\dfrac{2}{s+6}+4\right)=(s+6)\left(\dfrac{2}{s+6}\right)$

$2+(s+6)4=2$

$2+4s+24=2$

$4s+26=2$

$4s=-24$

$s=-6$

If $s=-6$, the denominators = 0. $\varnothing$

33) $\dfrac{3b}{b+7}-6=\dfrac{3}{b+7}$

$(b+7)\left(\dfrac{3b}{b+7}-6\right)=(b+7)\left(\dfrac{3}{b+7}\right)$

$3b-6(b+7)=3$

$3b-6b-42=3$

$-3b=45$

$b=-15 \qquad \{-15\}$

35) $\frac{8}{r}-1=\frac{6}{r}$

$r\left(\frac{8}{r}-1\right)=r\left(\frac{6}{r}\right)$

$8-r=6$

$-r=-2$

$r=2 \qquad \{2\}$

37) $z+\frac{12}{z}=-8$

$z\left(z+\frac{12}{z}\right)=z(-8)$

$z^2+12=-8z$

$z^2+8z+12=0$

$(z+6)(z+2)=0$

$z=-6$

$z=-2 \qquad \{-6,-2\}$

39) $\frac{15}{b}=8-b$

$b\left(\frac{15}{b}\right)=b(8-b)$

$15=8b-b^2$

$b^2-8b+15=0$

$(b-5)(b-3)=0$

$b=5$

$b=3 \qquad \{3,5\}$

41) $\frac{8}{c+2}-\frac{12}{c-4}=\frac{2}{c+2}$

$(c+2)(c-4)\left(\frac{8}{c+2}-\frac{12}{c-4}\right)=(c+2)(c-4)\left(\frac{2}{c+2}\right)$

$8(c-4)-12(c+2)=2(c-4)$

$8c-32-12c-24=2c-8$

$-4c-56=2c-8$

$-6c=48$

$c=-8 \qquad \{-8\}$

43) $\dfrac{9}{c-8}-\dfrac{15}{c}=1$

$$(c-8)c\left(\frac{9}{c-8}-\frac{15}{c}\right)=(c-8)c(1)$$

$$9c-15(c-8)=c^2-8c$$

$$9c-15c+120=c^2-8c$$

$$-6c+120=c^2-8c$$

$$c^2-2c-120=0$$

$$(c-12)(c+10)=0$$

$$c=12$$

$$c=-10 \qquad \{-12,10\}$$

45) $\dfrac{3}{p-4}+\dfrac{8}{p+4}=\dfrac{13}{p^2-16}$

$$\frac{3}{p-4}+\frac{8}{p+4}=\frac{13}{(p+4)(p-4)}$$

$$(p+4)(p-4)\left(\frac{3}{p-4}+\frac{8}{p+4}\right)=(p+4)(p-4)\left(\frac{13}{(p+4)(p-4)}\right)$$

$$3(p+4)+8(p-4)=13$$

$$3p+12+8p-32=13$$

$$11p-20=13$$

$$11p=33$$

$$p=3 \qquad \{3\}$$

47)
$$\frac{9}{k+5}-\frac{4}{k+1}=\frac{10}{k^2+6k+5}$$
$$\frac{9}{k+5}-\frac{4}{k+1}=\frac{10}{(k+5)(k+1)}$$
$$(k+5)(k+1)\left(\frac{9}{k+5}-\frac{4}{k+1}\right)=(k+5)(k+1)\left(\frac{10}{(k+5)(k+1)}\right)$$
$$9(k+1)-4(k+5)=10$$
$$9k+9-4k-20=10$$
$$5k-11=10$$
$$5k=21$$
$$k=\frac{21}{5} \qquad \left\{\frac{21}{5}\right\}$$

49)
$$\frac{12}{g^2-9}+\frac{2}{g+3}=\frac{7}{g-3}$$
$$\frac{12}{(g+3)(g-3)}+\frac{2}{g+3}=\frac{7}{g-3}$$
$$(g+3)(g-3)\left(\frac{12}{(g+3)(g-3)}+\frac{2}{g+3}\right)=(g+3)(g-3)\left(\frac{7}{g-3}\right)$$
$$12+2(g-3)=7(g+3)$$
$$12+2g-6=7g+21$$
$$2g+6=7g+21$$
$$-15=5g$$
$$-3=g$$
If $g=-3$, two denominators $=0$. $\varnothing$

51)

$$\frac{5}{p-3}-\frac{7}{p^2-7p+12}=\frac{8}{p-4}$$
$$\frac{5}{p-3}-\frac{7}{(p-4)(p-3)}=\frac{8}{p-4}$$
$$(p-4)(p-3)\left(\frac{5}{p-3}-\frac{7}{(p-4)(p-3)}\right)=(p-4)(p-3)\left(\frac{8}{p-4}\right)$$
$$5(p-4)-7=8(p-3)$$
$$5p-20-7=8p-24$$
$$5p-27=8p-24$$
$$-3=3p$$
$$-1=p \qquad \{-1\}$$

53)

$$\frac{k^2}{3}=\frac{k^2+3k}{4}$$
$$4k^2=3(k^2+3k)$$
$$4k^2=3k^2+9k$$
$$k^2-9k=0$$
$$k(k-9)=0$$
$$k=0 \text{ or } k-9=0$$
$$k=9 \qquad \{0,9\}$$

55)

$$\frac{5}{m^2-36}=\frac{4}{m^2+6m}$$
$$\frac{5}{(m+6)(m-6)}=\frac{4}{m(m+6)}$$
$$m(m+6)(m-6)\frac{5}{(m+6)(m-6)}=m(m+6)(m-6)\frac{4}{m(m+6)}$$
$$5m=4(m-6)$$
$$5m=4m-24$$
$$m=-24 \qquad \{-24\}$$

57)
$$\frac{3y-2}{y+2}=\frac{y}{4}+\frac{1}{4y+8}$$
$$\frac{3y-2}{y+2}=\frac{y}{4}+\frac{1}{4(y+2)}$$
$$4(y+2)\left(\frac{3y-2}{y+2}\right)=4(y+2)\left(\frac{y}{4}+\frac{1}{4(y+2)}\right)$$
$$4(3y-2)=y(y+2)+1$$
$$12y-8=y^2+2y+1$$
$$0=y^2-10y+9$$
$$0=(y-1)(y-9)$$
$y-1=0$ or $y-9=0$

$y=1$ $\quad y=9$ $\quad \{1,9\}$

59)
$$\frac{2}{c-6}-\frac{24}{c^2-36}=-\frac{3}{c+6}$$
$$\frac{2}{c-6}-\frac{24}{(c-6)(c+6)}=-\frac{3}{c+6}$$
$$(c-6)(c+6)\left(\frac{2}{c-6}-\frac{24}{(c-6)(c+6)}\right)=(c-6)(c+6)\left(-\frac{3}{c+6}\right)$$
$$2(c+6)-24=-3(c-6)$$
$$2c+12-24=-3c+18$$
$$2c-12=-3c+18$$
$$5c=30$$
$$c=6$$
If $c=6$, two denominators = 0. $\quad \varnothing$

61)
$$\frac{u}{7}=\frac{2}{9-u}$$
$$u(9-u)=2(7)$$
$$9u-u^2=14$$
$$0=u^2-9u+14$$
$$0=(u-7)(u-2)$$
$u-7=0$ or $u-2=0$

$u=7$ $\quad u=2$ $\quad \{2,7\}$

63) $$\frac{6}{x-1}+\frac{x}{x+3}=\frac{2x+28}{x^2+2x-3}$$

$$(x-1)(x+3)\left(\frac{6}{x-1}+\frac{x}{x+3}\right)=(x-1)(x+3)\left(\frac{2x+28}{(x-1)(x+3)}\right)$$

$$6(x+3)+x(x-1)=2x+28$$

$$6x+18+x^2-x=2x+28$$

$$x^2+5x+18=2x+28$$

$$x^2+3x-10=0$$

$$(x+5)(x-2)=0$$

$x+5=0$ or $x-2=0$

$x=-5$ $\quad x=2 \quad \{-5,2\}$

65) $$\frac{3}{2n^2+10n+8}=\frac{n}{2n+2}+\frac{1}{n+1}$$

$$2(n+4)(n+1)\left(\frac{3}{2(n+4)(n+1)}\right)=2(n+4)(n+1)\left(\frac{n}{2(n+1)}+\frac{1}{n+1}\right)$$

$$3=n(n+4)+2(n+4)$$

$$3=n^2+4n+2n+8$$

$$0=n^2+6n+5$$

$$0=(n+5)(n+1)$$

$$n+5=0 \; n=-5$$

$$n+1=0 \; n=-1$$

If $n=-1$, two denominators = 0. The solution set is $\{-5\}$.

67)
$$\frac{11}{c+9}=\frac{c}{c-4}-\frac{36-8c}{c^2+5c-36}$$
$$(c+9)(c-4)\left(\frac{11}{c+9}\right)=(c+9)(c-4)\left(\frac{c}{c-4}-\frac{36-8c}{(c+9)(c-4)}\right)$$
$$11(c-4)=c(c+9)-36+8c$$
$$11c-44=c^2+9c-36+8c$$
$$11c-44=c^2+17c-36$$
$$0=c^2+6c+8$$
$$0=(c+4)(c+2)$$
$c+4=0$ or $c+2=0$

$c=-4$ $\quad c=-2$ $\quad \{-4,-2\}$

69)
$$\frac{h}{h^2+2h-8}+\frac{4}{h^2+8h-20}=\frac{4}{h^2+14h+40}$$
$$\frac{h}{(h+4)(h-2)}+\frac{4}{(h+10)(h-2)}=\frac{4}{(h+10)(h+4)}$$
$$(h+10)(h+4)(h-2)\left(\frac{h}{(h+4)(h-2)}+\frac{4}{(h+10)(h-2)}\right)=(h+10)(h+4)(h-2)\left(\frac{4}{(h+10)(h+4)}\right)$$
$$h(h+10)+4(h+4)=4(h-2)$$
$$h^2+10h+4h+16=4h-8$$
$$h^2+14h+16=4h-8$$
$$h^2+10h+24=0$$
$$(h+6)(h+4)=0$$
$h+6=0$ or $h+4=0$

$h=-6$ $\quad h=-4$

If $h=-4$, two denominators = 0. The solution set is $\{-6\}$.

71)

$$\frac{5}{t^2+5t-6}-\frac{t}{t^2+10t+24}=\frac{1}{t^2+3t-4}$$

$$\frac{5}{(t+6)(t-1)}-\frac{t}{(t+6)(t+4)}=\frac{1}{(t+4)(t-1)}$$

$$(t+6)(t+4)(t-1)\left(\frac{5}{(t+6)(t-1)}-\frac{t}{(t+6)(t+4)}\right)=(t+6)(t+4)(t-1)\left(\frac{1}{(t+4)(t-1)}\right)$$

$$5(t+4)-t(t-1)=(t+6)$$

$$5t+20-t^2+t=t+6$$

$$6t+20-t^2=t+6$$

$$0=t^2-5t-14$$

$$0=(t+2)(t-7)$$

$t+2=0$ or $t-7=0$

$t=-2$ $\quad t=7$ $\quad \{-2,7\}$

73)

$$\frac{q}{q^2+4q-32}+\frac{2}{q^2-14q+40}=\frac{6}{q^2-2q-80}$$

$$\frac{q}{(q+8)(q-4)}+\frac{2}{(q-10)(q-4)}=\frac{6}{(q-10)(q+8)}$$

$$(q+8)(q-4)(q-10)\left(\frac{q}{(q+8)(q-4)}+\frac{2}{(q-10)(q-4)}\right)=(q+8)(q-4)(q-10)\left(\frac{6}{(q-10)(q+8)}\right)$$

$$q(q-10)+2(q+8)=6(q-4)$$

$$q^2-10q+2q+16=6q-24$$

$$q^2-8q+16=6q-24$$

$$q^2-14q+40=0$$

$$(q-10)(q-4)=0$$

$q-10=0$ or $q-4=0$

$q=10$ $\quad q=4$

If $q=10$ or 4, the denominators =0. $\varnothing$

75) $V = \frac{nRT}{\boxed{P}}$

$\boxed{P}V = nRT$

$\boxed{P} = \frac{nRT}{V}$

77) $y = \frac{kx}{\boxed{z}}$

$\boxed{z}y = kx$

$\boxed{z} = \frac{kx}{y}$

79) $Q = \frac{n-k}{5\boxed{r}}$

$5\boxed{r}Q = n-k$

$\boxed{r} = \frac{n-k}{5Q}$

81) $z = \frac{a}{\boxed{b}+c}$

$\left(\boxed{b}+c\right)z = a$

$\boxed{b}z + cz = a$

$\boxed{b}z = a - cz$

$\boxed{b} = \frac{a-cz}{z}$

83) $A = \frac{4r}{q-\boxed{t}}$

$\left(q-\boxed{t}\right)A = 4r$

$Aq - A\boxed{t} = 4r$

$Aq - 4r = A\boxed{t}$

$\frac{Aq-4r}{A} = \boxed{t}$

85) $w = \frac{na}{k\boxed{c}+b}$

$\left(k\boxed{c}+b\right)w = na$

$wk\boxed{c} + wb = na$

$wk\boxed{c} = na - wb$

$\boxed{c} = \frac{na-wb}{wk}$

87) $\frac{1}{R_1} + \frac{1}{\boxed{R_2}} = \frac{1}{R_3}$

$\left(R_1\boxed{R_2}R_3\right)\left(\frac{1}{R_1} + \frac{1}{\boxed{R_2}}\right) = \left(R_1\boxed{R_2}R_3\right)\left(\frac{1}{R_3}\right)$

$\boxed{R_2}R_3 + R_1R_3 = R_1\boxed{R_2}$

$\boxed{R_2}R_3 - R_1\boxed{R_2} = R_1R_3$

$\boxed{R_2}\left(R_3 - R_1\right) = R_1R_3$

$\boxed{R_2} = \frac{R_1R_3}{R_3 - R_1}$

89) $\frac{2}{A} + \frac{1}{\boxed{C}} = \frac{3}{B}$

$\left(A\boxed{C}B\right)\left(\frac{2}{A} + \frac{1}{\boxed{C}}\right) = \left(A\boxed{C}B\right)\left(\frac{3}{B}\right)$

$2\boxed{C}B + AB = 3A\boxed{C}$

$AB = 3A\boxed{C} - 2\boxed{C}B$

$AB = \boxed{C}(3A - 2B)$

$\frac{AB}{3A-2B} = \boxed{C}$

Section 8.7: Exercises

1) $\frac{8}{15} = \frac{32}{x}$

$8x = 15 \cdot 32$

$8x = 480$

$x = \frac{480}{8} = 60 \qquad \{60\}$

3) $\frac{4}{7} = \frac{n}{n+9}$

$4(n+9) = 7n$

$4n + 36 = 7n$

$36 = 3n$

$\frac{36}{3} = n$

$12 = n \qquad \{12\}$

5) l = length of the room

$\frac{2.5}{10} = \frac{3}{l}$

$2.5l = 3(10)$

$2.5l = 30$

$l = \frac{30}{2.5} = 12$

The length of the room is 12 feet.

7) n = number of employees that do not have direct deposit

$n + 14$ = number of employees that have direct deposit

$\frac{9}{2} = \frac{n+14}{n}$

$9n = 2(n+14)$

$9n = 2n + 28$

$7n = 28$

$n = \frac{28}{7} = 4$

$n + 14 = 4 + 14 = 18$

The number of employees that do not have direct deposit is 4.

9) n = number of students who graduated in more than four years.

$n - 1200$ = number of students who graduated in four years.

$\frac{2}{5} = \frac{n-1200}{n}$

$2n = 5(n-1200)$

$2n = 5n - 6000$

$-3n = -6000$

$n = \frac{-6000}{-3} = 2000$

$n - 1200 = 2000 - 1200 = 800$

The number of students who graduated in four years is 800.

11) l = length of the floor

$l - 18$ = width of the floor

$\frac{8}{5} = \frac{l}{l-18}$

$8(l-18) = 5l$

$8l - 144 = 5l$

$3l = 144$

$l = 48$

$l - 18 = 48 - 18 = 30$

The length is 48 feet, and the width is 30 feet.

13) $n =$ number of students who used the tutoring service.

$n+15 =$ number of students who did not use the tutoring service.

$$\frac{3}{8}=\frac{n}{n+15}$$

$$3(n+15)=8\cdot n$$

$$3n+45=8n$$

$$45=5n$$

$$n=\frac{45}{5}=9$$

$$n+15=9+15=24$$

The number of students who used tutoring service is 9.

The number of students who did not use the tutoring service is 24.

15) a) Speed against current $=8-2$
$=6$ mph

b) Speed with current $=8+2$
$=10$ mph

17) a) Speed with wind $= x+40$ mph

b) Speed against wind $= x-30$ mph

19) $s =$ speed of boat in still water

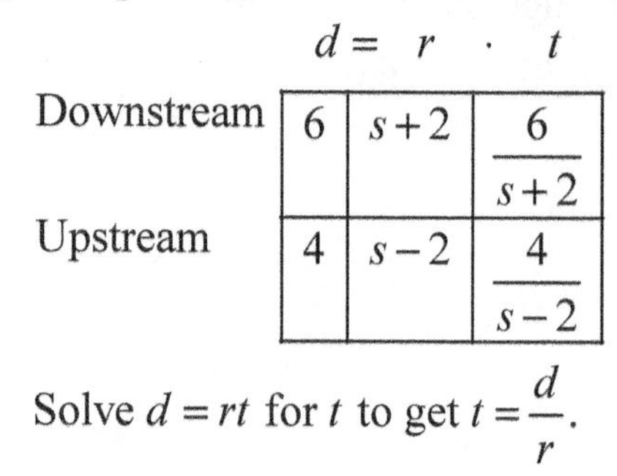

	$d =$	r	$\cdot\ t$
Downstream	6	$s+2$	$\frac{6}{s+2}$
Upstream	4	$s-2$	$\frac{4}{s-2}$

Solve $d=rt$ for t to get $t=\frac{d}{r}$.

$$\text{Time with the current} = \text{Time against the current}$$

$$\frac{6}{s+2}=\frac{4}{s-2}$$

$$6(s-2)=4(s+2)$$

$$6s-12=4s+8$$

$$2s=20$$

$$s=10$$

The speed of the boat in still water is 10 mph.

21) $s =$ speed of the plane

	$d =$	r	$\cdot\ t$
with wind	500	$s+25$	$\frac{500}{s+25}$
against wind	400	$s-25$	$\frac{400}{s-25}$

Solve $d=rt$ for t to get $t=\frac{d}{r}$.

$$\text{Time with wind} = \text{Time against wind}$$

$$\frac{500}{s+25}=\frac{400}{s-25}$$

$$500(s-25)=400(s+25)$$

$$500s-12{,}500=400s+10{,}000$$

$$100s=22{,}500$$

$$s=225$$

The speed of the plane is 225 mph.

23) $s =$ speed of the current

	$d =$	r	$\cdot\ t$
downstream	32	$28+s$	$\frac{32}{28+s}$
upstream	24	$28-s$	$\frac{24}{28-s}$

Solve $d=rt$ for t to get $t=\frac{d}{r}$.

$$\frac{\text{Time with}}{\text{the current}} = \frac{\text{Time against}}{\text{the current}}$$

Time with the current = Time against the current

$$\frac{32}{28+s} = \frac{24}{28-s}$$

$$32(28-s) = 24(28+s)$$

$$896 - 32s = 672 + 24s$$

$$224 = 56s$$

$$4 = s$$

The speed of the current is 4 mph.

25) $s =$ speed of the wind

	$d =$	r	$\cdot\ t$
with wind	800	$280+s$	$\frac{800}{280+s}$
against wind	600	$280-s$	$\frac{600}{280-s}$

Solve $d = rt$ for t to get $t = \frac{d}{r}$.

Time with wind = Time against wind

$$\frac{800}{280+s} = \frac{600}{280-s}$$

$$800(280-s) = 600(280+s)$$

$$224{,}000s - 800s = 168{,}000s + 600s$$

$$-1400s = -56{,}000$$

$$s = \frac{-56000}{-1400} = 40$$

The speed of the wind is 40 mph.

27) $s =$ average speed of Bill

	$d =$	r	$\cdot\ t$
San Diego to LA	120	s	$\frac{120}{s}$
LA to Las Vegas	240	s	$\frac{240}{s}$

Solve $d = rt$ for t to get $t = \frac{d}{r}$.

Time for San Diego to LA = (Time LA to Las Vegas) − 2

$$\frac{120}{s} = \frac{240}{s} - 2$$

$$120 = 240 - 2s$$

$$-120 = -2s$$

$$s = \frac{-120}{-2}$$

$$s = 60$$

Average speed of Bill is 60 mph.

29) $\text{rate} = \frac{1 \text{ homework}}{3 \text{ hours}} = \frac{1}{3}$ homework/hour

31) $\text{rate} = \frac{1 \text{ job}}{t \text{ hours}} = \frac{1}{t}$ job/hour

33) $t =$ hours to paint together

fractional part Rupinderjeet + fractional part Sana = 1 job

$$\frac{1}{4}t \quad + \quad \frac{1}{5}t \quad = 1$$

$$\frac{1}{4}t + \frac{1}{5}t = 1$$

$$20\left(\frac{1}{4}t + \frac{1}{5}t\right) = 20(1)$$

$$5t + 4t = 20$$

$$9t = 20$$

$$t = \frac{20}{9} = 2\frac{2}{9}$$

They could paint the room together in $2\frac{2}{9}$ hr.

35) t = the number of hours to clean the carpets together

$$\frac{\text{fractional}}{\text{part Wayne}} + \frac{\text{fractional}}{\text{part Garth}} = 1 \text{ job}$$

$$\frac{1}{4}t \quad + \quad \frac{1}{6}t \quad = 1$$

$$\frac{1}{4}t + \frac{1}{6}t = 1$$

$$24\left(\frac{1}{4}t + \frac{1}{6}t\right) = 24(1)$$

$$6t + 4t = 24$$

$$10t = 24$$

$$t = \frac{24}{10} = \frac{12}{5} = 2\frac{2}{5}$$

It would take $2\frac{2}{5}$ hours to clean the carpets together.

37) t = number of minutes it would take to fill a tub that has a leaky drain.

$$\frac{\text{fractional}}{\text{part faucet}} + \frac{\text{fractional}}{\text{part drain}} = 1 \text{ job}$$

$$\frac{1}{12}t \quad + \left(-\frac{1}{30}t\right) \quad = 1$$

$$\frac{1}{12}t - \frac{1}{30}t = 1$$

$$120\left(\frac{1}{12}t - \frac{1}{30}t\right) = 120(1)$$

$$10t - 4t = 120$$

$$6t = 120$$

$$t = 20$$

39) t = hours for old machine to do a job

$$\frac{\text{fractional}}{\text{part new machine}} + \frac{\text{fractional}}{\text{part old machine}} = 1 \text{ job}$$

$$\frac{1}{5}(3) \quad + \quad \frac{1}{t}(3) \quad = 1$$

$$\frac{3}{5} + \frac{3}{t} = 1$$

$$5t\left(\frac{3}{5} + \frac{3}{t}\right) = 1 \cdot 5t$$

$$3t + 15 = 5t$$

$$15 = 2t$$

$$t = \frac{15}{2} = 7.5$$

It would take the old machine 7.5 hours to do the job by itself.

41) t = number of hours for Ting to make decorations

$2t$ = number of hours for Mei.

$$\frac{\text{fractional part}}{\text{Ting}} + \frac{\text{fractional part}}{\text{Mei}} = 1 \text{ job}$$

$$\frac{1}{t}(40) \quad + \quad \frac{1}{2t}(40) \quad = 1$$

$$\frac{40}{t} + \frac{40}{2t} = 1$$

$$2t\left(\frac{40}{t} + \frac{40}{2t}\right) = 2t \cdot 1$$

$$80 + 40 = 2t$$

$$120 = 2t$$

$$60 = t$$

$$t = 1 \text{ hour}$$

It would take Ting 1 hour to make the decorations by herself.
It would take Mei 2 hours to make the decorations by herself.

43) increases

45) direct

47) inverse

49) $A = kw$

51) $x = \frac{k}{g}$

53) $C = kd^3$

55) $b = \frac{k}{z^2}$

57) a) $z = kx$

$63 = k(7)$

$9 = k$

b) $z = 9x$

c) $z = 9(6) = 54$

59) a) $T = \frac{k}{n}$

$10 = \frac{k}{6}$

$60 = k$

b) $T = \frac{60}{n}$

c) $T = \frac{60}{5} = 12$

61) a) $u = \frac{k}{v^2}$

$36 = \frac{k}{4^2}$

$36 = \frac{k}{16}$

$576 = k$

b) $u = \frac{576}{v^2}$

c) $u = \frac{576}{3^2} = \frac{576}{9} = 64$

63) $N = kd$

$28 = k(4)$

$7 = k$

$N = 7d$

$N = 7(11) = 77$

65) $b = \frac{k}{a}$

$18 = \frac{k}{5}$

$90 = k$

$b = \frac{90}{a}$

$b = \frac{90}{10} = 9$

67) $Q = \frac{k}{T^3}$

$216 = \frac{k}{2^3}$

$1728 = k$

$Q = \frac{1728}{T^3}$

$Q = \frac{1728}{3^3} = \frac{1728}{27} = 64$

69) $h = kv^2$

$175 = k(5)^2$

$175 = 25k$

$7 = k$

$h = 7v^2$

$h = 7(2)^2 = 7(4) = 28$

71) $E = kh$

$576.00 = k(32)$

$18 = k$

$E = 18h$

$E = 18(40) = \$720.00$

73) $w = \dfrac{k}{l}$

$12 = \dfrac{k}{24}$

$288 = k$

$w = \dfrac{288}{l}$

$w = \dfrac{288}{36} = 8 \text{ cm}$

75) $W = kr^3$

$3.24 = k(3)^3$

$3.24 = 27k$

$0.12 = k$

$W = 0.12r^3$

$W = 0.12(2)^3 = 0.12(8) = 0.96 \text{ lb}$

77) $h = \dfrac{k}{b}$

$10 = \dfrac{k}{18}$

$180 = k$

$h = \dfrac{180}{b}$

$h = \dfrac{180}{12} = 15 \text{ in.}$

79) $F = kd$

$120 = k(4)$

$30 = k$

$F = 30d$

$F = 30(6) = 180 \text{ lb}$

81) $I = \dfrac{k}{d^2}$

$40 = \dfrac{k}{(5)^2}$

$40 = \dfrac{k}{25}$

$1000 = k$

$I = \dfrac{1000}{d^2}$

$I = \dfrac{1000}{4^2} = \dfrac{1000}{16} = 62.5 \text{ lumens}$

Chapter 8 Review

1) a) $\dfrac{(5)^2-3(5)-10}{3(5)+2}=\dfrac{25-15-10}{15+2}$

$=\dfrac{0}{17}=0$

b) $\dfrac{(-2)^2-3(-2)-10}{3(-2)+2}=\dfrac{4+6-10}{-6+2}$

$=\dfrac{0}{-4}=0$

3) a) $s=0$ b) $4s+11=0$

$4s=-11$

$s=-\dfrac{11}{4}$

5) a) never equals zero

b) $4t^2-9=0$

$(2t+3)(2t-3)=0$

$2t+3=0$ or $2t-3=0$

$2t=-3$ $2t=3$

$t=-\dfrac{3}{2}$ $t=\dfrac{3}{2}$

7) a) $3m^2-m-10=0$

$(3m+5)(m-2)=0$

$3m+5=0$ or $m-2=0$

$3m=-5$ $m=2$

$m=-\dfrac{5}{3}$

b)
never undefined- any real number can be substituted for m.

9) $\dfrac{77k^9}{7k^3}=11k^6$

11) $\dfrac{r^2-14r+48}{4r^2-24r}=\dfrac{\cancel{(r-6)}(r-8)}{4r\cancel{(r-6)}}$

$=\dfrac{r-8}{4r}$

13) $\dfrac{3z-5}{6z^2-7z-5}=\dfrac{\cancel{3z-5}}{(2z+1)\cancel{(3z-5)}}$

$=\dfrac{1}{2z+1}$

15) $\dfrac{11-x}{x^2-121}=\dfrac{\overset{-1}{\cancel{11-x}}}{(x+11)\cancel{(x-11)}}$

$=-\dfrac{1}{x+11}$

17) $\dfrac{-4n-1}{5-3n},\dfrac{-(4n+1)}{5-3n},\dfrac{4n+1}{3n-5},$

$\dfrac{4n+1}{-5+3n},\dfrac{4n+1}{-(5-3n)}$

19) $w=\dfrac{2b^2+13b+21}{2b+7}$

$=\dfrac{(2b+7)(b+3)}{2b+7}=b+3$

21) $\dfrac{64}{45}\cdot\dfrac{27}{56}=\dfrac{\overset{8}{\cancel{64}}}{\underset{5}{\cancel{45}}}\cdot\dfrac{\overset{3}{\cancel{27}}}{\underset{7}{\cancel{56}}}=\dfrac{24}{35}$

23) $\dfrac{t+6}{4}\cdot\dfrac{2(t+2)}{(t+6)^2}=\dfrac{\cancel{t+6}}{\underset{2}{\cancel{4}}}\cdot\dfrac{\cancel{2}(t+2)}{(t+6)^{\cancel{2}}}$

$=\dfrac{t+2}{2(t+6)}$

25) $\dfrac{3x^2+11x+8}{15x+40} \div \dfrac{9x+9}{x-3}$

$= \dfrac{3x^2+11x+8}{15x+40} \cdot \dfrac{x-3}{9x+9}$

$= \dfrac{\cancel{(3x+8)}\cancel{(x+1)}}{5\cancel{(3x+8)}} \cdot \dfrac{x-3}{9\cancel{(x+1)}}$

$= \dfrac{x-3}{45}$

27) $\dfrac{r^2-16r+63}{2r^3-18r^2} \div (r-7)^2$

$= \dfrac{r^2-16r+63}{2r^3-18r^2} \cdot \dfrac{1}{(r-7)^2}$

$= \dfrac{\cancel{(r-7)}\cancel{(r-9)}}{2r^2\cancel{(r-9)}} \cdot \dfrac{1}{(r-7)\cancel{(r-7)}}$

$= \dfrac{1}{2r^2(r-7)}$

29) $\dfrac{3p^5}{20q^2} \cdot \dfrac{4q^3}{21p^7} = \dfrac{\cancel{3p^5}}{\underset{5}{\cancel{20q^2}}} \cdot \dfrac{\overset{q}{\cancel{4q^3}}}{\underset{7p^2}{\cancel{21p^7}}} = \dfrac{q}{35p^2}$

31) $\dfrac{\frac{3s+8}{12}}{\frac{3s+8}{4}} = \dfrac{3s+8}{12} \div \dfrac{3s+8}{4}$

$= \dfrac{\cancel{3s+8}}{\underset{3}{\cancel{12}}} \cdot \dfrac{\cancel{4}}{\cancel{3s+8}} = \dfrac{1}{3}$

33) $\dfrac{\frac{9}{8}}{\frac{15}{4}} = \dfrac{9}{8} \div \dfrac{15}{4} = \dfrac{\overset{3}{\cancel{9}}}{\underset{2}{\cancel{8}}} \cdot \dfrac{\cancel{4}}{\underset{5}{\cancel{15}}} = \dfrac{3}{10}$

35) LCD $= 30$

37) LCD $= k^5$

39) LCD $= (4x+9)(x-7)$

41) LCD $= w-5$ or $5-w$

43) $c^2+9c+20 = (c+4)(c+5)$

$c^2-2c-35 = (c+5)(c-7)$

LCD $= (c+4)(c+5)(c-7)$

45) $\dfrac{3}{5y} \cdot \dfrac{4y^2}{4y^2} = \dfrac{12y^2}{20y^3}$

47) $\dfrac{6}{2z+5} \cdot \dfrac{z}{z} = \dfrac{6z}{z(2z+5)}$

49) $\dfrac{t-3}{3t+1} \cdot \dfrac{t+4}{t+4} = \dfrac{t^2+t-12}{(3t+1)(t+4)}$

51) $\dfrac{8c}{c^2+5c-24} = \dfrac{8c}{(c+8)(c-3)}$

$\dfrac{5}{c^2-6c+9} = \dfrac{5}{(c-3)^2}$

LCD $= (c-3)^2(c+8)$

$\dfrac{8c}{(c+8)(c-3)} \cdot \dfrac{(c-3)}{(c-3)} = \dfrac{8c^2-24c}{(c-3)^2(c+8)}$

$\dfrac{5}{(c-3)^2} \cdot \dfrac{(c+8)}{(c+8)} = \dfrac{5c+40}{(c-3)^2(c+8)}$

53) $\dfrac{7}{2q^2-12q}=\dfrac{7}{2q(q-6)}$

$\dfrac{3q}{36-q^2}=\dfrac{3q}{(6+q)(6-q)}$

$\dfrac{q-5}{2q^2+12q}=\dfrac{q-5}{2q(q+6)}$

$\text{LCD}=2q(q+6)(q-6)$

$\dfrac{7}{2q(q-6)}\cdot\dfrac{(q+6)}{(q+6)}=\dfrac{7q+42}{2q(q+6)(q-6)}$

$\dfrac{3q}{(6+q)(6-q)}\cdot\dfrac{-2q}{-2q}=-\dfrac{6q^2}{2q(q+6)(q-6)}$

$\dfrac{q-5}{2q(q+6)}\cdot\dfrac{(q-6)}{(q-6)}=\dfrac{q^2-11q+30}{2q(q+6)(q-6)}$

55) $\dfrac{5}{9c}+\dfrac{7}{9c}=\dfrac{12}{9c}=\dfrac{4}{3c}$

57) $\dfrac{9}{10u^2v^2}-\dfrac{1}{8u^3v}=\dfrac{36u}{40u^3v^2}-\dfrac{5v}{40u^3v^2}=\dfrac{36u-5v}{40u^3v^2}$

59) $\dfrac{n}{3n-5}-\dfrac{4}{n}=\dfrac{n^2}{n(3n-5)}-\dfrac{4(3n-5)}{n(3n-5)}$

$=\dfrac{n^2-12n+20}{n(3n-5)}$

61) $\dfrac{8}{t+2}+\dfrac{8}{t}=\dfrac{8t}{t(t+2)}+\dfrac{8(t+2)}{t(t+2)}$

$=\dfrac{8t+8t+16}{t(t+2)}=\dfrac{16t+16}{t(t+2)}$

63) $\dfrac{9}{y+2}-\dfrac{5}{y-3}$

$=\dfrac{9(y-3)}{(y+2)(y-3)}-\dfrac{5(y+2)}{(y+2)(y-3)}$

$=\dfrac{9y-27-5y-10}{(y+2)(y-3)}$

$=\dfrac{4y-37}{(y+2)(y-3)}$

65) $\dfrac{k-3}{k^2+14k+49}-\dfrac{2}{k^2+7k}$

$=\dfrac{k-3}{(k+7)^2}-\dfrac{2}{k(k+7)}$

$=\dfrac{k(k-3)}{k(k+7)^2}-\dfrac{2(k+7)}{k(k+7)^2}$

$=\dfrac{k^2-3k-2k-14}{k(k+7)^2}$

$=\dfrac{k^2-5k-14}{k(k+7)^2}=\dfrac{(k-7)(k+2)}{k(k+7)^2}$

67) $\dfrac{t+9}{t-18}-\dfrac{11}{18-t}=\dfrac{t+9}{t-18}+\dfrac{11}{t-18}$

$=\dfrac{t+20}{t-18}$

69) $\dfrac{4w}{w^2+11w+24}-\dfrac{3w-1}{2w^2-w-21}=\dfrac{4w}{(w+3)(w+8)}-\dfrac{3w-1}{(w+3)(2w-7)}$

$=\dfrac{4w(2w-7)}{(w+3)(w+8)(2w-7)}-\dfrac{(3w-1)(w+8)}{(w+3)(w+8)(2w-7)}$

$=\dfrac{8w^2-28w-3w^2-24w+w+8}{(w+3)(w+8)(2w-7)}$

$=\dfrac{5w^2-51w+8}{(w+3)(w+8)(2w-7)}$

71) $\dfrac{b}{9b^2-4}+\dfrac{b+1}{6b^2-4b}-\dfrac{1}{6b+4}$

$=\dfrac{b}{(3b-2)(3b+2)}+\dfrac{b+1}{2b(3b-2)}-\dfrac{1}{2(3b+2)}$

$=\dfrac{2b\cdot b}{2b(3b-2)(3b+2)}+\dfrac{(3b+2)(b+1)}{2b(3b-2)(3b+2)}-\dfrac{b(3b-2)}{2b(3b-2)(3b+2)}$

$=\dfrac{2b^2+3b^2+3b+2b+2-3b^2+2b}{2b(3b-2)(3b+2)}=\dfrac{2b^2+7b+2}{2b(3b-2)(3b+2)}$

73) a) $A=\left(\dfrac{x}{x+2}\right)\left(\dfrac{2}{x^2}\right)=\dfrac{2\not{x}}{x^{\not{2}}(x+2)}$

$=\dfrac{2}{x(x+2)}$ sq. units

b) $P=2\left(\dfrac{x}{x+2}\right)+2\left(\dfrac{2}{x^2}\right)$

$=\dfrac{2x}{x+2}+\dfrac{4}{x^2}$

$=\dfrac{x^2\cdot 2x}{x^2(x+2)}+\dfrac{4(x+2)}{x^2(x+2)}$

$=\dfrac{2x^3+4x+8}{x^2(x+2)}$ units

75) $\dfrac{\frac{x}{y}}{\frac{x^3}{y^2}}=\dfrac{y^2\left(\frac{x}{y}\right)}{y^2\left(\frac{x^3}{y^2}\right)}=\dfrac{xy}{x^3}=\dfrac{y}{x^2}$

77) $\dfrac{p+\frac{4}{p}}{\frac{9}{p}+p}=\dfrac{p\left(p+\frac{4}{p}\right)}{p\left(\frac{9}{p}+p\right)}=\dfrac{p^2+4}{9+p^2}$

$=\dfrac{p^2+4}{p^2+9}$

79) $$\frac{\frac{4}{5}-\frac{2}{3}}{\frac{1}{2}+\frac{1}{6}}=\frac{30\left(\frac{4}{5}-\frac{2}{3}\right)}{30\left(\frac{1}{2}+\frac{1}{6}\right)}=\frac{24-20}{15+5}$$
$$=\frac{4}{20}=\frac{1}{5}$$

81) $$\frac{1-\frac{1}{y-8}}{\frac{2}{y+4}+1}=\frac{(y+4)(y-8)\left(1-\frac{1}{y-8}\right)}{(y+4)(y-8)\left(\frac{2}{y+4}+1\right)}$$
$$=\frac{(y+4)(y-8)-(y+4)}{2(y-8)+(y+4)(y-8)}$$
$$=\frac{y^2-4y-32-y-4}{2y-16+y^2-4y-32}$$
$$=\frac{y^2-5y-36}{y^2-2y-48}$$
$$=\frac{(y+4)(y-9)}{(y-8)(y+6)}$$

83) $$\frac{1+\frac{1}{r-t}}{\frac{1}{r^2-t^2}+\frac{1}{r+t}}$$
$$=\frac{(r+t)(r-t)\left(1+\frac{1}{r-t}\right)}{(r+t)(r-t)\left(\frac{1}{(r+t)(r-t)}+\frac{1}{r+t}\right)}$$
$$=\frac{(r+t)(r-t)+(r+t)}{1+r-t}$$
$$=\frac{(r+t)(\cancel{r-t+1})}{\cancel{r-t+1}}$$
$$=r+t$$

85) $$\frac{5a+4}{15}=\frac{a}{5}+\frac{4}{5}$$
$$15\left(\frac{5a+4}{15}\right)=15\left(\frac{a}{5}+\frac{4}{5}\right)$$
$$5a+4=3a+12$$
$$2a=8$$
$$a=4 \qquad \{4\}$$

87) $$\frac{m}{7}=\frac{5}{m+2}$$
$$m(m+2)=5\cdot 7$$
$$m^2+2m=35$$
$$m^2+2m-35=0$$
$$(m+7)(m-5)=0$$
$$m+7=0 \text{ or } m-5=0$$
$$m=-7 \qquad m=5 \qquad \{-7,5\}$$

89) $$\frac{r}{r+5}+4=\frac{5}{r+5}$$
$$(r+5)\left(\frac{r}{r+5}+4\right)=(r+5)\left(\frac{5}{r+5}\right)$$
$$r+4(r+5)=5$$
$$r+4r+20=5$$
$$5r=-15$$
$$r=-3 \qquad \{-3\}$$

91) $\dfrac{5}{t^2+10t+24}+\dfrac{5}{t^2+3t-18}=\dfrac{t}{t^2+t-12}$

$$\frac{5}{(t+6)(t+4)}+\frac{5}{(t+6)(t-3)}=\frac{t}{(t+4)(t-3)}$$

$$(t+6)(t+4)(t-3)\left(\frac{5}{(t+6)(t+4)}+\frac{5}{(t+6)(t-3)}\right)=(t+6)(t+4)(t-3)\left(\frac{t}{(t+4)(t-3)}\right)$$

$$5(t-3)+5(t+4)=t(t+6)$$

$$5t-15+5t+20=t^2+6t$$

$$10t+5=t^2+6t$$

$$t^2-4t-5=0$$

$$(t+1)(t-5)=0$$

$$t=-1 \quad \text{or} \quad t=5 \qquad \{-1,5\}$$

93) $\dfrac{3}{x+1}=\dfrac{6x}{(x+1)(x-1)}-\dfrac{4}{x-1}$

$$(x+1)(x-1)\left(\frac{3}{x+1}\right)=(x+1)(x-1)\left(\frac{6x}{(x+1)(x-1)}-\frac{4}{x-1}\right)$$

$$3(x-1)=6x-4(x+1)$$

$$3x-3=6x-4x-4$$

$$3x-3=2x-4$$

$$x=-1$$

If $x=-1$, two denominators $=0$. $\varnothing$

95) $R=\dfrac{s+T}{\boxed{D}}$

$$\boxed{D}R=s+T$$

$$\boxed{D}=\frac{s+T}{R}$$

97) $w=\dfrac{N}{c-a\boxed{k}}$

$$w\left(c-a\boxed{k}\right)=N$$

$$wc-wa\boxed{k}=N$$

$$-wa\boxed{k}=N-wc$$

$$\boxed{k}=\frac{N-wc}{-wa}$$

$$\boxed{k}=\frac{wc-N}{wa}$$

99) $$\frac{1}{\boxed{R_1}}+\frac{1}{R_2}=\frac{1}{R_3}$$

$$\left(\boxed{R_1}R_2R_3\right)\left(\frac{1}{\boxed{R_1}}+\frac{1}{R_2}\right)=\left(\boxed{R_1}R_2R_3\right)\left(\frac{1}{R_3}\right)$$

$$R_2R_3+\boxed{R_1}R_3=\boxed{R_1}R_2$$

$$R_2R_3=\boxed{R_1}R_2-\boxed{R_1}R_3$$

$$R_2R_3=\boxed{R_1}\left(R_2-R_3\right)$$

$$\frac{R_2R_3}{R_2-R_3}=\boxed{R_1}$$

101) s = speed of the current

	$d =$	r	$\cdot\ t$
downstream	8	$14+s$	$\frac{8}{14+s}$
upstream	6	$14-s$	$\frac{6}{14-s}$

Solve $d=rt$ for t to get $t=\frac{d}{r}$.

$$\text{Time dowstream} = \text{Time upstream}$$

$$\frac{8}{14+s}=\frac{6}{14-s}$$

$$8(14-s)=6(14+s)$$

$$112-8s=84+6s$$

$$28=14s$$

$$2=s$$

The speed of the current is 2 mph.

103) t = the number of hours to assemble the notebooks together

$$\text{fractional part Crayton} + \text{fractional part Flow} = 1 \text{ job}$$

$$\frac{1}{5}t + \frac{1}{8}t = 1$$

$$40\left(\frac{1}{5}t+\frac{1}{8}t\right)=40\cdot 1$$

$$8t+5t=40$$

$$13t=40$$

$$t=\frac{40}{13}=3\frac{1}{13}$$

It would take $3\frac{1}{13}$ hours to assemble the notebooks together.

105) $c=km$

$$96=k(12)$$

$$8=k$$

$$c=8m$$

$$c=8(3)=24$$

107) $S=kl^2$

$$54=k(3)^2$$

$$54=9k$$

$$6=k$$

$$S=6l^2$$

$$S=6(6)^2=216\text{ cm}^2$$

109) $$\frac{5n}{2n-1}-\frac{2n+3}{n+2}$$

$$=\frac{5n(n+2)}{(2n-1)(n+2)}-\frac{(2n+3)(2n-1)}{(2n-1)(n+2)}$$

$$=\frac{5n^2+10n-\left(4n^2-2n+6n-3\right)}{(2n-1)(n+2)}$$

$$=\frac{5n^2+10n-4n^2-4n+3}{(2n-1)(n+2)}$$

$$=\frac{n^2+6n+3}{(2n-1)(n+2)}$$

111) $\dfrac{2a^2+9a+10}{4a-7}\div(2a+5)^2$

$$=\frac{2a^2+9a+10}{4a-7}\cdot\frac{1}{(2a+5)^2}$$

$$=\frac{\cancel{(2a+5)}(a+2)}{(4a-7)}\cdot\frac{1}{(2a+5)\cancel{(2a+5)}}$$

$$=\frac{a+2}{(4a-7)(2a+5)}$$

113) $\dfrac{c^2}{c^2-d^2}+\dfrac{c}{d-c}$

$$=\frac{c^2}{(c+d)(c-d)}-\frac{c}{c-d}$$

$$=\frac{c^2}{(c+d)(c-d)}-\frac{c(c+d)}{(c+d)(c-d)}$$

$$=\frac{c^2-c^2-cd}{(c+d)(c-d)}=\frac{cd}{(c+d)(c-d)}$$

115) $\dfrac{h}{5}=\dfrac{h-3}{h+1}+\dfrac{12}{5n+5}$

$$\frac{h}{5}=\frac{h-3}{h+1}+\frac{12}{5(n+1)}$$

$$5(h+1)\frac{h}{5}=5(h+1)\left(\frac{h-3}{h+1}+\frac{12}{5(h+1)}\right)$$

$$(h+1)h=5(h-3)+12$$

$$h^2+h=5h-15+12$$

$$h^2+h=5h-3$$

$$h^2-4h+3=0$$

$$(h-3)(h-1)=0$$

$h-3=0$ or $h-1=0$

$h=3$ $\quad$ $h=1$ $\quad$ $\{1,3\}$

117)

$$\frac{8}{3g^2-7g-6}-\frac{g}{3g+2}=-\frac{4}{g-3}$$

$$\frac{8}{(3g+2)(g-3)}-\frac{8}{3g+2}=-\frac{4}{g-3}$$

$$(3g+2)(g-3)\left(\frac{8}{(3g+2)(g-3)}-\frac{8}{3g+2}\right)=(3g+2)(g-3)\left(-\frac{4}{g-3}\right)$$

$$8-8(g-3)=-4(3g+2)$$

$$8-8g+24=-12g-8$$

$$32-8g=-12g-8$$

$$32+4g=-8$$

$$4g=-40$$

$$g=-10 \quad \{-10\}$$

Chapter 8 Test

1) $\frac{5(-4)+8}{(-4)^2+16}=\frac{-20+8}{16+16}=\frac{-12}{32}=-\frac{3}{8}$

3) a) $n^2-5n-36=0$

$$(k-9)(k+4)=0$$

$$k-9=0 \text{ or } k+4=0$$

$$k=9 \qquad k=-4$$

b) It never equals zero.

5) $\frac{3h^2-25h+8}{9h^2-1}=\frac{(h-8)\cancel{(3h-1)}}{\cancel{(3h-1)}(3h+1)}$

$$=\frac{h-8}{3h+1}$$

7) z and $z+6$ are different factors. The LCD will be the product of these factors. LCD $=z(z+6)$

9) $\frac{28a^9}{b^2}\div\frac{20a^{15}}{b^3}=\frac{\overset{7}{\cancel{28a^9}}}{\cancel{b^2}}\cdot\frac{\overset{b}{\cancel{b^3}}}{\underset{5a^6}{\cancel{20a^{15}}}}$

$$=\frac{7b}{5a^6}$$

11) $\frac{6}{c+2}+\frac{c}{3c+5}=\frac{6(3c+5)+c(c+2)}{(c+2)(3c+5)}$

$$=\frac{18c+30+c^2+2c}{(c+2)(3c+5)}$$

$$=\frac{c^2+20c+30}{(c+2)(3c+5)}$$

13) $\dfrac{8d^2+24d}{20} \div (d+3)^2$

$= \dfrac{8d(d+3)}{20} \div (d+3)^2$

$= \dfrac{\overset{2}{\cancel{8}}\, d\,\cancel{(d+3)}}{\underset{5}{\cancel{20}}} \cdot \dfrac{1}{(d+3)\,\cancel{(d+3)}}$

$= \dfrac{2d}{5(d+3)}$

15) $\dfrac{3}{2v^2-7v+6} - \dfrac{v+4}{v^2+7v-18}$

$= \dfrac{3}{(2v-3)(v-2)} - \dfrac{v+4}{(v+9)(v-2)}$

$= \dfrac{3(v+9)-(v+4)(2v-3)}{(2v-3)(v-2)(v+9)}$

$= \dfrac{3v+27-(2v^2-3v+8v-12)}{(2v-3)(v-2)(v+9)}$

$= \dfrac{3v+27-2v^2-5v+12}{(2v-3)(v-2)(v+9)}$

$= \dfrac{-2v^2-2v+39}{(2v-3)(v-2)(v+9)}$

17) $\dfrac{\dfrac{5x+5y}{x^2y^2}}{\dfrac{20}{xy}} = \dfrac{x^2y^2\left(\dfrac{5x+5y}{x^2y^2}\right)}{x^2y^2\left(\dfrac{20}{xy}\right)}$

$= \dfrac{5(x+y)}{20xy} = \dfrac{x+y}{4xy}$

19) $\dfrac{28}{w^2-4} = \dfrac{7}{w-2} - \dfrac{5}{w+2}$

$\dfrac{28}{(w+2)(w-2)} = \dfrac{7}{w-2} - \dfrac{5}{w+2}$

$28 = 7(w+2) - 5(w-2)$

$28 = 7w + 14 - 5w + 10$

$28 = 2w + 24$

$4 = 2w$

$2 = w$

If $w = 2$, two denominators equal 0. $\varnothing$

21) $\dfrac{1}{a} + \dfrac{1}{\boxed{b}} = \dfrac{1}{c}$

$a\boxed{b}c\left(\dfrac{1}{a} + \dfrac{1}{\boxed{b}}\right) = a\boxed{b}c\left(\dfrac{1}{c}\right)$

$\boxed{b}c + ac = a\boxed{b}$

$ac = \boxed{b}a - \boxed{b}c$

$ac = \boxed{b}(a-c)$

$\dfrac{ac}{a-c} = \boxed{b}$

23) $s =$ speed of boat in still water

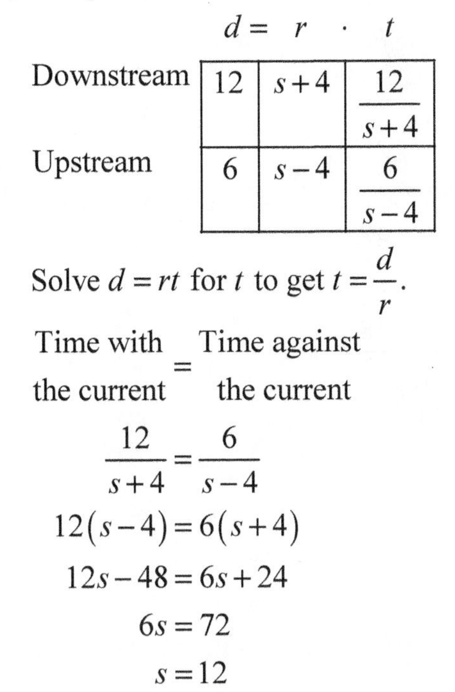

	$d =$	r	$\cdot\ t$
Downstream	12	$s+4$	$\frac{12}{s+4}$
Upstream	6	$s-4$	$\frac{6}{s-4}$

Solve $d = rt$ for t to get $t = \frac{d}{r}$.

Time with the current = Time against the current

$$\frac{12}{s+4} = \frac{6}{s-4}$$
$$12(s-4) = 6(s+4)$$
$$12s - 48 = 6s + 24$$
$$6s = 72$$
$$s = 12$$

The speed of the boat in still water is 12 mph.

25) $V = \frac{k}{P}$

$$6.25 = \frac{k}{2}$$
$$12.5 = k$$
$$V = \frac{12.5}{P}$$
$$V = \frac{12.5}{1.25} = 10\text{L}$$

Cumulative Review: Chapters 1-8

1) $A = \frac{1}{2}bh$

$$= \frac{1}{2}(18)(5)$$
$$= 45 \text{ cm}^2$$

3) $(2p^3)^5 = 2^5 p^{15} = 32p^{15}$

5) Let $l =$ the length of the rectangle.
Let $w =$ the width of the rectangle.

$$l = w + 4$$
$$2l + 2w = 28$$

Use substitution.

$$2(w+4) + 2w = 28$$
$$2w + 8 + 2w = 28$$
$$4w + 8 = 28$$
$$4w = 20$$
$$w = 5$$

$l = w + 4;\quad w = 5$

$$l = 5 + 4$$
$$l = 9$$

width: 5 ft, length: 9 ft

7) $4 \le \frac{3}{5}t + 4 \le 13$

$$20 \le 3t + 20 \le 65$$
$$0 \le 3t \le 45$$
$$0 \le t \le 15$$

$[0, 15]$

9) $(x_1, y_1) = (4, 1)$
$(x_2, y_2) = (-2, 9)$

$$m = \frac{y_2 - y_1}{x_2 - x_1} = \frac{9-1}{-2-4} = \frac{8}{-6} = -\frac{4}{3}$$

11) $(2n-3)^2 = 4n^2 - 12n + 9$

13) $\dfrac{45h^4-25h^3+15h^2-10}{15h^2}$

$$=\frac{\overset{3h^2}{\cancel{45h^4}}}{\cancel{15h^2}}-\frac{\overset{5h}{\cancel{25h^3}}}{\underset{3}{\cancel{15h^2}}}+\frac{\cancel{15h^2}}{\cancel{15h^2}}-\frac{\overset{2}{\cancel{10}}}{\underset{3h^2}{\cancel{15h^2}}}$$

$$=3h^2-\frac{5}{3}h+1-\frac{2}{3h^2}$$

15) $4d^2+4d-15=(2d+5)(2d-3)$

17) $rt+8t-r-8=t(r+8)-1(r+8)$

$$=(r+8)(t-1)$$

19) a) $a^2-6a=0$

$$a(a-6)=0$$

$$a=0 \text{ or } a-6=0$$

$$a=6$$

b) $7a+2=0$

$$7a=-2$$

$$a=-\frac{2}{7}$$

21) $\dfrac{10n^2}{n^2-8n+16}\cdot\dfrac{3n^2-14n+8}{10n-15n^2}$

$$=\frac{\overset{2}{\cancel{10}}\,\overset{n}{\cancel{n^2}}}{(n-4)\cancel{(n-4)}}\cdot\frac{\cancel{(n-4)}\overset{-1}{\cancel{(3n-2)}}}{\cancel{5}\,\cancel{n}\cancel{(2-3n)}}$$

$$=-\frac{2n}{n-4} \text{ or } \frac{2n}{4-n}$$

23) $\dfrac{\frac{2}{r-8}+1}{1-\frac{3}{r-8}}=\dfrac{(r-8)\left(\frac{2}{r-8}+1\right)}{(r-8)\left(1-\frac{3}{r-8}\right)}$

$$=\frac{2+r-8}{r-8-3}$$

$$=\frac{r-6}{r-11}$$

25) $h=\dfrac{k}{p^2}$

$$12=\frac{k}{2^2}$$

$$48=k$$

$$h=\frac{48}{p^2}$$

$$h=\frac{48}{4^2}=3$$

Chapter 9: Roots and Radicals

Section 9.1: Exercises

1) False; the $\sqrt{\ }$ symbol means to find only the positive square root of 100. $\sqrt{100}=10$

3) False; the square root of a negative number is not a real number.

5) 9 and -9

7) 2 and -2

9) 30 and -30

11) $\frac{1}{6}$ and $-\frac{1}{6}$

13) 0.5 and -0.5

15) $\sqrt{144}=12$

17) $\sqrt{9}=3$

19) $\sqrt{-64}$ is not real

21) $-\sqrt{36}=-6$

23) $\sqrt{\frac{64}{81}}=\frac{8}{9}$

25) $\sqrt{\frac{4}{9}}=\frac{2}{3}$

27) $-\sqrt{\frac{1}{16}}=-\frac{1}{4}$

29) $\sqrt{0.49}=0.7$

31) $-\sqrt{0.0144}=-0.12$

33) Since 11 is between 9 and 16,

$$\sqrt{9}<\sqrt{11}<\sqrt{16}$$
$$3<\sqrt{11}<4$$
$$\sqrt{11}\approx 3.3$$

$\sqrt{11}$

0 1 2 3 4 5 6 7 8 9

35) Since 2 is between 1 and 4,

$$\sqrt{1}<\sqrt{2}<\sqrt{4}$$
$$1<\sqrt{2}<2$$
$$\sqrt{2}\approx 1.4$$

$\sqrt{2}$

0 1 2 3 4 5 6 7 8 9

37) Since 33 is between 25 and 36,

$$\sqrt{25}<\sqrt{33}<\sqrt{36}$$
$$5<\sqrt{33}<6$$
$$\sqrt{33}\approx 5.7$$

$\sqrt{33}$

2 3 4 5 6 7

39) Since 55 is between 49 and 64,

$$\sqrt{49}<\sqrt{55}<\sqrt{64}$$
$$7<\sqrt{55}<8$$
$$\sqrt{55}\approx 7.4$$

$\sqrt{55}$

3 4 5 6 7 8 9

41) $c^2=6^2+8^2$

$c^2=36+64$

$c^2=100$

$c=\sqrt{100}=10$

43) $9^2=4^2+b^2$

$81=16+b^2$

$81-16=b^2$

$65=b^2$

$\sqrt{65}=b$

45) $13^2=a^2+5^2$

$169=a^2+25$

$169-25=a^2$

$144=a^2$

$\sqrt{144}=a$

$12=a$

47) $c^2=1^2+1^2$

$c^2=1+1$

$c^2=2$

$c=\sqrt{2}$

49) Let x be the length of the rectangle.

diagonal^2=length^2+width^2

$8^2=x^2+3^2$

$64=x^2+9$

$64-9=x^2$

$55=x^2$

$\sqrt{55}=x$

Length is $\sqrt{55}$ in.

51) Let x be the height that the ladder reaches

diagonal^2=height^2+width^2

$13^2=x^2+5^2$

$169=x^2+25$

$169-25=x^2$

$144=x^2$

$\sqrt{144}=x$

$12=x$

Height is 12 ft.

53) Let x be the length of the fencing needed.

diagonal^2=length^2+width^2

$x^2=15^2+8^2$

$x^2=225+64$

$x^2=289$

$x=\sqrt{289}=17$

Fencing needed is 17 ft.

55) Let x be the width of the laptop

diagonal^2=length^2+width^2

$13^2=11.2^2+x^2$

$169=125.44+x^2$

$169-125.44=x^2$

$43.56=x^2$

$\sqrt{43.56}=x$

$x\approx 6.6\text{in}$

Width is 6.6 in.

57) Ivan is right. You must combine the like terms in the radicand before evaluating the square root.

59) $\sqrt{25+144}=\sqrt{169}=13$

61) $\sqrt{(1)^2+(8)^2}=\sqrt{1+64}=\sqrt{65}$

63) $\sqrt{(3)^2+(-10)^2}=\sqrt{9+100}=\sqrt{109}$

65) $\sqrt{(-40)^2+(-30)^2}=\sqrt{1600+900}=\sqrt{2500}=50$

67) $d=\sqrt{(x_2-x_1)^2+(y_2-y_1)^2}$

$d=\sqrt{(8-4)^2+(5-2)^2}$

$=\sqrt{(4)^2+(3)^2}$

$=\sqrt{16+9}$

$=\sqrt{25}=5$

69) $d=\sqrt{(x_2-x_1)^2+(y_2-y_1)^2}$

$d=\sqrt{(-2-(-5))^2+(-8-(-6))^2}$

$=\sqrt{(3)^2+(-2)^2}$

$=\sqrt{9+4}$

$=\sqrt{13}$

71) $d=\sqrt{(x_2-x_1)^2+(y_2-y_1)^2}$

$d=\sqrt{(2-7)^2+(-3-3)^2}$

$=\sqrt{(-5)^2+(-6)^2}$

$=\sqrt{25+36}$

$=\sqrt{61}$

73) $d=\sqrt{(x_2-x_1)^2+(y_2-y_1)^2}$

$d=\sqrt{(-6-0)^2+(0-11)^2}$

$=\sqrt{(-6)^2+(-11)^2}$

$=\sqrt{36+121}$

$=\sqrt{157}$

75) $d=\sqrt{(x_2-x_1)^2+(y_2-y_1)^2}$

$d=\sqrt{\left(\frac{1}{4}-\frac{5}{8}\right)^2+\left(\frac{5}{6}-\frac{1}{3}\right)^2}$

$=\sqrt{\left(-\frac{3}{8}\right)^2+\left(\frac{3}{6}\right)^2}$

$=\sqrt{\frac{9}{64}+\frac{1}{4}}$

$=\sqrt{\frac{25}{64}}=\frac{5}{8}$

77) True

79) False: the odd root of a negative number is a negative number.

81) $\sqrt[3]{64}$ is the number you cube to get 64. $\sqrt[3]{64}=4$

83) No; the even root of a negative number is not a real number.

85) $\sqrt[3]{8}=2$

87) $\sqrt[3]{125}=5$

89) $\sqrt[3]{-1}=-1$

91) $\sqrt[4]{81}=3$

93) $\sqrt[4]{-1}$ is not real

95) $-\sqrt[4]{16}=-2$

97) $\sqrt[5]{-32}=-2$

99) $-\sqrt[3]{-27}=-(-3)=3$

101) $\sqrt[6]{-64}$ is not real

103) $\sqrt[3]{\frac{8}{125}}=\frac{2}{5}$

105) $\sqrt{60-11}=\sqrt{49}=7$

107) $\sqrt[3]{100+25}=\sqrt[3]{125}=5$

109) $\sqrt{1-9}=\sqrt{-8}$; not real

Section 9.2: Exercises

1) $\sqrt{2}\cdot\sqrt{3}=\sqrt{2\cdot 3}=\sqrt{6}$

3) $\sqrt{13}\cdot\sqrt{10}=\sqrt{13\cdot 10}=\sqrt{130}$

5) $\sqrt{15}\cdot\sqrt{n}=\sqrt{15\cdot n}=\sqrt{15n}$

7) False; 24 contains the factor 4 which is a perfect square.

9) $\sqrt{60}=\sqrt{4\cdot 15}$ $\boxed{\text{Factor}}$

$=\boxed{\sqrt{4}\cdot\sqrt{15}}$ Product Rule

$=\boxed{2\sqrt{15}}$ Simplify

11) $\sqrt{20}=\sqrt{4\cdot 5}=\sqrt{4}\cdot\sqrt{5}=2\sqrt{5}$

13) $\sqrt{90}=\sqrt{9\cdot 10}=\sqrt{9}\cdot\sqrt{10}=3\sqrt{10}$

15) $\sqrt{21}$; simplified

17) $-\sqrt{75}=-\sqrt{25\cdot 3}=-\sqrt{25}\cdot\sqrt{3}=-5\sqrt{3}$

19) $3\sqrt{80}=3\sqrt{16\cdot 5}=3\sqrt{16}\cdot\sqrt{5}=12\sqrt{5}$

21) $-\sqrt{1600}=-40$

23) $\sqrt{6}\sqrt{2}=\sqrt{12}=\sqrt{4}\sqrt{3}=2\sqrt{3}$

25) $\sqrt{12}\sqrt{3}=\sqrt{36}=6$

27) $\sqrt{20}\sqrt{3}=\sqrt{60}=\sqrt{4}\sqrt{15}=2\sqrt{15}$

29) $7\sqrt{6}\sqrt{12}=7\sqrt{72}=7\sqrt{36\cdot 2}=42\sqrt{2}$

31) $\sqrt{20}\cdot\sqrt{18}=\sqrt{360}=\sqrt{36\cdot 10}=6\sqrt{10}$

33) $\sqrt{\frac{9}{16}}=\frac{\sqrt{9}}{\sqrt{16}}=\frac{3}{4}$

35) $\frac{\sqrt{32}}{\sqrt{2}}=\sqrt{\frac{32}{2}}=\sqrt{16}=4$

37) $\sqrt{\frac{60}{5}}=\sqrt{12}=\sqrt{4\cdot 3}=\sqrt{4}\cdot\sqrt{3}=2\sqrt{3}$

39) $-\sqrt{\frac{3}{64}}=-\frac{\sqrt{3}}{\sqrt{64}}=-\frac{\sqrt{3}}{8}$

41) $\sqrt{\frac{52}{49}}=\frac{\sqrt{52}}{\sqrt{49}}=\frac{2\sqrt{13}}{7}$

43) $\frac{16\sqrt{63}}{2\sqrt{3}}=8\sqrt{\frac{63}{3}}=8\sqrt{21}$

45) $\frac{10\sqrt{80}}{15\sqrt{2}}=\frac{2}{3}\sqrt{\frac{80}{2}}=\frac{2}{3}\sqrt{40}=\frac{2}{3}\sqrt{4\cdot 10}=\frac{4\sqrt{10}}{3}$

47) $\sqrt{\frac{1}{2}}\sqrt{\frac{5}{8}}=\frac{\sqrt{5}}{\sqrt{16}}=\frac{\sqrt{5}}{4}$

49) $\sqrt{\frac{2}{7}}\sqrt{\frac{50}{7}}=\frac{\sqrt{100}}{7}=\frac{10}{7}$

51) $\sqrt{c^2}=c^{2/2}=c$

53) $\sqrt{t^6}=t^{6/2}=t^3$

55) $\sqrt{121b^{10}}=\sqrt{121}\cdot\sqrt{b^{10}}=11b^{10/2}=11b^5$

57) $\sqrt{36q^{14}}=\sqrt{36}\cdot\sqrt{q^{14}}=6q^{14/2}=6q^7$

59) $\sqrt{28r^4}=\sqrt{28}\cdot\sqrt{r^4}=\sqrt{4}\cdot\sqrt{7}\cdot r^{4/2}$
$=2\sqrt{7}\cdot r^2=2r^2\sqrt{7}$

61) $\sqrt{18z^{12}}=\sqrt{18}\cdot\sqrt{z^{12}}=\sqrt{9}\cdot\sqrt{2}\cdot z^{12/2}$
$=3\sqrt{2}\cdot z^6=3z^6\sqrt{2}$

63) $\sqrt{\frac{y^8}{169}}=\frac{\sqrt{y^8}}{\sqrt{169}}=\frac{y^4}{13}$

65) $\frac{\sqrt{99}}{\sqrt{w^2}}=\frac{\sqrt{9}\cdot\sqrt{11}}{w^{2/2}}=\frac{3\sqrt{11}}{w}$

67) $\sqrt{r^4s^{12}}=\sqrt{r^4}\cdot\sqrt{s^{12}}=r^2s^6$

69) $\sqrt{36x^{10}y^2}=\sqrt{36}\cdot\sqrt{x^{10}}\cdot\sqrt{y^2}$
$=\sqrt{36}\cdot x^{10/2}\cdot y^{2/2}=6x^5y$

71) $\sqrt{\frac{m^{18}n^8}{49}}=\frac{\sqrt{m^{18}n^8}}{\sqrt{49}}=\frac{\sqrt{m^{18}}\cdot\sqrt{n^8}}{7}=\frac{m^9n^4}{7}$

73) $\sqrt{p^9}=\sqrt{p^8\cdot p^1}$ Factor
$=\sqrt{p^8}\cdot\sqrt{p^1}$ Product Rule
$=p^4\sqrt{p}$ Simplify

75) $\sqrt{h^3}=\sqrt{h^2}\cdot\sqrt{h}=h^{2/2}\cdot\sqrt{h}=h\sqrt{h}$

77) $\sqrt{g^{13}}=\sqrt{g^{12}}\cdot\sqrt{g}=g^6\sqrt{g}$

79) $\sqrt{100w^5}=\sqrt{100}\cdot\sqrt{w^5}$
$=10\cdot\sqrt{w^4}\cdot\sqrt{w}$
$=10\cdot w^{4/2}\cdot\sqrt{w}=10w^2\sqrt{w}$

81) $\sqrt{75t^{11}}=\sqrt{75}\cdot\sqrt{t^{11}}$
$=\sqrt{25}\cdot\sqrt{3}\cdot\sqrt{t^{10}}\cdot\sqrt{t}$
$=5\sqrt{3}\cdot t^5\sqrt{t}=5t^5\sqrt{3t}$

83) $\sqrt{13q^7}=\sqrt{13}\cdot\sqrt{q^7}$
$=\sqrt{13}\cdot\sqrt{q^6}\cdot\sqrt{q}$
$=\sqrt{13}\cdot q^3\sqrt{q}=q^3\sqrt{13q}$

85) $\sqrt{x^2y^9}=\sqrt{x^2}\cdot\sqrt{y^9}=x\cdot\sqrt{y^8}\cdot\sqrt{y}$
$=xy^4\sqrt{y}$

87) $\sqrt{4t^9u^5}=\sqrt{4}\cdot\sqrt{t^9}\cdot\sqrt{u^5}$
$=2\cdot\sqrt{t^8t}\cdot\sqrt{u^4}\cdot\sqrt{u}$
$=2t^4u^2\sqrt{tu}$

89) i) The radicand will not contain any factors that are perfect cubes.
ii) There will be no radical in the denominator of a fraction.
iii) The radicand will not contain fractions.

91) $\sqrt[3]{2}\cdot\sqrt[3]{6}=\sqrt[3]{12}$

93) $\sqrt[4]{9}\cdot\sqrt[4]{n}=\sqrt[4]{9n}$

95) $\sqrt[3]{24}=\sqrt[3]{8}\cdot\sqrt[3]{3}=2\sqrt[3]{3}$

97) $\sqrt[3]{72}=\sqrt[3]{8}\cdot\sqrt[3]{9}=2\sqrt[3]{9}$

99) $\sqrt[3]{108}=\sqrt[3]{27}\cdot\sqrt[3]{4}=3\sqrt[3]{4}$

101) $\sqrt[4]{48}=\sqrt[4]{16}\cdot\sqrt[4]{3}=2\sqrt[4]{3}$

103) $\sqrt[4]{162}=\sqrt[4]{81}\cdot\sqrt[4]{2}=3\sqrt[4]{2}$

105) $\sqrt[3]{\frac{1}{64}}=\frac{\sqrt[3]{1}}{\sqrt[3]{64}}=\frac{1}{4}$

107) $\sqrt[4]{\frac{1}{16}}=\frac{\sqrt[4]{1}}{\sqrt[4]{16}}=\frac{1}{2}$

109) $\frac{\sqrt[3]{500}}{\sqrt[3]{2}}=\sqrt[3]{\frac{500}{2}}=\sqrt[3]{250}$
$=\sqrt[3]{125}\cdot\sqrt[3]{2}=5\sqrt[3]{2}$

111) $\frac{\sqrt[3]{8000}}{\sqrt[3]{4}}=\sqrt[3]{\frac{8000}{4}}=\sqrt[3]{2000}$
$=\sqrt[3]{1000}\cdot\sqrt[3]{2}=10\sqrt[3]{2}$

113) $\sqrt[3]{r^3}=r^{3/3}=r$

115) $\sqrt[3]{w^{12}}=w^{12/3}=w^4$

117) $\sqrt[3]{27d^{15}}=\sqrt[3]{3^3d^{15}}=3^{3/3}d^{15/3}=3d^5$

119) $\sqrt[3]{125a^{18}b^6}=\sqrt[3]{5^3a^{18}b^6}=5^{3/3}a^{18/3}b^{6/3}=5a^6b^2$

121) $\sqrt[3]{\frac{t^9}{8}}=\frac{\sqrt[3]{t^9}}{\sqrt[3]{8}}=\frac{t^{9/3}}{2}=\frac{t^3}{2}$

123) $\sqrt[3]{t^4}=\sqrt[3]{t^3}\cdot\sqrt[3]{t}=t\sqrt[3]{t}$

125) $\sqrt[3]{x^{11}}=\sqrt[3]{x^9}\cdot\sqrt[3]{x^2}=x^3\sqrt[3]{x^2}$

127) $\sqrt[3]{b^{17}}=\sqrt[3]{b^{15}}\cdot\sqrt[3]{b^2}=b^5\sqrt[3]{b^2}$

129) $\sqrt[3]{125g^6}=\sqrt[3]{125}\cdot\sqrt[3]{g^6}=\sqrt[3]{5^3}\cdot\sqrt[3]{g^6}=5g^2$

131) $\sqrt[3]{27x^3y^{12}}=\sqrt[3]{27}\cdot\sqrt[3]{x^3}\cdot\sqrt[3]{y^{12}}$
$=\sqrt[3]{3^3}\cdot\sqrt[3]{x^3}\cdot\sqrt[3]{y^{12}}$
$=3\cdot x^{3/3}\cdot y^{12/3}=3xy^4$

133) $\sqrt[3]{24k^7}=\sqrt[3]{24}\cdot\sqrt[3]{k^7}$
$=\sqrt[3]{2^3\cdot3}\cdot\sqrt[3]{k^6}\cdot\sqrt[3]{k}$
$=2\cdot k^{6/3}\sqrt[3]{3k}$
$=2k^2\sqrt[3]{k}$

135) $\sqrt[4]{n^4}=n^{4/4}=n$

137) $\sqrt[4]{m^5}=\sqrt[4]{m^4}\cdot\sqrt[4]{m}=m^{4/4}\cdot\sqrt[4]{m}=m\sqrt[4]{m}$

139) Let $A=2\sqrt{3}\text{ in}^2$ and $s=\sqrt{\frac{4\sqrt{3}A}{3}}$.

$s=\sqrt{\frac{4\sqrt{3}\left(2\sqrt{3}\right)}{3}}=\sqrt{\frac{8\sqrt{3}\cdot\sqrt{3}}{3}}$
$=\sqrt{\frac{8\cdot3}{3}}=\sqrt{\frac{24}{3}}=\sqrt{8}=2\sqrt{2}$

Each side of the equilateral triangle is $2\sqrt{2}$ in.

Section 9.3: Exercises

1) They have the same index and the same radicand.

3) $4\sqrt{11}+5\sqrt{11}=9\sqrt{11}$

5) $4\sqrt{2}-9\sqrt{2}=-5\sqrt{2}$

7) $9\sqrt[3]{5}-2\sqrt[3]{5}=7\sqrt[3]{5}$

9) $11\sqrt[3]{2}+7\sqrt[3]{2}=18\sqrt[3]{2}$

11) $4-\sqrt{13}+8-6\sqrt{13}=12-7\sqrt{13}$

13) $\sqrt{24}+\sqrt{6}=\sqrt{4\cdot 6}+\sqrt{6}$ Factor.
$=\sqrt{4}\cdot\sqrt{6}+\sqrt{6}$ Product Rule.
$=2\sqrt{6}+\sqrt{6}$ Simplify.
$=3\sqrt{6}$ Add like radicals.

15) $6\sqrt{3}-\sqrt{12}=6\sqrt{3}-2\sqrt{3}=4\sqrt{3}$

17) $\sqrt{75}+\sqrt{3}=5\sqrt{3}+\sqrt{3}=6\sqrt{3}$

19) $\sqrt{28}-3\sqrt{63}=2\sqrt{7}-3\left(3\sqrt{7}\right)$
$=2\sqrt{7}-9\sqrt{7}=-7\sqrt{7}$

21) $3\sqrt{72}-4\sqrt{8}=3\left(6\sqrt{2}\right)-8\sqrt{2}$
$=18\sqrt{2}-8\sqrt{2}=10\sqrt{2}$

23) $\sqrt{32}-3\sqrt{18}=4\sqrt{2}-3\left(3\sqrt{2}\right)$
$=4\sqrt{2}-9\sqrt{2}=-5\sqrt{2}$

25) $\frac{5}{2}\sqrt{40}+\frac{1}{3}\sqrt{90}=5\sqrt{10}+\sqrt{10}=6\sqrt{10}$

27) $\frac{4}{3}\sqrt{18}-\frac{5}{8}\sqrt{128}=4\sqrt{2}-5\sqrt{2}=-\sqrt{2}$

29) $\sqrt{50}-\sqrt{2}+\sqrt{98}=5\sqrt{2}-\sqrt{2}+7\sqrt{2}=11\sqrt{2}$

31) $\sqrt{96}+3\sqrt{24}-\sqrt{54}=4\sqrt{6}+6\sqrt{6}-3\sqrt{6}=7\sqrt{6}$

33) $2\sqrt[3]{11}+5\sqrt[3]{88}=2\sqrt[3]{11}+10\sqrt[3]{11}=12\sqrt[3]{11}$

35) $8\sqrt[3]{3}-3\sqrt[3]{81}=8\sqrt[3]{3}-9\sqrt[3]{3}=-\sqrt[3]{3}$

37) $11\sqrt[3]{16}+10\sqrt[3]{2}=22\sqrt[3]{2}+10\sqrt[3]{2}=32\sqrt[3]{2}$

39) $12\sqrt{c}+3\sqrt{c}=15\sqrt{c}$

41) $5\sqrt{3a}-9\sqrt{3a}=-4\sqrt{3a}$

43) $\sqrt{5b}+\sqrt{45b}=\sqrt{5b}+3\sqrt{5b}=4\sqrt{5b}$

45) $\sqrt{50n}-\sqrt{18n}=5\sqrt{2n}-3\sqrt{2n}=2\sqrt{2n}$

47) $11\sqrt{3z}+2\sqrt{12z}=11\sqrt{3z}+4\sqrt{3z}=15\sqrt{3z}$

49) $5\sqrt{63v}+6\sqrt{7v}=15\sqrt{7v}+6\sqrt{7v}=21\sqrt{7v}$

51) $\sqrt{4h}-8\sqrt{8p}+\sqrt{4h}+6\sqrt{8p}$
$=2\sqrt{4h}-2\sqrt{8p}=4\sqrt{h}-4\sqrt{2p}$

53) $9z\sqrt{12}-\sqrt{3z^2}=18z\sqrt{3}-z\sqrt{3}=17z\sqrt{3}$

55) $8\sqrt{7r^2}+3r\sqrt{28}=8r\sqrt{7}+6r\sqrt{7}=14r\sqrt{7}$

$=-34x\sqrt{2}$

57) $6q\sqrt{q}+7\sqrt{q^3}=6q\sqrt{q}+7q\sqrt{q}$
$=13q\sqrt{q}$

59) $16m^3\sqrt{m}-13\sqrt{m^7}$
$=16m^3\sqrt{m}-13m^3\sqrt{m}=3m^3\sqrt{m}$

61) $8w\sqrt{w^5}+4\sqrt{w^7}$
$=8w^3\sqrt{w}+4w^3\sqrt{w}=12w^3\sqrt{w}$

63) $\sqrt{xy^3}+6y\sqrt{xy}$
$=y\sqrt{xy}+6y\sqrt{xy}=7y\sqrt{xy}$

65) $9v\sqrt{6u^3}-2u\sqrt{54uv^2}$
$=9uv\sqrt{6u}-6uv\sqrt{6u}=3uv\sqrt{6u}$

67) $3\sqrt{75m^3n}+m\sqrt{12mn}$
$=15m\sqrt{3mn}+2m\sqrt{3mn}=17m\sqrt{3mn}$

69) $15\sqrt[3]{t^2}-25\sqrt[3]{t^2}=-10\sqrt[3]{t^2}$

71) $5\sqrt[3]{27x^2}+\sqrt[3]{8x^2}=5\sqrt[3]{3^3}\cdot\sqrt[3]{x^2}+\sqrt[3]{2^3}\cdot\sqrt[3]{x^2}$
$=15\sqrt[3]{x^2}+2\sqrt[3]{x^2}=17\sqrt[3]{x^2}$

73) $7r^5\sqrt[3]{r}-14\sqrt[3]{r^{16}}=7r^5\sqrt[3]{r}-14\sqrt[3]{r^{15}}\sqrt[3]{r}$
$=7r^5\sqrt[3]{r}-14r^{15/3}\sqrt[3]{r}=7r^5\sqrt[3]{r}-14r^5\sqrt[3]{r}$
$=-7r^5\sqrt[3]{r}$

75) $7\sqrt[3]{81a^5}+4a\sqrt[3]{3a^2}$
$=7\left(3a\sqrt[3]{3a^2}\right)+4a\sqrt[3]{3a^2}$
$=21a\sqrt[3]{3a^2}+4a\sqrt[3]{3a^2}=25a\sqrt[3]{3a^2}$

77) $2c^2\sqrt[3]{108c}-12\sqrt[3]{32c^7}$
$=2c^2\left(3\sqrt[3]{4c}\right)-12\left(2c^2\sqrt[3]{4c}\right)$
$=6c^2\sqrt[3]{4c}-24c^2\sqrt[3]{4c}$
$=-18c^2\sqrt[3]{4c}$

79) $3p^2\sqrt[3]{88pq^5}+8p^2q\sqrt[3]{11pq^2}$
$=3p^2\left(2q\sqrt[3]{11pq^2}\right)+8p^2q\left(\sqrt[3]{11pq^2}\right)$
$=6p^2q\sqrt[3]{11pq^2}+8p^2q\sqrt[3]{11pq^2}$
$=14p^2q\sqrt[3]{11pq^2}$

81) $3\sqrt[4]{x^4}+\sqrt[4]{16x^4}=3x+\sqrt[4]{2^4}\cdot\sqrt[4]{x^4}$
$=3x+2x=5x$

83) $-3\sqrt[4]{c^{15}}+8c^3\sqrt[4]{c^3}=-3\sqrt[4]{c^{12}}\cdot\sqrt[4]{c^3}+8c^3\sqrt[4]{c^3}$
$=-3c^3\sqrt[4]{c^3}+8c^3\sqrt[4]{c^3}$
$=5c^3\sqrt[4]{c^3}$

85) $\sqrt[3]{y^2}+11\sqrt[5]{y^2}-9\sqrt[3]{y^2}+\sqrt[5]{y^2}$
$=\sqrt[3]{y^2}-9\sqrt[3]{y^2}+11\sqrt[5]{y^2}+\sqrt[5]{y^2}$
$=-8\sqrt[3]{y^2}+12\sqrt[5]{y^2}$

87) $15cd\sqrt[4]{9cd}-\sqrt[4]{9c^5d^5}$
$=15cd\sqrt[4]{9cd}-\sqrt[4]{9c^5d^5}$
$=15cd\sqrt[4]{9cd}-\sqrt[4]{9cd}\cdot\sqrt[4]{c^4}\cdot\sqrt[4]{d^4}$
$=15cd\sqrt[4]{9cd}-cd\sqrt[4]{9cd}$
$=14cd\sqrt[4]{9cd}$

89) $P=A+B+C$
$=5\sqrt{3}+3\sqrt{2}+7\sqrt{3}$
$=12\sqrt{3}+3\sqrt{2}$
$P=\left(12\sqrt{3}+3\sqrt{2}\right)$ cm

Section 9.4 Exercises

1) $5(m+3)=5m+15$

3) $7\left(\sqrt{2}+6\right)=7\sqrt{2}+42$

5) $\sqrt{2}\left(\sqrt{5}-8\right)=\sqrt{10}-8\sqrt{2}$

7) $-6\left(\sqrt{32}+\sqrt{2}\right)=-6\left(4\sqrt{2}+\sqrt{2}\right)$
$=-6\left(5\sqrt{2}\right)=-30\sqrt{2}$

9) $6\left(\sqrt{45}-\sqrt{20}\right)=6\left(3\sqrt{5}-2\sqrt{5}\right)$
$=6\left(\sqrt{5}\right)=6\sqrt{5}$

11) $\sqrt{5}\left(\sqrt{54}-\sqrt{96}\right)=\sqrt{5}\left(3\sqrt{6}-4\sqrt{6}\right)$
$=\sqrt{5}\left(-\sqrt{6}\right)$
$=-\sqrt{30}$

13) $\sqrt{2}\left(7+\sqrt{6}\right)=\sqrt{2}\left(7+\sqrt{6}\right)=7\sqrt{2}+\sqrt{12}$
$=7\sqrt{2}+2\sqrt{3}$

15) $\sqrt{a}\left(\sqrt{a}-\sqrt{64b}\right)=\sqrt{a}\left(\sqrt{a}-8\sqrt{b}\right)$
$=a-8\sqrt{ab}$

17) $\sqrt{xy}\left(\sqrt{6x}+\sqrt{27y}\right)$
$=\sqrt{xy}\left(\sqrt{6x}+3\sqrt{3y}\right)$
$=\sqrt{6x^2y}+3\sqrt{3xy^2}=x\sqrt{6y}+3y\sqrt{3x}$

19) Both are examples of multiplication of two binomials. They can be multiplied using FOIL.

21) $\left(4+\sqrt{5}\right)\left(3+\sqrt{5}\right)$
$=4\cdot3+4\sqrt{5}+3\sqrt{5}+\sqrt{5}\cdot\sqrt{5}$ Use FOIL.
$=12+4\sqrt{5}+3\sqrt{5}+5$ Multiply.
$=\boxed{17+7\sqrt{5}}$ Combine like terms.

23) $(k+3)(k+6)=k^2+3k+6k+18$
$=k^2+9k+18$

25) $\left(6+\sqrt{7}\right)\left(2+\sqrt{7}\right)=12+6\sqrt{7}+2\sqrt{7}+\sqrt{49}$
$=12+8\sqrt{7}+7=19+8\sqrt{7}$

27) $\left(\sqrt{2}+7\right)\left(\sqrt{2}-4\right)$
$=2-4\sqrt{2}+7\sqrt{2}-28=-26+3\sqrt{2}$

29) $\left(\sqrt{3}-4\sqrt{5}\right)\left(2\sqrt{3}-\sqrt{5}\right)$
$=2(3)-\sqrt{15}-8\sqrt{15}+4(5)$
$=6-9\sqrt{15}+20=26-9\sqrt{15}$

31) $\left(3\sqrt{6}-2\sqrt{2}\right)\left(\sqrt{2}+5\sqrt{6}\right)$
$=3\sqrt{12}+15(6)-2(2)-10\sqrt{12}$
$=6\sqrt{3}+90-4-20\sqrt{3}$
$=86-14\sqrt{3}$

33) $\left(3+2\sqrt{5}\right)\left(\sqrt{7}+\sqrt{2}\right)$
$=3\sqrt{2}+2\sqrt{35}+3\sqrt{7}+2\sqrt{10}$

35) $\left(\sqrt{m}+\sqrt{7n}\right)\left(\sqrt{m}+5\sqrt{7n}\right)$
$=m+5\sqrt{7mn}+\sqrt{7mn}+5(7n)$
$=m+35n+6\sqrt{7mn}$

37) $\left(\sqrt{6p}-2\sqrt{q}\right)\left(8\sqrt{q}+5\sqrt{6p}\right)$
$=8\sqrt{6pq}+5(6p)-16(q)-10\sqrt{6pq}$
$=30p-16q-2\sqrt{6pq}$

39) Both are examples of squaring a binomial. Multiply them using the formula: $(a-b)^2=a^2-2ab+b^2$

41) $(2b-11)^2$
$=(2b)^2-2(2b)(11)+(11)^2$
$=4b^2-44b+121$

43) $\left(\sqrt{3}+1\right)^2$
$=\left(\sqrt{3}\right)^2+2\left(\sqrt{3}\right)(1)+(1)^2$
$=3+2\sqrt{3}+1=4+2\sqrt{3}$

45) $\left(\sqrt{13}-\sqrt{5}\right)^2$
$=\left(\sqrt{13}\right)^2-2\left(\sqrt{13}\right)\left(\sqrt{5}\right)+\left(\sqrt{5}\right)^2$
$=13-2\sqrt{65}+5=18-2\sqrt{65}$

47) $\left(2\sqrt{7}+\sqrt{10}\right)^2$
$=\left(2\sqrt{7}\right)^2+2\left(2\sqrt{7}\right)\left(\sqrt{10}\right)+\left(\sqrt{10}\right)^2$
$=28+4\sqrt{70}+10=38+4\sqrt{70}$

49) $\left(\sqrt{6}-4\sqrt{2}\right)^2$
$=\left(\sqrt{6}\right)^2-2\left(\sqrt{6}\right)\left(4\sqrt{2}\right)+\left(4\sqrt{2}\right)^2$
$=6-8\sqrt{12}+32=38-16\sqrt{3}$

51) $\left(\sqrt{k}+\sqrt{11}\right)^2$
$=\left(\sqrt{k}\right)^2+2\left(\sqrt{k}\right)\left(\sqrt{11}\right)+\left(\sqrt{11}\right)^2$
$=k+2\sqrt{11k}+11$

53) $\left(\sqrt{x}-\sqrt{y}\right)^2$
$=\left(\sqrt{x}\right)^2-2\left(\sqrt{x}\right)\left(\sqrt{y}\right)+\left(\sqrt{y}\right)^2$
$=x-2\sqrt{xy}+y$

55) The product of the sum and difference of two terms formula can be used.
$(a+b)(a-b)=a^2-b^2$

57) $(a+9)(a-9)=a^2-(9)^2=a^2-81$

59) $\left(\sqrt{5}+3\right)\left(\sqrt{5}-3\right)=\left(\sqrt{5}\right)^2-(3)^2=5-9=-4$

61) $\left(6-\sqrt{2}\right)\left(6+\sqrt{2}\right)=6^2-\left(\sqrt{2}\right)^2$
$=36-2=34$

63) $\left(4\sqrt{3}+\sqrt{2}\right)\left(4\sqrt{3}-\sqrt{2}\right)$
$=\left(4\sqrt{3}\right)^2-\left(\sqrt{2}\right)^2$
$=16(3)-2=48-2=46$

65) $\left(\sqrt{11}+5\sqrt{2}\right)\left(\sqrt{11}-5\sqrt{2}\right)$
$=\left(\sqrt{11}\right)^2-\left(5\sqrt{2}\right)^2$
$=11-25(2)=11-50=-39$

67) $\left(\sqrt{c}+\sqrt{d}\right)\left(\sqrt{c}-\sqrt{d}\right)$
$=\left(\sqrt{c}\right)^2-\left(\sqrt{d}\right)^2=c-d$

69) $\left(6-\sqrt{t}\right)\left(6+\sqrt{t}\right)$

$=(6)^2-\left(\sqrt{t}\right)^2=36-t$

71) $\left(8\sqrt{f}-\sqrt{g}\right)\left(8\sqrt{f}+\sqrt{g}\right)$

$=\left(8\sqrt{f}\right)^2-\left(\sqrt{g}\right)^2=64f-g$

73) $\left(7+\sqrt{2}\right)\left(11+\sqrt{2}\right)$

$=77+7\sqrt{2}+11\sqrt{2}+2$

$=79+18\sqrt{2}$

75) $\sqrt{n}\left(\sqrt{98m}+\sqrt{5n}\right)$

$=\sqrt{98mn}+\sqrt{5n^2}=7\sqrt{2mn}+n\sqrt{5mn}$

77) $\left(3\sqrt{5}+4\right)\left(3\sqrt{5}-4\right)$

$=\left(3\sqrt{5}\right)^2-(4)^2$

$=45-16=29$

79) $\left(2-\sqrt{7}\right)^2=(2)^2-2\sqrt{7}-2\sqrt{7}+\left(\sqrt{7}\right)^2$

$=4-4\sqrt{7}+7=11-4\sqrt{7}$

81) $\left(\sqrt{5a}-\sqrt{b}\right)\left(3\sqrt{5a}-\sqrt{b}\right)$

$=3(5a)-\sqrt{5ab}-3\sqrt{5ab}+\sqrt{b^2}$

$=15a+b-4\sqrt{5ab}$

83) $\left(\sqrt{5}+\sqrt{10}\right)\left(\sqrt{2}-3\sqrt{3}\right)$

$=\sqrt{10}-3\sqrt{15}+\sqrt{20}-3\sqrt{30}$

$=\sqrt{10}-3\sqrt{15}+2\sqrt{5}-3\sqrt{30}$

85) $\left(\sqrt[3]{2}-4\right)\left(\sqrt[3]{2}+4\right)$

$=\left(\sqrt[3]{2}\right)^2-(4)^2=\sqrt[3]{2^2}-16=\sqrt[3]{4}-16$

87) $\left(1+2\sqrt[3]{5}\right)\left(1-2\sqrt[3]{5}+4\sqrt[3]{25}\right)$

$=(1)^2-2\sqrt[3]{5}+4\sqrt[3]{25}+2\sqrt[3]{5}-4\sqrt[3]{25}+8\sqrt[3]{125}$

$=1+8(5)=1+40=41$

89) $\left[\left(\sqrt{5}+\sqrt{2}\right)+\sqrt{7}\right]\left[\left(\sqrt{5}+\sqrt{2}\right)-\sqrt{7}\right]$

$=\left(\sqrt{5}+\sqrt{2}\right)^2-\left(\sqrt{7}\right)^2$

$=\left(\sqrt{5}\right)^2+2\left(\sqrt{5}\right)\left(\sqrt{2}\right)+\left(\sqrt{2}\right)^2-7$

$=5+2\sqrt{10}+2-7=2\sqrt{10}$

Section 9.5: Exercises

1) Eliminate the radical of the denominator.

3) $\dfrac{1}{\sqrt{6}}=\dfrac{1}{\sqrt{6}}\cdot\dfrac{\sqrt{6}}{\sqrt{6}}=\dfrac{\sqrt{6}}{6}$

5) $\dfrac{7}{\sqrt{3}}=\dfrac{7}{\sqrt{3}}\cdot\dfrac{\sqrt{3}}{\sqrt{3}}=\dfrac{7\sqrt{3}}{3}$

7) $\dfrac{8}{\sqrt{2}}=\dfrac{8}{\sqrt{2}}\cdot\dfrac{\sqrt{2}}{\sqrt{2}}=\dfrac{8\sqrt{2}}{2}=4\sqrt{2}$

9) $\dfrac{45}{\sqrt{10}}=\dfrac{45}{\sqrt{10}}\cdot\dfrac{\sqrt{10}}{\sqrt{10}}=\dfrac{45\sqrt{10}}{10}=\dfrac{9\sqrt{10}}{2}$

11) $\dfrac{4}{\sqrt{72}}=\dfrac{4}{6\sqrt{2}}\cdot\dfrac{\sqrt{2}}{\sqrt{2}}=\dfrac{4\sqrt{2}}{6\sqrt{4}}=\dfrac{4\sqrt{2}}{12}=\dfrac{\sqrt{2}}{3}$

13) $-\frac{20}{\sqrt{8}}=-\frac{20}{2\sqrt{2}}\cdot\frac{\sqrt{2}}{\sqrt{2}}=-\frac{20\sqrt{2}}{4}=-5\sqrt{2}$

15) $\frac{\sqrt{11}}{\sqrt{7}}=\frac{\sqrt{11}}{\sqrt{7}}\cdot\frac{\sqrt{7}}{\sqrt{7}}=\frac{\sqrt{77}}{7}$

17) $\frac{2\sqrt{10}}{\sqrt{3}}=\frac{2\sqrt{10}}{\sqrt{3}}\cdot\frac{\sqrt{3}}{\sqrt{3}}=\frac{2\sqrt{30}}{3}$

19) $\frac{\sqrt{5}}{\sqrt{18}}=\frac{\sqrt{5}}{3\sqrt{2}}\cdot\frac{\sqrt{2}}{\sqrt{2}}=\frac{\sqrt{10}}{3\sqrt{4}}=\frac{\sqrt{10}}{6}$

21) $\frac{6\sqrt{8}}{\sqrt{27}}=\frac{12\sqrt{2}}{3\sqrt{3}}\cdot\frac{\sqrt{3}}{\sqrt{3}}$
$=\frac{12\sqrt{6}}{9}=\frac{4\sqrt{6}}{3}$

23) $\frac{35\sqrt{6}}{14\sqrt{5}}=\frac{5\sqrt{6}}{2\sqrt{5}}=\frac{5\sqrt{6}}{2\sqrt{5}}\cdot\frac{\sqrt{5}}{\sqrt{5}}$
$=\frac{5\sqrt{30}}{10}=\frac{\sqrt{30}}{2}$

25) $\sqrt{\frac{7}{14}}=\sqrt{\frac{1}{2}}=\frac{\sqrt{1}}{\sqrt{2}}\cdot\frac{\sqrt{2}}{\sqrt{2}}=\frac{\sqrt{2}}{2}$

27) $\sqrt{\frac{12}{80}}=\sqrt{\frac{3}{20}}=\frac{\sqrt{3}}{\sqrt{20}}=\frac{\sqrt{3}}{2\sqrt{5}}\cdot\frac{\sqrt{5}}{\sqrt{5}}$
$=\frac{\sqrt{15}}{10}$

$=\frac{\sqrt{21}}{30}$

29) $\sqrt{\frac{50}{2000}}=\sqrt{\frac{1}{40}}=\frac{\sqrt{1}}{\sqrt{40}}=\frac{1}{2\sqrt{10}}\cdot\frac{\sqrt{10}}{\sqrt{10}}$
$=\frac{\sqrt{10}}{20}$

31) $\sqrt{\frac{10}{7}}\cdot\sqrt{\frac{7}{3}}=\sqrt{\frac{10}{7}\cdot\frac{7}{3}}=\sqrt{\frac{10}{3}}=\frac{\sqrt{10}}{\sqrt{3}}$
$=\frac{\sqrt{10}}{\sqrt{3}}\cdot\frac{\sqrt{3}}{\sqrt{3}}=\frac{\sqrt{30}}{3}$

33) $\sqrt{\frac{8}{15}}\cdot\sqrt{\frac{3}{2}}=\sqrt{\frac{8}{15}\cdot\frac{3}{2}}=\sqrt{\frac{4}{5}}=\frac{\sqrt{4}}{\sqrt{5}}$
$=\frac{2}{\sqrt{5}}\cdot\frac{\sqrt{5}}{\sqrt{5}}=\frac{2\sqrt{5}}{5}$

35) $\sqrt{\frac{1}{13}}\cdot\sqrt{\frac{8}{3}}=\sqrt{\frac{1}{13}\cdot\frac{8}{3}}=\sqrt{\frac{8}{39}}=\frac{\sqrt{8}}{\sqrt{39}}$
$=\frac{2\sqrt{2}}{\sqrt{39}}\cdot\frac{\sqrt{39}}{\sqrt{39}}=\frac{2\sqrt{78}}{39}$

37) $\sqrt{\frac{3}{8}}\cdot\sqrt{\frac{27}{2}}=\sqrt{\frac{3}{8}\cdot\frac{27}{2}}=\sqrt{\frac{81}{16}}=\frac{\sqrt{81}}{\sqrt{16}}=\frac{9}{4}$

39) $\frac{5}{\sqrt{k}}=\frac{5}{\sqrt{k}}\cdot\frac{\sqrt{k}}{\sqrt{k}}=\frac{5\sqrt{k}}{k}$

41) $\frac{\sqrt{6}}{\sqrt{d}}=\frac{\sqrt{6}}{\sqrt{d}}\cdot\frac{\sqrt{d}}{\sqrt{d}}=\frac{\sqrt{6d}}{d}$

43) $\sqrt{\frac{16}{a}}=\frac{\sqrt{16}}{\sqrt{a}}=\frac{4}{\sqrt{a}}\cdot\frac{\sqrt{a}}{\sqrt{a}}=\frac{4\sqrt{a}}{a}$

45) $\sqrt{\frac{x}{y}}=\frac{\sqrt{x}}{\sqrt{y}}=\frac{\sqrt{x}}{\sqrt{y}}\cdot\frac{\sqrt{y}}{\sqrt{y}}=\frac{\sqrt{xy}}{y}$

47) $\sqrt{\dfrac{40p^3}{q}} = \dfrac{\sqrt{40p^3}}{\sqrt{q}} = \dfrac{2p\sqrt{10p}}{\sqrt{q}} \cdot \dfrac{\sqrt{q}}{\sqrt{q}} = \dfrac{2p\sqrt{10pq}}{q}$

49) $\sqrt{\dfrac{64v^7}{5w}} = \dfrac{8v^3\sqrt{v}}{\sqrt{5w}} = \dfrac{8v^3\sqrt{v}}{\sqrt{5w}} \cdot \dfrac{\sqrt{5w}}{\sqrt{5w}} = \dfrac{8v^3\sqrt{5vw}}{5w}$

51) $\dfrac{\sqrt{18h^5}}{\sqrt{2h}} = \dfrac{3h^2\sqrt{2h}}{\sqrt{2h}} = 3h^2$

53) $\dfrac{\sqrt{15x^6}}{\sqrt{9y}} = \dfrac{x^3\sqrt{15}}{3\sqrt{y}} \cdot \dfrac{\sqrt{y}}{\sqrt{y}} = \dfrac{x^3\sqrt{15y}}{3y}$

55) $\sqrt{\dfrac{2mn^2}{mn^3}} = \sqrt{\dfrac{2}{n}} = \dfrac{\sqrt{2}}{\sqrt{n}} \cdot \dfrac{\sqrt{n}}{\sqrt{n}} = \dfrac{\sqrt{2n}}{n}$

57) $\sqrt{\dfrac{8u^5v^3}{24u^6v}} = \sqrt{\dfrac{v^2}{3u}} = \dfrac{\sqrt{v^2}}{\sqrt{3u}} = \dfrac{v}{\sqrt{3u}} \cdot \dfrac{\sqrt{3u}}{\sqrt{3u}} = \dfrac{v\sqrt{3u}}{3u}$

59) $\sqrt{\dfrac{500}{p^3}} = \dfrac{\sqrt{500}}{\sqrt{p^3}} = \dfrac{10\sqrt{5}}{p\sqrt{p}} \cdot \dfrac{\sqrt{p}}{\sqrt{p}} = \dfrac{10\sqrt{5p}}{p^2}$

61) 2^2 or 4

63) 3

65) c^2

67) He multiplied incorrectly. The correct way is:
$\dfrac{9}{\sqrt[3]{2}} \cdot \dfrac{\sqrt[3]{4}}{\sqrt[3]{4}} = \dfrac{9\sqrt[3]{4}}{\sqrt[3]{8}} = \dfrac{9\sqrt[3]{4}}{2}$

69) $\dfrac{4}{\sqrt[3]{3}} = \dfrac{4}{\sqrt[3]{3}} \cdot \dfrac{\sqrt[3]{3^2}}{\sqrt[3]{3^2}} = \dfrac{4\sqrt[3]{9}}{3}$

71) $\dfrac{1}{\sqrt[3]{4}} = \dfrac{1}{\sqrt[3]{2^2}} \cdot \dfrac{\sqrt[3]{2}}{\sqrt[3]{2}} = \dfrac{\sqrt[3]{2}}{2}$

73) $\dfrac{35}{\sqrt[3]{5}} = \dfrac{35}{\sqrt[3]{5}} \cdot \dfrac{\sqrt[3]{5^2}}{\sqrt[3]{5^2}} = \dfrac{35\sqrt[3]{25}}{\sqrt[3]{5^3}} = 7\sqrt[3]{25}$

75) $\sqrt[3]{\dfrac{8}{9}} = \dfrac{\sqrt[3]{8}}{\sqrt[3]{9}} = \dfrac{2}{\sqrt[3]{3^2}} \cdot \dfrac{\sqrt[3]{3}}{\sqrt[3]{3}} = \dfrac{2\sqrt[3]{3}}{3}$

77) $\sqrt[3]{\dfrac{11}{25}} = \dfrac{\sqrt[3]{11}}{\sqrt[3]{25}} = \dfrac{\sqrt[3]{11}}{\sqrt[3]{5^2}} \cdot \dfrac{\sqrt[3]{5}}{\sqrt[3]{5}} = \dfrac{\sqrt[3]{55}}{5}$

79) $\dfrac{6}{\sqrt[3]{x}} = \dfrac{6}{\sqrt[3]{x}} \cdot \dfrac{\sqrt[3]{x^2}}{\sqrt[3]{x^2}} = \dfrac{6\sqrt[3]{x^2}}{x}$

81) $\dfrac{7}{\sqrt[3]{m^2}} = \dfrac{7}{\sqrt[3]{m^2}} \cdot \dfrac{\sqrt[3]{m}}{\sqrt[3]{m}} = \dfrac{7\sqrt[3]{m}}{m}$

83) $\sqrt[3]{\dfrac{1000}{c}} = \dfrac{10}{\sqrt[3]{c}} \cdot \dfrac{\sqrt[3]{c^2}}{\sqrt[3]{c^2}} = \dfrac{10\sqrt[3]{c}}{c}$

85) $\sqrt[3]{\dfrac{4}{w^2}} = \dfrac{\sqrt[3]{4}}{\sqrt[3]{w^2}} \cdot \dfrac{\sqrt[3]{w}}{\sqrt[3]{w}} = \dfrac{\sqrt[3]{4w}}{\sqrt[3]{w^3}} = \dfrac{\sqrt[3]{4w}}{w}$

87) $\dfrac{16}{\sqrt[3]{2t}} = \dfrac{16}{\sqrt[3]{2t}} \cdot \dfrac{\sqrt[3]{(2t)^2}}{\sqrt[3]{(2t)^2}} = \dfrac{16\sqrt[3]{4t^2}}{2t} = \dfrac{8\sqrt[3]{4t^2}}{t}$

89) $\dfrac{\sqrt[3]{5a}}{\sqrt[3]{36b}}=\dfrac{\sqrt[3]{5a}}{\sqrt[3]{6^2b}}\cdot\dfrac{\sqrt[3]{6b^2}}{\sqrt[3]{6b^2}}=\dfrac{\sqrt[3]{30ab^2}}{6b}$

91) $\dfrac{7}{\sqrt[4]{2}}=\dfrac{7}{\sqrt[4]{2}}\cdot\dfrac{\sqrt[4]{2^3}}{\sqrt[4]{2^3}}=\dfrac{7\sqrt[4]{8}}{2}$

93) $\dfrac{2}{\sqrt[4]{x^3}}=\dfrac{2}{\sqrt[4]{x^3}}\cdot\dfrac{\sqrt[4]{x}}{\sqrt[4]{x}}=\dfrac{2\sqrt[4]{x}}{x}$

Change the sign between the two terms.

97) $\left(4-\sqrt{3}\right)$;

$\left(4+\sqrt{3}\right)\left(4-\sqrt{3}\right)=(4)^2-\left(\sqrt{3}\right)^2$

$=16-3=13$

99) $\left(\sqrt{5}+\sqrt{11}\right)$;

$\left(\sqrt{5}-\sqrt{11}\right)\left(\sqrt{5}+\sqrt{11}\right)=\left(\sqrt{5}\right)^2-\left(\sqrt{11}\right)^2$

$=5-11=-6$

101) $\left(\sqrt{p}+6\right)$;

$\left(\sqrt{p}-6\right)\left(\sqrt{p}+6\right)=\left(\sqrt{p}\right)^2-(6)^2$

$=p-36$

103) $\dfrac{5}{4-\sqrt{3}}=\dfrac{5}{4-\sqrt{3}}\cdot\dfrac{4+\sqrt{3}}{4+\sqrt{3}}$ Multiply by the conjugate.

$=\dfrac{5\left(4+\sqrt{3}\right)}{(4)^2-\left(\sqrt{3}\right)^2}$ $\quad (a+b)(a-b)=a^2-b^2$

$=\dfrac{20+5\sqrt{3}}{16-3}$ Multiply terms in numerator, square terms in denominator

$=\dfrac{20+5\sqrt{3}}{13}$ Simplify.

105) $\dfrac{6}{3+\sqrt{2}}=\dfrac{6}{3+\sqrt{2}}\cdot\dfrac{3-\sqrt{2}}{3-\sqrt{2}}$

$=\dfrac{6\left(3-\sqrt{2}\right)}{(3)^2-\left(\sqrt{2}\right)^2}=\dfrac{6\left(3-\sqrt{2}\right)}{9-2}$

$=\dfrac{18-6\sqrt{2}}{7}$

107) $\dfrac{5}{7-\sqrt{3}}=\dfrac{5}{7-\sqrt{3}}\cdot\dfrac{7+\sqrt{3}}{7+\sqrt{3}}$

$=\dfrac{5\left(7+\sqrt{3}\right)}{(7)^2-\left(\sqrt{3}\right)^2}=\dfrac{5\left(7+\sqrt{3}\right)}{49-3}$

$=\dfrac{35+5\sqrt{3}}{46}$

109) $\dfrac{10}{\sqrt{5}-6}=\dfrac{10}{\sqrt{5}-6}\cdot\dfrac{\sqrt{5}+6}{\sqrt{5}+6}$

$=\dfrac{10\left(\sqrt{5}+6\right)}{\left(\sqrt{5}\right)^2-(6)^2}=\dfrac{10\left(\sqrt{5}+6\right)}{5-36}$

$=\dfrac{10\sqrt{5}+60}{-31}=-\dfrac{10\sqrt{5}+60}{31}$

111) $\dfrac{\sqrt{2}}{\sqrt{10}-3}=\dfrac{\sqrt{2}}{\sqrt{10}-3}\cdot\dfrac{\sqrt{10}+3}{\sqrt{10}+3}$

$=\dfrac{\sqrt{2}\left(\sqrt{10}+3\right)}{\left(\sqrt{10}\right)^2-(3)^2}=\dfrac{\sqrt{20}+3\sqrt{2}}{10-9}$

$=2\sqrt{5}+3\sqrt{2}$

113) $\dfrac{5+\sqrt{3}}{4+\sqrt{3}}=\dfrac{5+\sqrt{3}}{4+\sqrt{3}}\cdot\dfrac{4-\sqrt{3}}{4-\sqrt{3}}$
$=\dfrac{\left(5+\sqrt{3}\right)\left(4-\sqrt{3}\right)}{(4)^2-\left(\sqrt{3}\right)^2}$
$=\dfrac{20-5\sqrt{3}+4\sqrt{3}-3}{16-3}$
$=\dfrac{17-\sqrt{3}}{13}$

115) $\dfrac{1-2\sqrt{3}}{\sqrt{2}-\sqrt{3}}=\dfrac{1-2\sqrt{3}}{\sqrt{2}-\sqrt{3}}\cdot\dfrac{\sqrt{2}+\sqrt{3}}{\sqrt{2}+\sqrt{3}}$

117) $\dfrac{12}{\sqrt{w}+3}=\dfrac{12}{\sqrt{w}+3}\cdot\dfrac{\sqrt{w}-3}{\sqrt{w}-3}$
$=\dfrac{12\left(\sqrt{w}-3\right)}{\left(\sqrt{w}\right)^2-(3)^2}=\dfrac{12\sqrt{w}-36}{w-9}$

119) $\dfrac{9}{5-\sqrt{n}}=\dfrac{9}{5-\sqrt{n}}\cdot\dfrac{5+\sqrt{n}}{5+\sqrt{n}}$
$=\dfrac{9\left(5+\sqrt{n}\right)}{(5)^2-\left(\sqrt{n}\right)^2}=\dfrac{45+9\sqrt{n}}{25-n}$

121) $\dfrac{\sqrt{m}}{\sqrt{m}+\sqrt{n}}=\dfrac{\sqrt{m}}{\sqrt{m}+\sqrt{n}}\cdot\dfrac{\sqrt{m}-\sqrt{n}}{\sqrt{m}-\sqrt{n}}$
$=\dfrac{\sqrt{m}\left(\sqrt{m}-\sqrt{n}\right)}{\left(\sqrt{m}\right)^2-\left(\sqrt{n}\right)^2}=\dfrac{m-\sqrt{mn}}{m-n}$

123) $\dfrac{9+36\sqrt{5}}{9}=\dfrac{9\left(1+4\sqrt{5}\right)}{9}=1+4\sqrt{5}$

125) $\dfrac{20-44\sqrt{3}}{12}=\dfrac{4\left(5-11\sqrt{3}\right)}{12}$
$=\dfrac{5-11\sqrt{3}}{3}$

127) $\dfrac{\sqrt{45}+6}{9}=\dfrac{3\sqrt{5}+6}{9}=\dfrac{3\left(\sqrt{5}+2\right)}{9}$
$=\dfrac{\sqrt{5}+2}{3}$

129) $\dfrac{-24-\sqrt{800}}{4}=\dfrac{-24-20\sqrt{2}}{4}$
$=\dfrac{4\left(-6-5\sqrt{2}\right)}{4}=-6-5\sqrt{2}$

131) $\dfrac{\sqrt{11}}{\sqrt{50}}=\dfrac{\sqrt{11}}{5\sqrt{2}}\cdot\dfrac{\sqrt{2}}{\sqrt{2}}=\dfrac{\sqrt{22}}{10}$

133) $\sqrt[3]{\dfrac{4}{3}}=\dfrac{\sqrt[3]{4}}{\sqrt[3]{3}}=\dfrac{\sqrt[3]{4}}{\sqrt[3]{3}}\cdot\dfrac{\sqrt[3]{3^2}}{\sqrt[3]{3^2}}=\dfrac{\sqrt[3]{36}}{3}$

135) $\sqrt{\dfrac{125a^3}{6b}}=\dfrac{\sqrt{125a^3}}{\sqrt{6b}}=\dfrac{5a\sqrt{5a}}{\sqrt{6b}}\cdot\dfrac{\sqrt{6b}}{\sqrt{6b}}$
$=\dfrac{5a\sqrt{30ab}}{6b}$

137) $\dfrac{3+\sqrt{10}}{\sqrt{10}-9}=\dfrac{3+\sqrt{10}}{\sqrt{10}-9}\cdot\dfrac{\sqrt{10}+9}{\sqrt{10}+9}$
$=\dfrac{\left(3+\sqrt{10}\right)\left(\sqrt{10}+9\right)}{\left(\sqrt{10}\right)^2-(9)^2}$
$=\dfrac{3\sqrt{10}+27+10+9\sqrt{10}}{10-81}$
$=-\dfrac{37+12\sqrt{10}}{71}$

139) $\sqrt{\dfrac{108}{63}}=\sqrt{\dfrac{12}{7}}=\dfrac{2\sqrt{3}}{\sqrt{7}}\cdot\dfrac{\sqrt{7}}{\sqrt{7}}=\dfrac{2\sqrt{21}}{7}$

141) $\dfrac{\sqrt{256}}{\sqrt{2y}}=\dfrac{16}{\sqrt{2y}}\cdot\dfrac{\sqrt{2y}}{\sqrt{2y}}=\dfrac{16\sqrt{2y}}{2y}=\dfrac{8\sqrt{2y}}{y}$

143) $r=\sqrt{\frac{A}{\pi}}$

$r=\sqrt{\frac{12\pi}{\pi}}=\sqrt{12}=2\sqrt{3}$

When the area of the circle is 12π in^2, the radius is $2\sqrt{3}$ in.

Section 9.6: Exercises

1) Sometimes there are extraneous solutions.

3) $\sqrt{d}=12$

$\left(\sqrt{d}\right)^2=12^2$

$d=144$

Check $\sqrt{144}=12$ $\{144\}$

5) $\sqrt{x}=\frac{1}{3}$

$\left(\sqrt{x}\right)^2=\left(\frac{1}{3}\right)^2$

$x=\frac{1}{9}$ $\left\{\frac{1}{9}\right\}$

Check is left to the student.

7) $\sqrt{b-11}-3=0$

$\sqrt{b-11}=3$

$\left(\sqrt{b-11}\right)^2=(3)^2$

$b-11=9$

$b=20$

Check $\sqrt{20-11}-3=0$

$3-3=0$

$\{20\}$

9) $\sqrt{c-6}+3=4$

$\sqrt{c-6}=1$

$\left(\sqrt{c-6}\right)^2=(1)^2$

$c-6=1$

$c=7$

Check is left to the student. $\{7\}$

11) $-9=\sqrt{3k+1}-11$

$2=\sqrt{3k+1}$

$(2)^2=\left(\sqrt{3k+1}\right)^2$

$4=3k+1$

$3=3k$

$1=k$

Check is left to the student. $\{1\}$

13) $\sqrt{5p+8}+7=5$

$\sqrt{5p+8}=-2$

$\left(\sqrt{5p+8}\right)^2=(-2)^2$

$5p+8=4$

$5p=-4$

$p=-\frac{4}{5}$

$-\frac{4}{5}$ is an extraneous solution. $\varnothing$

15) $\sqrt{5-2r}-13=-7$

$$\sqrt{5-2r}=6$$
$$\left(\sqrt{5-2r}\right)^2=(6)^2$$
$$5-2r=36$$
$$-2r=31$$
$$r=-\frac{31}{2}$$

Check is left to the student. $\left\{-\frac{31}{2}\right\}$

17) $g=\sqrt{g^2+8g-24}$

$$g^2=\left(\sqrt{g^2+8g-24}\right)^2$$
$$g^2=g^2+8g-24$$
$$0=8g-24$$
$$-8g=-24$$
$$g=3$$

Check is left to the student. $\{3\}$

19) $\sqrt{v^2-7v-42}=v$

$$\left(\sqrt{v^2-7v-42}\right)^2=v^2$$
$$v^2-7v-42=v^2$$
$$-7v-42=0$$
$$-7v=42$$
$$v=-6$$

-6 is an extraneous solution. $\varnothing$

21) $2p=\sqrt{4p^2-3p+6}$

$$(2p)^2=\left(\sqrt{4p^2-3p+6}\right)^2$$
$$4p^2=4p^2-3p+6$$
$$0=-3p+6$$
$$-6=-3p$$
$$2=p$$

Check is left to the student. $\{2\}$

23) $\sqrt{5y+4}=\sqrt{y+8}$

$$\left(\sqrt{5y+4}\right)^2=\left(\sqrt{y+8}\right)^2$$
$$5y+4=y+8$$
$$4y=4$$
$$y=1$$

Check is left to the student. $\{1\}$

25) $\sqrt{z}=3\sqrt{7}$

$$\left(\sqrt{z}\right)^2=\left(3\sqrt{7}\right)^2$$
$$z=9(7)$$
$$z=63$$

Check is left to the student. $\{63\}$

27) $\sqrt{3-11h}-3\sqrt{h+7}=0$

$$\left(\sqrt{3-11h}\right)^2=\left(3\sqrt{h+7}\right)^2$$
$$3-11h=9(h+7)$$
$$3-11h=9h+63$$
$$-60=20h$$
$$-3=h$$

Check is left to the student. $\{-3\}$

29) $\sqrt{2p-1}+2\sqrt{p+4}=0$

$$\left(\sqrt{2p-1}\right)^2=\left(-2\sqrt{p+4}\right)^2$$
$$2p-1=4p+16$$
$$2p-1-4p=16$$
$$-2p-1=16$$
$$-2p=17$$

$\frac{-17}{2}$ is an extraneous solution. $\varnothing$

31) $\sqrt{x}=x-6$

$$\left(\sqrt{x}\right)^2=(x-6)^2$$
$$x=x^2-12x+36$$
$$0=x^2-13x+36$$
$$0=(x-9)(x-4)$$
$$x-9=0 \qquad x-4=0$$
$$x=9 \qquad x=4$$
$$\sqrt{9}\stackrel{?}{=}9-6 \qquad \sqrt{4}\stackrel{?}{=}4-6$$
$$3=3 \qquad 2\neq-2$$

$\{9\}$

33) $m+4=5\sqrt{m}$

$$(m+4)^2=\left(5\sqrt{m}\right)^2$$
$$m^2+8m+16=25m$$
$$m^2-17m+16=0$$
$$(m-1)(m-16)=0$$
$$m-1=0 \qquad m-16=0$$
$$m=1 \qquad m=16$$
$$1+4\stackrel{?}{=}5\sqrt{1} \qquad 16+4\stackrel{?}{=}5\sqrt{16}$$
$$5=5 \qquad 20=20$$

$\{1,16\}$

35) $c=\sqrt{5c-9}+3$

$$c-3=\sqrt{5c-9}$$
$$(c-3)^2=\left(\sqrt{5c-9}\right)^2$$
$$c^2-6c+9=5c-9$$
$$c^2-11c+18=0$$
$$(c-9)(c-2)=0$$
$$c-9=0 \qquad c-2=0$$
$$c=9 \qquad c=2$$
$$9\stackrel{?}{=}\sqrt{5(9)-9}+3 \qquad 2\stackrel{?}{=}\sqrt{5(2)-9}+3$$
$$9=\sqrt{45-9}+3 \qquad 2=\sqrt{10-9}+3$$
$$9=\sqrt{36}+3 \qquad 2=\sqrt{1}+3$$
$$9=6+3 \qquad 2\neq1+3$$

$\{9\}$

37) $6+\sqrt{w^2+3w-9}=w$

$$\sqrt{w^2+3w-9}=w-6$$
$$\left(\sqrt{w^2+3w-9}\right)^2=(w-6)^2$$
$$w^2+3w-9=w^2-12w+36$$
$$0=-15k-45$$
$$15k=-45$$
$$k=-3$$

$$6+\sqrt{(-3)^2+3(-3)-9}\stackrel{?}{=}-1$$
$$\sqrt{9-9-9}=-7$$
$$\sqrt{-9}\neq-7$$

$\varnothing$

39) $\sqrt{7u-5}-u=1$

$$\sqrt{7u-5}=u+1$$
$$\left(\sqrt{7u-5}\right)^2=(u+1)^2$$
$$7u-5=u^2+2u+1$$
$$0=u^2-5u+6$$
$$0=(u-3)(u-2)$$
$$u-3=0 \qquad u-2=0$$
$$u=3 \qquad u=2$$
$$\sqrt{7(3)-5}-3\overset{?}{=}1 \qquad \sqrt{7(2)-5}-2\overset{?}{=}1$$
$$\sqrt{21-5}-3=1 \qquad \sqrt{14-5}-2=1$$
$$\sqrt{16}-3=1 \qquad \sqrt{9}-2=1$$
$$4-3=1 \qquad 3-2=1$$

$\{2,3\}$

41) $a-2\sqrt{a+1}=-1$

$$-2\sqrt{a+1}=-a-1$$
$$\left(-2\sqrt{a+1}\right)^2=(-a-1)^2$$
$$4(a+1)=a^2+2a+1$$
$$4a+4=a^2+2a+1$$
$$0=a^2-2a-3$$
$$0=(a-3)(a+1)$$
$$a-3=0 \qquad a+1=0$$
$$a=3 \qquad a=-1$$
$$3-2\sqrt{3+1}\overset{?}{=}-1 \qquad -1-2\sqrt{-1+1}\overset{?}{=}-1$$
$$3-2\sqrt{4}=-1 \qquad -1-2\sqrt{0}=-1$$
$$3-4=-1 \qquad -1-0=-1$$

$\{-1,3\}$

43) $(y+7)^2=(y)^2+2(y)(7)+(7)^2$

$$=y^2+14y+49$$

45) $\left(\sqrt{n}+7\right)^2=\left(\sqrt{n}\right)^2+2\left(\sqrt{n}\right)(7)+(7)^2$

$$=n+14\sqrt{n}+49$$

47) $\left(4-\sqrt{c+3}\right)^2$

$$=(4)^2-2(4)\left(\sqrt{c+3}\right)+\left(\sqrt{c+3}\right)^2$$
$$=16-8\sqrt{c+3}+c+3$$
$$=19-8\sqrt{c+3}+c$$

49) $\left(2\sqrt{3p-4}+1\right)^2$

$$=\left(2\sqrt{3p-4}\right)^2+2\left(2\sqrt{3p-4}\right)(1)+1^2$$
$$=4(3p-4)+4\sqrt{3p-4}+1$$
$$=12p-16+4\sqrt{3p-4}+1$$
$$=12p+4\sqrt{3p-4}-15$$

51) $\left(\sqrt{w+5}\right)^2=\left(5-\sqrt{w}\right)^2$

$$w+5=25-10\sqrt{w}+w$$
$$0=20-10\sqrt{w}$$
$$10\sqrt{w}=20$$
$$\sqrt{w}=2$$
$$\left(\sqrt{w}\right)^2=(2)^2$$
$$w=4$$

$\{4\}$

53) $\sqrt{y}-\sqrt{y-13}=1$

$$\sqrt{y}-1=\sqrt{y-13}$$
$$\left(\sqrt{y-13}\right)^2=\left(\sqrt{y}-1\right)^2$$
$$y-13=y-2\sqrt{y}+1$$
$$0=-2\sqrt{y}+14$$
$$-14=-2\sqrt{y}$$
$$7=\sqrt{y}$$
$$(7)^2=\left(\sqrt{y}\right)^2$$
$$49=y$$
$$\{49\}$$

55) $\sqrt{3t+10}-\sqrt{2t}=2$

$$\left(\sqrt{3t+10}\right)^2=\left(\sqrt{2t}+2\right)^2$$
$$3t+10=2t+4\sqrt{2t}+4$$
$$t+6-4\sqrt{2t}=0$$
$$(t+6)^2=\left(4\sqrt{2t}\right)^2$$
$$t^2+12t+36=16(2t)$$
$$t^2+12t+36=32t$$
$$t^2-20t+36=0$$
$$(t-18)(t-2)=0$$
$$t-18=0 \qquad t-2=0$$
$$t=18 \qquad t=2$$

Check is left to the student. $\{2,18\}$

57) $\sqrt{4c+5}+\sqrt{2c-1}=4$

$$\left(\sqrt{4c+5}\right)^2=\left(4-\sqrt{2c-1}\right)^2$$
$$4c+5=16-8\sqrt{2c-1}+2c-1$$
$$8\sqrt{2c-1}=10-2c$$
$$\left(8\sqrt{2c-1}\right)^2=(10-2c)^2$$
$$64(2c-1)=100-40c+4c^2$$
$$128c-64=100-40c+4c^2$$
$$0=4c^2-168c+164$$
$$0=c^2-42c+41$$
$$(c-41)(c-1)=0$$
$$c-41=0 \qquad c-1=0$$
$$c=41 \qquad c=1$$

41 is an extraneous solution. $\{1\}$

59) $3-\sqrt{3h+1}=\sqrt{3h-14}$

$$\left(3-\sqrt{3h+1}\right)^2=\left(\sqrt{3h-14}\right)^2$$
$$9-6\sqrt{3h+1}+3h+1=3h-14$$
$$24=6\sqrt{3h+1}$$
$$4=\sqrt{3h+1}$$
$$(4)^2=\left(\sqrt{3h+1}\right)^2$$
$$16=3h+1$$
$$15=3h$$
$$5=h$$

5 is an extraneous solution. $\varnothing$

61) $\sqrt{5a+19}-\sqrt{a+12}=1$

$$\left(\sqrt{5a+19}\right)^2=\left(1+\sqrt{a+12}\right)^2$$
$$5a+19=1+2\sqrt{a+12}+a+12$$
$$4a+6=2\sqrt{a+12}$$
$$(4a+6)^2=\left(2\sqrt{a+12}\right)^2$$
$$16a^2+48a+36=4(a+12)$$
$$16a^2+44a-12=0$$
$$4a^2+11a-3=0$$
$$(4a-1)(a+3)=0$$
$$4a-1=0 \qquad a+3=0$$
$$a=\frac{1}{4} \qquad a=-3$$

-3 is an extraneous solution. $\left\{\frac{1}{4}\right\}$

63) $3=\sqrt{5-2y}+\sqrt{y+2}$

$$\left(3-\sqrt{y+2}\right)^2=\left(\sqrt{5-2y}\right)^2$$
$$6+3y=6\sqrt{y+2}$$
$$(6+3y)^2=\left(6\sqrt{y+2}\right)^2$$
$$36+36y+9y^2=36(y+2)$$
$$9y^2+36y+36=36y+72$$
$$9y^2-36=0$$
$$y^2-4=0$$
$$(y+2)(y-2)=0$$
$$y+2=0 \qquad y-2=0$$
$$y=-2 \qquad y=2$$

Check is left to the student. $\{-2,2\}$

65) $\sqrt{2n-5}=\sqrt{2n+3}-2$

$$\left(\sqrt{2n-5}\right)^2=\left(\sqrt{2n+3}-2\right)^2$$
$$2n-5=2n+3-4\sqrt{2n+3}+4$$
$$2n-5=2n-4\sqrt{2n+3}+7$$
$$-12=-4\sqrt{2n+3}$$
$$3=\sqrt{2n+3}$$
$$(3)^2=\left(\sqrt{2n+3}\right)^2$$
$$9=2n+3$$
$$6=2n$$
$$3=n$$

Check is left to the student. $\{3\}$

67) $\sqrt{3b+4}-\sqrt{2b+9}=-1$

$$\sqrt{3b+4}=-1+\sqrt{2b+9}$$
$$\left(\sqrt{3b+4}\right)^2=\left(-1+\sqrt{2b+9}\right)^2$$
$$3b+4=1-2\sqrt{2b+9}+2b+9$$
$$2\sqrt{2b+9}=6-b$$
$$\left(2\sqrt{2b+9}\right)^2=(6-b)^2$$
$$4(2b+9)=36-12b+b^2$$
$$8b+36=36-12b+b^2$$
$$0=b^2-20b$$
$$0=b(b-20)$$
$$b=0 \qquad b-20=0$$
$$b=20$$

20 is an extraneous solution. $\{0\}$

69) Raise both sides of the equation to the third power.

71) $\sqrt[3]{n}=-5$

$$\left(\sqrt[3]{n}\right)^3=(-5)^3$$

$$n=-125$$

Check is left to the student. $\{-125\}$

73) $\sqrt[3]{6j-2}=\sqrt[3]{j-7}$

$$\left(\sqrt[3]{6j-2}\right)^3=\left(\sqrt[3]{j-7}\right)^3$$

$$6j-2=j-7$$

$$5j=-5$$

$$j=-1$$

Check is left to the student.

$\{-1\}$

75) $\sqrt[3]{5a+4}-\sqrt[3]{2a-3}=0$

$$\left(\sqrt[3]{5a+4}\right)^3=\left(\sqrt[3]{2a-3}\right)^3$$

$$5a+4=2a-3$$

$$3a=-7$$

$$a=-\frac{7}{3}$$

Check is left to the student. $\left\{-\frac{7}{3}\right\}$

77) $\sqrt[3]{x^2}-\sqrt[3]{5x+14}=0$

$$\left(\sqrt[3]{x^2}\right)^3=\left(\sqrt[3]{5x+14}\right)^3$$

$$x^2=5x+14$$

$$x^2-5x-14=0$$

$$(x-7)(x+2)=0$$

$$x-7=0 \qquad x+2=0$$

$$x=7 \qquad x=-2$$

Check is left to the student. $\{-2,7\}$

81) $\sqrt[4]{y+10}=3$

$$\left(\sqrt[4]{y+10}\right)^4=(3)^4$$

$$y+10=81$$

$$y=71$$

Check is left to the student. $\{71\}$

83) $2=\sqrt[4]{b^2-6b}$

$$(2)^4=\left(\sqrt[4]{b^2-6b}\right)^4$$

$$16=b^2-6b$$

$$0=b^2-6b-16$$

$$0=(b-8)(b+2)$$

$$b-8=0 \qquad b+2=0$$

$$b=8 \qquad b=-2$$

Check is left to the student. $\{-2,8\}$

85) $\sqrt[5]{4x}=-2$

$$\left(\sqrt[5]{4x}\right)^5=(-2)^5$$

$$4x=-32$$

$$x=-8$$

Check is left to the student. $\{-8\}$

87) $7+\sqrt{x}=12$

$$\sqrt{x}=5$$

$$\left(\sqrt{x}\right)^2=(5)^2$$

$$x=25$$

89) $3\sqrt{x}=4+\sqrt{x}$

$$2\sqrt{x}=4$$

$$\sqrt{x}=2$$

$$\left(\sqrt{x}\right)^2=(2)^2$$

$$x=4$$

91) Let $A = 36$, and solve for l.

$l = \sqrt{A}$

$l = \sqrt{36} = 6$

The rug measures 6 ft × 6 ft.

93) a) Use formula, $S = \sqrt{30fd}$ and let $f = 0.90$ and $d = 142$ft to find S.

$S = \sqrt{30(0.90)(142)}$

$= \sqrt{3834}$

≈ 61.9

62 mph; the car was not speeding.

95) Use formula, $d = 1.2\sqrt{h}$ and let $d = 48$ miles to find h.

$48 = 1.2\sqrt{h}$

$40 = \sqrt{h}$

$(40)^2 = \left(\sqrt{h}\right)^2$

$1600 \approx h$

1600 ft

Section 9.7: Exercises

1) The denominator of 2 becomes the index of the radical. $36^{1/2} = \sqrt{36}$

3) $16^{1/2} = \sqrt{16} = 4$

5) $1000^{1/3} = \sqrt[3]{1000} = 10$

7) $81^{1/4} = \sqrt[4]{81} = 3$

9) $-64^{1/3} = -\sqrt[3]{64} = -4$

11) $\left(\frac{9}{4}\right)^{1/2} = \sqrt{\frac{9}{4}} = \frac{3}{2}$

13) $\left(\frac{125}{64}\right)^{1/3} = \sqrt[3]{\frac{125}{64}} = \frac{5}{4}$

15) $-\left(\frac{100}{49}\right)^{1/2} = -\sqrt{\frac{100}{49}} = -\frac{10}{7}$

17) The denominator of 4 becomes the index of the radical. The numerator of 3 is the power to which we raise the radical expression.

$16^{3/4} = \left(\sqrt[4]{16}\right)^3$

19) $8^{4/3} = \left(8^{1/3}\right)^4 = \left(\sqrt[3]{8}\right)^4 = 2^4 = 16$

21) $81^{3/4} = \left(81^{1/4}\right)^3 = \left(\sqrt[4]{81}\right)^3 = 3^3 = 27$

23) $64^{5/6} = \left(64^{1/6}\right)^5 = \left(\sqrt[6]{64}\right)^5 = 2^5 = 32$

25) $-36^{3/2} = -\left(\sqrt{36}\right)^3 = -(6)^3 = -216$

27) $\left(\frac{16}{81}\right)^{5/4} = \left(\sqrt[4]{\frac{16}{81}}\right)^5 = \left(\frac{2}{3}\right)^5 = \frac{32}{243}$

29) $-\left(\frac{8}{27}\right)^{4/3} = -\left(\sqrt[3]{\frac{8}{27}}\right)^4$

$= -\left(\frac{2}{3}\right)^4 = -\frac{16}{81}$

31) False; the negative exponent does not make the result negative.

$121^{-1/2} = -11$

33) $16^{-1/2}=\left(\frac{1}{16}\right)^{1/2}=\sqrt{\frac{1}{16}}=\frac{1}{4}$

35) $100^{-1/2}=\left(\frac{1}{100}\right)^{1/2}=\sqrt{\frac{1}{100}}=\frac{1}{10}$

37) $125^{-1/3}=\left(\frac{1}{125}\right)^{1/3}=\sqrt[3]{\frac{1}{125}}=\frac{1}{5}$

39) $-81^{-1/4}=-\left(\frac{1}{81}\right)^{1/4}=-\sqrt[4]{\frac{1}{81}}=-\frac{1}{3}$

41) $64^{-5/6}=\left(\frac{1}{64}\right)^{5/6}=\left(\sqrt[6]{\frac{1}{64}}\right)^5$
$=\left(\frac{1}{2}\right)^5=\frac{1}{32}$

43) $64^{-2/3}=\left(\frac{1}{64}\right)^{2/3}=\left(\sqrt[3]{\frac{1}{64}}\right)^2$
$=\left(\frac{1}{4}\right)^2=\frac{1}{16}$

45) $-27^{-2/3}=-\left(\frac{1}{27}\right)^{2/3}=-\left(\sqrt[3]{\frac{1}{27}}\right)^2$
$=-\left(\frac{1}{3}\right)^2=-\frac{1}{9}$

47) $-16^{-5/4}=-\left(\frac{1}{16}\right)^{5/4}=-\left(\sqrt[4]{\frac{1}{16}}\right)^5$
$=-\left(\frac{1}{2}\right)^5=-\frac{1}{32}$

49) $5^{2/3}\cdot 5^{7/3}=5^{2/3+7/3}=5^{9/3}=5^3=125$

51) $\left(16^{1/4}\right)^2=16^{\frac{1}{4}\cdot 2}=16^{1/2}=\sqrt{16}=4$

53) $7^{7/5}\cdot 7^{-3/5}=7^{7/5+(-3/5)}=7^{4/5}$

55) $\frac{9^{10/3}}{9^{4/3}}=9^{10/3-4/3}=9^{6/3}=9^2=81$

57) $\frac{5^{7/2}}{5^{13/2}}=5^{7/2-13/2}=5^{-6/2}=5^{-3}=\frac{1}{5^3}=\frac{1}{125}$

59) $\frac{7^{-1}}{7^{1/2}\cdot 7^{-5/2}}=\frac{7^{-1}}{7^{1/2+(-5/2)}}=\frac{7^{-1}}{7^{-4/2}}=\frac{7^{-1}}{7^{-2}}$
$=7^{-1-(-2)}=7^{-1+2}=7^1=7$

61) $\frac{6^{4/9}\cdot 6^{1/9}}{6^{2/9}}=\frac{6^{4/9+1/9}}{6^{2/9}}=\frac{6^{5/9}}{6^{2/9}}=$
$=6^{5/9-2/9}=6^{3/9}=6^{1/3}$

63) $k^{9/4}\cdot k^{5/4}=k^{9/4+5/4}=k^{14/4}=k^{7/2}$

65) $m^{-4/5}\cdot m^{1/10}=m^{-4/5+1/10}=m^{-8/10+1/10}$
$=m^{-7/10}=\frac{1}{m^{7/10}}$

67) $\left(-9v^{5/8}\right)\left(8v^{3/4}\right)=-72v^{5/8+3/4}$
$=-72v^{5/8+6/8}$
$=-72v^{11/8}$

69) $\frac{n^{5/9}}{n^{2/9}}=n^{5/9-2/9}=n^{3/9}=n^{1/3}$

71) $\frac{26c^{-2/3}}{72c^{5/6}}=\frac{13}{36}c^{-2/3-5/6}=\frac{13}{36}c^{-4/6-5/6}$
$=\frac{13}{36}c^{-9/6}=\frac{13}{36}c^{-3/2}=\frac{13}{36c^{3/2}}$

73) $\left(q^{3/5}\right)^{10}=q^{\frac{3}{5}\cdot 10}=q^{30/5}=q^6$

75) $\left(x^{-5/9}\right)^3 = x^{-\frac{5}{9}\cdot 3} = x^{-5/3} = \frac{1}{x^{5/3}}$

77) $\left(z^{1/5}\right)^{3/4} = z^{\frac{1}{5}\cdot\frac{3}{4}} = z^{3/20}$

79) $\sqrt[6]{49^3} = \left(49^3\right)^{1/6} = 49^{1/2} = \sqrt{49} = 7$

81) $\sqrt[9]{8^3} = \left(8^3\right)^{1/9} = 8^{1/3} = \sqrt[3]{8} = 2$

83) $\sqrt[6]{t^4} = t^{4/6} = t^{2/3} = \sqrt[3]{t^2}$

85) $\sqrt[4]{h^2} = h^{2/4} = h^{1/2} = \sqrt{h}$

87) $\sqrt{d^4} = d^{4/2} = d^2$

89) $T = 2\pi\left(\frac{L}{g}\right)^{1/2}$

$$
\begin{aligned}
T &= 2\pi\left(\frac{0.25}{9.8}\right)^{1/2} \\
&= 2\pi(0.0255)^{1/2} \\
&= 2\pi(0.1597) \\
&= 1.06 \approx 1 \text{ sec}
\end{aligned}
$$

91) a) $N_t = N_i \cdot 2^{t/d}$

$$
\begin{aligned}
&= 5000 \cdot 2^{30/48} \\
&= 5000 \cdot 2^{5/8} \\
&\approx 7711
\end{aligned}
$$

b) $N_t = N_i \cdot 2^{t/d}$

$$
\begin{aligned}
&= 5000 \cdot 2^{120/48} \\
&= 5000 \cdot 2^{5/2} \\
&\approx 28,284
\end{aligned}
$$

Chapter 9 Review Exercises

1) $\sqrt{49} = 7$

3) $-\sqrt{16} = -4$

5) $\sqrt[3]{-125} = -5$

7) $\sqrt[4]{81} = 3$

9) Pythagorean Theorem is $a^2 + b^2 = c^2$.
Let $b = 8$ and $c = 10$.

$$
\begin{aligned}
a^2 + b^2 &= c^2 \\
a^2 + (8)^2 &= (10)^2 \\
a^2 + 64 &= 100 \\
a^2 &= 36 \\
\sqrt{a^2} &= \sqrt{36} \\
a &= 6
\end{aligned}
$$

11) Distance formula is $d = \sqrt{(x_2 - x_1)^2 + (y_2 - y_1)^2}$.
Let $(x_1, y_1) = (4,1)$ and $(x_2, y_2) = (10,-7)$

$$
\begin{aligned}
d &= \sqrt{(10-4)^2 + (-7-1)^2} \\
d &= \sqrt{(6)^2 + (-8)^2} \\
d &= \sqrt{36+64} = \sqrt{100} = 10
\end{aligned}
$$

13) $\sqrt{45} = \sqrt{9 \cdot 5} = 3\sqrt{5}$

15) $\sqrt[3]{81} = \sqrt[3]{27 \cdot 3} = 3\sqrt[3]{3}$

17) $\sqrt[4]{80} = \sqrt[4]{16 \cdot 5} = 2\sqrt[4]{5}$

19) $\sqrt{\frac{48}{6}} = \sqrt{8} = 2\sqrt{2}$

21) $\sqrt[3]{-\frac{16}{2}} = \sqrt[3]{-8} = -2$

23) $\sqrt{h^8} = h^{8/2} = h^4$

25) $\sqrt[3]{b^{24}} = b^{24/3} = b^8$

27) $\sqrt{w^{15}} = \sqrt{w^{14}} \cdot \sqrt{w} = w^{14/2}\sqrt{w} = w^7\sqrt{w}$

29) $\sqrt{63d^4} = \sqrt{9 \cdot 7d^{4/2}} = 3d^2\sqrt{7}$

31) $\sqrt{44x^{12}y^5} = \sqrt{4 \cdot 11 \cdot x^{12} \cdot y^4 \cdot y}$
$= \sqrt{4} \cdot \sqrt{11} \cdot \sqrt{x^{12}} \cdot \sqrt{y^4} \cdot \sqrt{y}$
$= 2x^6y^2\sqrt{11y}$

33) $\sqrt[3]{48m^{17}} = \sqrt[3]{8 \cdot 6 \cdot m^{15} \cdot m^2}$
$= 2\sqrt[3]{6} \cdot \sqrt[3]{m^{15}} \cdot \sqrt[3]{m^2}$
$= 2m^5\sqrt[3]{6} \cdot \sqrt[3]{m^2} = 2m^5\sqrt[3]{6m^2}$

35) $\frac{40\sqrt{150}}{56\sqrt{2}} = \frac{40}{56} \cdot \frac{\sqrt{150}}{\sqrt{2}} = \frac{5}{7} \cdot \sqrt{\frac{150}{2}}$
$= \frac{5}{7}\sqrt{75} = \frac{5}{7} \cdot 5\sqrt{3} = \frac{25\sqrt{3}}{7}$

37) $\sqrt{6} \cdot \sqrt{5} = \sqrt{30}$

39) $\sqrt{5} \cdot \sqrt{10} = \sqrt{50} = 5\sqrt{2}$

41) $\sqrt[3]{3} \cdot \sqrt[3]{10} = \sqrt[3]{30}$

43) $\sqrt{80} + \sqrt{150} = 4\sqrt{5} + 5\sqrt{6}$

45) $4\sqrt[3]{9} - 9\sqrt[3]{72} = 4\sqrt[3]{9} - 9\left(2\sqrt[3]{9}\right)$
$= 4\sqrt[3]{9} - 18\sqrt[3]{9} = -14\sqrt[3]{9}$

47) $10n^2\sqrt{8n} - 45n\sqrt{2n^3}$
$= 10n^2\sqrt{8n} - 45n\left(n\sqrt{2n}\right)$
$= 20n^2\sqrt{2n} - 45n^2\sqrt{2n} = -25n^2\sqrt{2n}$

49) $\sqrt{7}\left(\sqrt{7} - \sqrt{5}\right) = 7 - \sqrt{35}$

51) $\left(5 - \sqrt{2}\right)\left(3 + \sqrt{2}\right)$
$= 15 + 5\sqrt{2} - 3\sqrt{2} - 2 = 13 + 2\sqrt{2}$

53) $\left(2\sqrt{6} - 5\right)^2$
$= \left(2\sqrt{6}\right)^2 - 2\left(2\sqrt{6}\right)(5) + (5)^2$
$= 4(6) - 20\sqrt{6} + 25$
$= 24 - 20\sqrt{6} + 25 = 49 - 20\sqrt{6}$

55) $\left(\sqrt{7} - \sqrt{6}\right)\left(\sqrt{7} + \sqrt{6}\right)$
$= \left(\sqrt{7}\right)^2 - \left(\sqrt{6}\right)^2 = 7 - 6 = 1$

57) $\frac{18}{\sqrt{3}} = \frac{18}{\sqrt{3}} \cdot \frac{\sqrt{3}}{\sqrt{3}} = \frac{18\sqrt{3}}{3} = 6\sqrt{3}$

59) $\sqrt{\frac{98}{h}} = \frac{\sqrt{98}}{\sqrt{h}} = \frac{7\sqrt{2}}{\sqrt{h}} \cdot \frac{\sqrt{h}}{\sqrt{h}} = \frac{7\sqrt{2h}}{h}$

61) $\frac{15}{\sqrt[3]{3}} = \frac{15}{\sqrt[3]{3}} \cdot \frac{\sqrt[3]{3^2}}{\sqrt[3]{3^2}} = \frac{15\sqrt[3]{3^2}}{\sqrt[3]{3^3}} = \frac{15\sqrt[3]{9}}{3} = 5\sqrt[3]{9}$

63) $\frac{7}{\sqrt[3]{2c}} = \frac{7}{\sqrt[3]{2c}} \cdot \frac{\sqrt[3]{2^2c^2}}{\sqrt[3]{2^2c^2}} = \frac{7\sqrt[3]{4c^2}}{2c}$
$= \frac{7\sqrt[3]{4c^2}}{2c}$

65) $\frac{2}{3+\sqrt{2}}=\frac{2}{3+\sqrt{2}}\cdot\frac{3-\sqrt{2}}{3-\sqrt{2}}$

$=\frac{2\left(3-\sqrt{2}\right)}{(3)^2-\left(\sqrt{2}\right)^2}$

$=\frac{2\left(3-\sqrt{2}\right)}{9-2}$

$=\frac{2\left(3-\sqrt{2}\right)}{7}=\frac{6-2\sqrt{2}}{7}$

67) $\frac{6}{9-\sqrt{x}}=\frac{6}{9-\sqrt{x}}\cdot\frac{9+\sqrt{x}}{9+\sqrt{x}}$

$=\frac{6\left(9+\sqrt{x}\right)}{(9)^2-\left(\sqrt{x}\right)^2}$

$=\frac{6\left(9+\sqrt{x}\right)}{81-x}=\frac{54+6\sqrt{x}}{81-x}$

69) $\frac{16-24\sqrt{3}}{8}=\frac{8\left(2-3\sqrt{3}\right)}{8}=2-3\sqrt{3}$

71) $\sqrt{r}-9=0$

$\sqrt{r}=9$

$\left(\sqrt{r}\right)^2=(9)^2$

$r=81\quad\{81\}$

73) $3\sqrt{2v+7}-15=0$

$3\sqrt{2v+7}=15$

$\left(3\sqrt{2v+7}\right)^2=15^2$

$9(2v+7)=225$

$18v+63=225$

$18v=162$

$v=9$

Check $3\sqrt{2(9)+7}-15\stackrel{?}{=}0$

$3\sqrt{25}-15\stackrel{?}{=}0$

$15-15=0\quad\{9\}$

75) $\sqrt{a+11}-2\sqrt{a+8}=0$

$\sqrt{a+11}=2\sqrt{a+8}$

$\left(\sqrt{a+11}\right)^2=\left(2\sqrt{a+8}\right)^2$

$a+11=4(a+8)$

$a+11=4a+32$

$-21=3a$

$-7=a$

Check is left to the student. $\{-7\}$

77) $1=\sqrt{5x-1}-x$

$x+1=\sqrt{5x-1}$

$(x+1)^2=\left(\sqrt{5x-1}\right)^2$

$x^2+2x+1=5x-1$

$x^2-3x+2=0$

$(x-1)(x-2)=0$

$x-1=0$ or $x-2=0$

$x=1\qquad x=2\qquad\{1,2\}$

Check is left to the student.

79) $\sqrt{3b+1}+\sqrt{b}=3$

$$\sqrt{3b+1}=3-\sqrt{b}$$
$$\left(\sqrt{3b+1}\right)^2=\left(3-\sqrt{b}\right)^2$$
$$3b+1=9-6\sqrt{b}+b$$
$$0=8-6\sqrt{b}-2b$$
$$2b-8=-6\sqrt{b}$$
$$b-4=-3\sqrt{b}$$
$$(b-4)^2=\left(3\sqrt{b}\right)^2$$
$$b^2-8b+16=9b$$
$$b^2-17b+16=0$$
$$(b-16)(b-1)=0$$
$$b-16=0 \text{ or } b-1=0$$
$$b=16 \qquad b=1$$

16 is an extraneous solution. $\{1\}$

81) The denominator of the fractional exponent becomes the index on the radical. The numerator is the power to which we raise the radical expression. $8^{2/3}=\left(\sqrt[3]{8}\right)^2$

83) $64^{1/2}=\sqrt{64}=8$

85) $\left(\frac{64}{125}\right)^{1/3}=\sqrt[3]{\frac{64}{125}}=\frac{4}{5}$

87) $-81^{1/4}=-\sqrt[4]{81}=-3$

89) $125^{2/3}=\left(\sqrt[3]{125}\right)^2=(5)^2=25$

91) $\left(\frac{1000}{27}\right)^{2/3}=\left(\sqrt[3]{\frac{1000}{27}}\right)^2=\left(\frac{10}{3}\right)^2=\frac{100}{9}$

93) $121^{-1/2}=\left(\frac{1}{121}\right)^{1/2}=\sqrt{\frac{1}{121}}=\frac{1}{11}$

95) $-81^{-3/4}=\left(-\frac{1}{81}\right)^{3/4}=\left(-\sqrt[4]{\frac{1}{81}}\right)^3$

$$=\left(-\frac{1}{3}\right)^3=-\frac{1}{27}$$

97) $4^{6/7}\cdot 4^{8/7}=4^{6/7+8/7}=4^{14/7}=4^2=16$

99) $\frac{64^2}{64^{8/3}}=64^{6/3-8/3}=64^{-2/3}=\left(\frac{1}{64}\right)^{2/3}$

$$=\left(\sqrt[3]{\frac{1}{64}}\right)^2=\left(\frac{1}{4}\right)^2=\frac{1}{16}$$

101) $\left(-8p^{5/6}\right)\left(3p^{-1/2}\right)=-24p^{5/6-3/6}=-24p^{2/6}$

$$=-24p^{1/3}$$

103) $\left(m^{4/3}\right)^{-3}=m^{\frac{4}{3}\cdot-3}=m^{-4}=\frac{1}{m^4}$

105) $\sqrt[10]{n^5}=\left(n^5\right)^{1/10}=n^{1/2}=\sqrt{n}$

107) $9d\sqrt{d}-4\sqrt{d^3}=9d\sqrt{d}-4d\sqrt{d}=5d\sqrt{d}$

109) $3\left(5\sqrt{8}+\sqrt{2}\right)=15\sqrt{8}+3\sqrt{2}$

$$=30\sqrt{2}+3\sqrt{2}=33\sqrt{2}$$

111) $t^{9/5}\cdot t^{6/5}=t^{9/5+6/5}=t^{15/5}=t^3$

113) $\dfrac{\sqrt{5}}{6+\sqrt{10}}=\dfrac{\sqrt{5}}{6+\sqrt{10}}\cdot\dfrac{6-\sqrt{10}}{6-\sqrt{10}}$
$=\dfrac{\sqrt{5}\left(6-\sqrt{10}\right)}{6^2-\left(\sqrt{10}\right)^2}=\dfrac{6\sqrt{5}-\sqrt{50}}{36-10}=\dfrac{6\sqrt{5}-5\sqrt{2}}{26}$

115) $32^{-4/5}=\left(\dfrac{1}{32}\right)^{4/5}=\left(\sqrt[5]{\dfrac{1}{32}}\right)^4=\left(\dfrac{1}{2}\right)^4=\dfrac{1}{16}$

117) $\left(4-\sqrt{3}\right)^2=4^2-4\sqrt{3}-4\sqrt{3}+\left(\sqrt{3}\right)^2$
$=16-8\sqrt{3}+3=19-8\sqrt{3}$

119) $\sqrt{128p^{15}}=\sqrt{64\cdot 2\cdot p^{14}\cdot p}=8p^7\sqrt{2p}$

Chapter 9 Test

1) $\sqrt{121}=11$

3) $\sqrt[3]{-1000}=-10$

5) Pythagorean Theorem is $a^2+b^2=c^2$.
Let $a=5$ and $c=8$.
$a^2+b^2=c^2$
$(5)^2+b^2=(8)^2$
$25+b^2=64$
$b^2=39$
$\sqrt{b^2}=\sqrt{39}$
$b=\sqrt{39}$ cm

7) $\sqrt{125}=\sqrt{25}\cdot\sqrt{5}=5\sqrt{5}$

9) $\sqrt{\dfrac{192}{6}}=\sqrt{32}=\sqrt{16}\cdot\sqrt{2}=4\sqrt{2}$

11) $\sqrt{h^7}=\sqrt{h^6}\cdot\sqrt{h}=h^3\sqrt{h}$

13) $\sqrt[3]{b^{14}}=\sqrt[3]{b^{12}}\cdot\sqrt[3]{b^2}=b^{12/3}\cdot\sqrt[3]{b^2}=b^4\sqrt[3]{b^2}$

15) $\sqrt[3]{\dfrac{x^{15}y^7}{8}}=\dfrac{\sqrt[3]{x^{15}}\cdot\sqrt[3]{y^7}}{\sqrt[3]{8}}$
$=\dfrac{\sqrt[3]{x^{15}}\cdot\sqrt[3]{y^6}\cdot\sqrt[3]{y}}{\sqrt[3]{8}}=\dfrac{x^5y^2\sqrt[3]{y}}{2}$

17) $\sqrt{3y}\cdot\sqrt{12y}=\sqrt{36y^2}=6y$

19) $2\sqrt{3}-\sqrt{18}+\sqrt{108}$
$=2\sqrt{3}-3\sqrt{2}+6\sqrt{3}=8\sqrt{3}-3\sqrt{2}$

21) $\sqrt{\dfrac{8}{3}}\cdot\sqrt{\dfrac{10}{27}}=\sqrt{\dfrac{80}{81}}=\dfrac{\sqrt{80}}{\sqrt{81}}=\dfrac{4\sqrt{5}}{9}$

23) $\left(4-3\sqrt{7}\right)\left(\sqrt{3}+9\right)$
$=4\sqrt{3}+36-3\sqrt{21}-27\sqrt{7}$

25) $\left(7-2\sqrt{t}\right)^2$
$=(7)^2-2(7)\left(2\sqrt{t}\right)+\left(2\sqrt{t}\right)^2$
$=49-28\sqrt{t}+4t$

27) $\dfrac{11}{\sqrt{2}+3}=\dfrac{11}{\sqrt{2}+3}\cdot\dfrac{\sqrt{2}-3}{\sqrt{2}-3}=\dfrac{11\left(\sqrt{2}-3\right)}{\left(\sqrt{2}\right)^2-(3)^2}$
$=\dfrac{11\left(\sqrt{2}-3\right)}{2-9}=\dfrac{11\left(\sqrt{2}-3\right)}{-7}$
$=\dfrac{11\sqrt{2}-33}{-7}=\dfrac{33-11\sqrt{2}}{7}$

29) $\dfrac{12}{\sqrt[3]{4}}=\dfrac{12}{\sqrt[3]{2^2}}\cdot\dfrac{\sqrt[3]{2}}{\sqrt[3]{2}}=\dfrac{12\sqrt[3]{2}}{\sqrt[3]{2^3}}=\dfrac{12\sqrt[3]{2}}{2}=6\sqrt[3]{2}$

31) There is no real number such that if you take the square root you will get a negative solution.

33) $y=2+\sqrt{y-2}$

$y-2=\sqrt{y-2}$

$(y-2)^2=\left(\sqrt{y-2}\right)^2$

$y^2-4y+4=y-2$

$y^2-5y+6=0$

$(y-2)(y-3)=0$

$y-2=0$ or $y-3=0$

$y=2$ $\quad y=3$ $\quad \{2,3\}$

Check is left to the student.

35) a) Let $V=64\pi$ and $h=4$, solve for r.

$r=\sqrt{\frac{V}{\pi h}}=\sqrt{\frac{64\pi}{\pi(4)}}=\sqrt{16}=4$ in.

37) $169^{-1/2}=\left(\frac{1}{169}\right)^{1/2}=\sqrt{\frac{1}{169}}=\frac{\sqrt{1}}{\sqrt{169}}=\frac{1}{13}$

39) $\left(-4g^{5/8}\right)\cdot\left(5g^{1/4}\right)=-20g^{5/8+2/8}=-20g^{7/8}$

41) $\left(x^{3/10}y^{-2/5}\right)^{-5}=x^{-15/10}y^{10/5}=x^{-3/2}y^2=\frac{y^2}{x^{3/2}}$

Cumulative Review: Chapters 1-9

1) $\frac{5}{8}\div\frac{7}{12}=\frac{5}{\cancel{8}_2}\cdot\frac{\cancel{12}^3}{7}=\frac{15}{14}$

3) $2^{-3}-4^{-2}=\frac{1}{2^3}-\frac{1}{4^2}=\frac{1}{8}-\frac{1}{16}=\frac{2}{16}-\frac{1}{16}=\frac{1}{16}$

5) $0.000941=9.41\times10^{-4}$

7) $-x+4y=8$

Put into slope-intercept form:

$-x+4y=8$

$4y=x+8$

$y=\frac{1}{4}x+2$

$m=\frac{1}{4}$; y-int $=(0,2)$

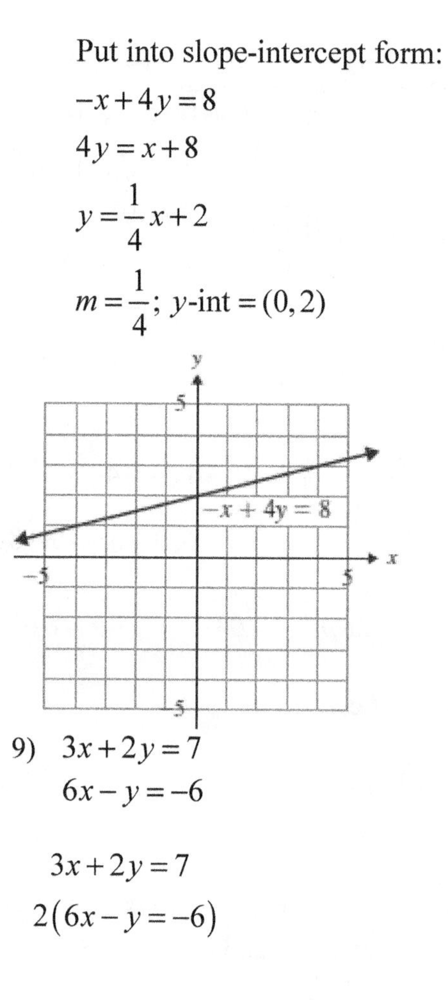

9) $3x+2y=7$

$6x-y=-6$

$3x+2y=7$

$2(6x-y=-6)$

$3x+2y=7$

$+12x-2y=-12$

$15x \quad =-5$

$x=-\frac{1}{3}$

$3\left(-\frac{1}{3}\right)+2y=7$

$-1+2y=7$

$2y=8$

$y=4$

$\left(-\frac{1}{3},4\right)$

11) $\left(7n^2+10n-1\right)-\left(8n^2+n-5\right)$

$=7n^2+10n-1-8n^2-n+5$

$=-n^2+9n+4$

13) $p^2-10p+25=(p-5)(p-5)=(p-5)^2$

15) $h^3-9h^2+4h-36=\left(h^3-9h^2\right)+(4h-36)$

$=h^2(h-9)+4(h-9)=\left(h^2+4\right)(h-9)$

17) $A=l\cdot w$

$w=l-3$

$40=lw$

$40=l(l-3)$

$40=l^2-3l$

$0=l^2-3l-40$

$0=(l-8)(l+5)$

$l-8=0$ or $l+5=0$

$l=8$ $\quad l=-5$

$w=8-3$ Length cannot be negative.

$w=5$

length = 8 in., width = 5 in.

19) $\dfrac{k^2-2k-63}{64-k^2}\cdot\dfrac{3k^2-24k}{k+7}$

$=\dfrac{\cancel{(k+7)}(k-9)}{-\cancel{(k-8)}(k+8)}\cdot\dfrac{3k\,\cancel{(k-8)}}{\cancel{(k+7)}}$

$=-\dfrac{(k-9)}{(k+8)}\cdot\dfrac{3k}{1}=-\dfrac{3k(k-9)}{k+8}$

21) $\dfrac{10\sqrt{80}}{15\sqrt{12}}=\dfrac{40\sqrt{5}}{30\sqrt{3}}=\dfrac{4\sqrt{5}}{3\sqrt{3}}\cdot\dfrac{\sqrt{3}}{\sqrt{3}}=\dfrac{4\sqrt{15}}{9}$

23) $\dfrac{\sqrt{2}}{\sqrt{6}-4}=\dfrac{\sqrt{2}}{\sqrt{6}-4}\cdot\dfrac{\sqrt{6}+4}{\sqrt{6}+4}=\dfrac{\sqrt{2}\left(\sqrt{6}+4\right)}{\left(\sqrt{6}\right)^2-(4)^2}$

$=\dfrac{\sqrt{12}+4\sqrt{2}}{6-16}=\dfrac{2\sqrt{3}+4\sqrt{2}}{-10}$

$=\dfrac{2\left(\sqrt{3}+2\sqrt{2}\right)}{-10}=-\dfrac{\sqrt{3}+2\sqrt{2}}{5}$

25) $2\sqrt{x}=x-3$

$\left(2\sqrt{x}\right)^2=(x-3)^2$

$4x=x^2-6x+9$

$0=x^2-10x+9$

$0=(x-1)(x-9)$

$x-1=0 \qquad x-9=0$

$x=1 \qquad x=9$

1 is an extraneous solution. $\{9\}$

Chapter 10: Quadratic Equations

Section 10.1: Exercises

1) Factoring and Square Root Property

Factoring:

$$x^2-81=0$$
$$(x+9)(x-9)=0$$
$$x+9=0 \text{ or } x-9=0$$
$$x=9 \text{ or } x=-9$$

Square Root Property:

$$x^2-81=0$$
$$x^2=81$$
$$x=\pm\sqrt{81}$$
$$x=\pm 9$$
$$\{-9,9\}$$

3) $b^2=16$

$b=\pm\sqrt{16}$

$b=\pm 4 \qquad \{-4,4\}$

5) $w^2=11$

$w=\pm\sqrt{11} \qquad \{-\sqrt{11},\sqrt{11}\}$

7) $p^2=-49$

$\varnothing$

9) $x^2=\dfrac{25}{9}$

$n=\pm\sqrt{\dfrac{25}{9}}=\pm\dfrac{5}{3} \qquad \left\{-\dfrac{5}{3},\dfrac{5}{3}\right\}$

11) $y^2=0.04$

$y=\pm\sqrt{0.04}$

$=\pm 0.2 \qquad \{-0.2,0.2\}$

13) $r^2-144=0$

$r^2=144$

$r=\pm\sqrt{144} \qquad \{-12,12\}$

15) $c^2-19=0$

$c^2=19$

$c=\pm\sqrt{19} \qquad \{-\sqrt{19},\sqrt{19}\}$

17) $v^2-54=0$

$v^2=54$

$v^2=\pm\sqrt{54}$

$v=\pm 3\sqrt{6} \qquad \{-3\sqrt{6},3\sqrt{6}\}$

19) $r^2-\dfrac{5}{64}=0$

$r^2=\dfrac{5}{64}$

$r=\pm\sqrt{\dfrac{5}{64}}=\pm\dfrac{\sqrt{5}}{8} \qquad \left\{-\dfrac{\sqrt{5}}{8},\dfrac{\sqrt{5}}{8}\right\}$

21) $z^2+5=19$

$z^2=19-5$

$z^2=14$

$z=\pm\sqrt{14} \qquad \{-\sqrt{14},\sqrt{14}\}$

23) $n^2+10=6$

$n^2=-4$

There is no solution $\varnothing$

25) $3d^2 + 14 = 41$
$3d^2 = 27$
$d^2 = 9$
$d = \pm\sqrt{9} = \pm 3$ $\quad \{-3, 3\}$

27) $4p^2 - 9 = 39$
$4p^2 = 48$
$p^2 = 12$
$p = \pm\sqrt{12} = \pm 2\sqrt{3}$ $\quad \{-2\sqrt{3}, 2\sqrt{3}\}$

29) $3 = 35 - 8h^2$
$8h^2 = 32$
$h^2 = 4$
$h = \pm\sqrt{4} = \pm 2$ $\quad \{-2, 2\}$

31) $10 = 14 + 2x^2$
$2x^2 = -4$
There is no real solution. $\varnothing$

33) $4y^2 + 15 = 24$
$4y^2 = 9$
$y^2 = \frac{9}{4}$
$y = \pm\sqrt{\frac{9}{4}} = \pm\frac{3}{2}$ $\quad \left\{-\frac{3}{2}, \frac{3}{2}\right\}$

35) $9w^2 - 5 = 5$
$9w^2 = 10$
$w^2 = \frac{10}{9}$
$w = \pm\sqrt{\frac{10}{9}} = \pm\frac{\sqrt{10}}{3}$ $\quad \left\{-\frac{\sqrt{10}}{3}, \frac{\sqrt{10}}{3}\right\}$

37) $-7 = 4 - 5b^2$
$5b^2 = 11$
$b^2 = \frac{11}{5}$
$b = \pm\sqrt{\frac{11}{5}} \cdot \frac{\sqrt{5}}{\sqrt{5}} = \pm\frac{\sqrt{55}}{5}$ $\quad -\frac{\sqrt{55}}{5}, \frac{\sqrt{55}}{5}$

39) $(r+6)^2 = 25$
$r + 6 = \pm\sqrt{25}$
$r + 6 = \pm 5$
$r = -6 \pm 5$
$r = -11, -1$ $\quad \{-11, -1\}$

41) $(q-8)^2 = 1$
$q - 8 = \pm 1$
$q = 8 \pm 1$ $\quad \{7, 9\}$

43) $(a+2)^2 = 13$
$a + 2 = \pm\sqrt{13}$
$a = -2 \pm \sqrt{13}$ $\quad \{-2 - \sqrt{13}, -2 + \sqrt{13}\}$

45) $(k-10)^2 = 45$

$k-10 = \pm\sqrt{45}$

$k = 10 \pm 3\sqrt{5}$ $\{10-3\sqrt{5}, 10+3\sqrt{5}\}$

47)

$(m+7)^2 = -18$

There is no real solution. $\varnothing$

49) $0 = (p+3)^2 - 68$

$(p+3)^2 = 68$

$p+3 = \pm\sqrt{68}$

$p+3 = \pm 2\sqrt{17}$

$p = -3 \pm 2\sqrt{17}$

$\{-3-2\sqrt{17}, -3+2\sqrt{17}\}$

51) $(2z-1)^2 = 9$

$2z-1 = \pm\sqrt{9} = \pm 3$

$2z-1 = -3$ or $2z-1 = 3$

$2z = -2$ or $2z = 4$

$z = -1$ or 2 $\{-1, 2\}$

53) $121 = (4q+5)^2$

$\pm 11 = 4q+5$ $\quad 4q = -5+11$

$4q = -5 \pm 11$ $\quad 4q = 6$

$4q = -5-11$ $\quad q = \frac{4}{6} = \frac{2}{3}$

$4q = -16$

$q = -4$

$\left\{-4, \frac{3}{2}\right\}$

55) $(3g-10)^2 = 24$

$3g-10 = \pm\sqrt{24}$

$3g = 10 \pm 2\sqrt{6}$

$g = \frac{10}{3} \pm \frac{2\sqrt{6}}{3}$ $\left\{\frac{10}{3} - \frac{2\sqrt{6}}{3}, \frac{10}{3} + \frac{2\sqrt{6}}{3}\right\}$

57) $125 = (5u+8)^2$

$\pm\sqrt{125} = 5u+8$

$5u = -8 \pm 5\sqrt{5}$

$u = \frac{-8 \pm 5\sqrt{5}}{5}$ $\left\{\frac{-8-5\sqrt{5}}{5}, \frac{-8+5\sqrt{5}}{5}\right\}$

59) $(2x+3)^2 - 54 = 0$

$(2x+3)^2 = 54$

$2x+3 = \pm\sqrt{54}$

$2x = -3 \pm 3\sqrt{6}$

$x = \frac{-3 \pm 3\sqrt{6}}{2}$ $\left\{\frac{-3-3\sqrt{6}}{2}, \frac{-3+3\sqrt{6}}{2}\right\}$

61) $(7h-8)^2 + 32 = 0$

$(7h-8)^2 = -32$

There is no real solution. $\varnothing$

63) $(5y-2)^2 + 6 = 22$

$(5y-2)^2 = 16$

$5y-2 = \pm 4$

$5y = 2 \pm 4$

$y = \frac{2 \pm 4}{5}$ $\left\{\frac{6}{5}, -\frac{2}{5}\right\}$

65) $1=(6r+7)^2-8$

$9=(6r+7)^2$

$6r+7=\pm 3$

$6r=-7\pm 3$

$r=\dfrac{-7\pm 3}{6}$

$r=-\dfrac{4}{6},-\dfrac{10}{6}$ $\left\{-\dfrac{2}{3},-\dfrac{5}{3}\right\}$

67) $(2z-11)^2+3=17$

$(2z-11)^2=14$

$2z-11=\pm\sqrt{14}$

$2z=11\pm\sqrt{14}$

$z=\dfrac{11\pm\sqrt{14}}{2}$

$\left\{\dfrac{11+\sqrt{14}}{2},\dfrac{11-\sqrt{14}}{2}\right\}$

69) $\left(1-\dfrac{1}{2}c\right)^2-6=-5$

$\left(1-\dfrac{1}{2}c\right)^2=1$

$1-\dfrac{1}{2}c=\pm\sqrt{1}$

$-\dfrac{1}{2}c=1\pm 1$

$c=-2(-1\pm 1)$

$c=\{0,4\}$

71) Area of circle, $A=\pi r^2$

$A=81\pi cm^2$

$81\pi=\pi r^2$

$r^2=81$

$r=\pm\sqrt{81}$

$r=\pm 9$

Choosing the positive solution, the radius is 9 cm.

73) Surface area of a sphere, $S=4\pi r^2$

$S=\pi\ m^2$

$\pi=4\pi r^2$

$r^2=\dfrac{1}{4}$

$r=\pm\sqrt{\dfrac{1}{4}};\quad r=\pm\dfrac{1}{2}$

Choosing the positive solution, the radius is $\dfrac{1}{2}$ *m.*

75) Illuminance, $E=\dfrac{I}{d^2}$

$E=300$ lux; $I=2700$ candela

$300=\dfrac{2700}{d^2}$

$d^2=9$

$d=\pm\sqrt{9};\quad d=\pm 3$

Choosing the positive solution, the distance is 3 m.

77) Kinetic Energy, $k=\dfrac{1}{2}mv^2$

$k=153{,}600$ joules; $m=1200$ kg

$153{,}600=\dfrac{1}{2}(1200)v^2$

$v^2=\dfrac{(153{,}600)2}{1200}$

$v^2=\pm\sqrt{256};\quad v=\pm 16$

Choosing the positive solution, the velocity is 16 m/sec.

Chapter 10: Quadratic Equations

Section 10.2: Exercises

1) A perfect square trinomial is a trinomial whose factored form is the square of a binomial. Some examples are

$x^2+8x+16=(x+4)^2$

$p^2-14p+49=(p-7)^2$

3) y^2+18y

$\frac{1}{2}(18)=9$ Find half of the coefficient of y

$9^2=81$ Square the result.

$y^2+18y+81$ Add the constant to the expression.

The perfect square trinomial is $y^2+18y+81$.

The factored form of the trinomial is $(y+9)^2$.

5) $a^2+12a+36=(a+6)^2$

7) $k^2-10k+25=(k-5)^2$

9) $g^2-24g+144=(g-12)^2$

11) $h^2+9h+\frac{81}{4}=\left(h+\frac{9}{2}\right)^2$

13) $x^2-x+\frac{1}{4}=\left(x-\frac{1}{2}\right)^2$

15) Answers may vary.

17)
$$x^2+6x+8=0$$
$$x^2+6x=-8$$
$$x^2+6x+9=-8+9$$
$$(x+3)^2=1$$
$$x+3=\pm\sqrt{1}$$
$$x+3=\pm1$$

$x+3=1$ or $x+3=-1$

$x=-2$ $\quad x=-4$ $\quad \{-4,-2\}$

19)
$$z^2-14z+45=0$$
$$z^2-14z=-45$$
$$z^2-14z+49=-45+49$$
$$(z-7)^2=4$$
$$z-7=\pm\sqrt{4}$$
$$z-7=\pm2$$

$z-7=2$ or $z-7=-2$

$z=9$ $\quad z=5$ $\quad \{5,9\}$

21) $p^2+8p+20=0$
$p^2+8p=-20$
$p^2+8p+16=-20+16$
$(p+4)^2=-4$
There is no real solution. $\varnothing$

23) $y^2-11=-4y$
$y^2+4y=11$
$y^2+4y+4=11+4$
$(y+2)^2=15$
$y+2=\pm\sqrt{15}$
$y=-2\pm\sqrt{15}$
$\{-2+\sqrt{15}\ ,\ -2-\sqrt{15}\}$

25) $x^2-10x=3$
$x^2-10x+25=3+25$
$(x-5)^2=28$
$x-5=\pm\sqrt{28}$
$x-5=\pm2\sqrt{7}$
$x=5\pm2\sqrt{7}$
$\{5-2\sqrt{7}\ ,\ 5+2\sqrt{7}\}$

27) $2a=22+a^2$
$a^2-2a=-22$
$a^2-2a+1=-22+1$
$(a-1)^2=-21$
There is no real solution. $\varnothing$

29) $m^2+3m-40=0$
$m^2+3m=40$
$m^2+3m+\frac{9}{4}=40+\frac{9}{4}$
$\left(m+\frac{3}{2}\right)^2=\frac{169}{4}$
$m+\frac{3}{2}=\pm\sqrt{\frac{169}{4}}$
$m+\frac{3}{2}=\pm\frac{13}{2}$
$m=-\frac{3}{2}-\frac{13}{2};\ m=-\frac{3}{2}+\frac{13}{2}$
$\{-8,5\}$

31) $c^2-56=c$
$c^2-c=56$
$c^2-c+\frac{1}{4}=56+\frac{1}{4}$
$\left(c-\frac{1}{2}\right)^2=\frac{225}{4}$
$c-\frac{1}{2}=\pm\sqrt{\frac{225}{4}}$
$h-\frac{1}{2}=\pm\frac{15}{2}$
$h=\frac{1}{2}-\frac{15}{2};\ m=\frac{1}{2}+\frac{15}{2}$
$\{-7,8\}$

33) $h^2+9h=-12$
$h^2+9h+\frac{81}{4}=-12+\frac{81}{4}$
$\left(h+\frac{9}{2}\right)^2=\frac{33}{4}$
$h+\frac{9}{2}=\pm\sqrt{\frac{33}{4}}$
$h=-\frac{9}{2}\pm\frac{\sqrt{33}}{2}$
$\left\{-\frac{9}{2}+\frac{\sqrt{33}}{2},-\frac{9}{2}-\frac{\sqrt{33}}{2}\right\}$

35) $b^2-5b+27=6$
$b^2-5b=-21$
$b^2-5b+\frac{25}{4}=-21+\frac{25}{4}$
$\left(b-\frac{5}{2}\right)^2=-\frac{59}{4}$
There is no solution $\varnothing$

37) No, because the coefficient of x^2 is not one.

39) $4r^2+32r+55=0$
$r^2+8r=-\frac{55}{4}$
$r^2+8r+16=16-\frac{55}{4}$
$(r+4)^2=\frac{9}{4}$
$r+4=\pm\frac{3}{2}$
$r=-4-\frac{3}{2}$ or $r=-4+\frac{3}{2}$
$\left\{-\frac{11}{2},-\frac{5}{2}\right\}$

41) $3x^2+39=30x$
$3x^2-30x=-39$
$x^2-10x=-13$
$x^2-10x+25=25-13$
$(x-5)^2=12$
$x-5=\pm\sqrt{12}$
$x=5\pm2\sqrt{3}$ $\{5-2\sqrt{3},\ 5+2\sqrt{3}\}$

43) $7k^2+84=49k$
$7k^2-49k=-84$
$k^2-7k=-12$
$k^2-7k+\frac{49}{4}=\frac{49}{4}-12$
$\left(k-\frac{7}{2}\right)^2=\frac{1}{4}$
$k-\frac{7}{2}=\pm\frac{1}{2}$
$k=\frac{7}{2}\pm\frac{1}{2}$ $\{3,4\}$

45) $54y-6y^2=72$
$9y-y^2=12$
$y^2-9y=-12$
$x-9x+\frac{81}{4}=-12+\frac{81}{4}$
$\left(y-\frac{9}{2}\right)^2=\frac{33}{4}$
$y-\frac{9}{2}=\pm\frac{\sqrt{33}}{2}$
$y=\frac{9}{2}\pm\frac{\sqrt{33}}{2}$
$\left\{\frac{9}{2}-\frac{\sqrt{33}}{2},\ \frac{9}{2}+\frac{\sqrt{33}}{2}\right\}$

47) $16z^2+3=16z$
$16z^2-16z=-3$
$z^2-z=-\frac{3}{16}$
$z^2-z+\frac{1}{4}=\frac{1}{4}-\frac{3}{16}$
$\left(z-\frac{1}{2}\right)^2=\frac{1}{16}$

$$z-\frac{1}{2}=\pm\frac{1}{4}$$

$$z=\frac{1}{2}\pm\frac{1}{4} \qquad \left\{\frac{1}{4},\frac{3}{4}\right\}$$

49) $3g^2+15g+37=0$

$$g^2+5g=-\frac{37}{3}$$

$$g^2+5g+\frac{25}{4}=\frac{25}{4}-\frac{37}{3}$$

$$\left(g+\frac{5}{2}\right)^2=\frac{75-148}{12}$$

$$\left(g+\frac{5}{2}\right)^2=-\frac{73}{12}$$

There is no real solution. $\varnothing$

51) $-v^2-2v+35=0$

$$v^2+2v=35$$

$$v^2+2v+1=1+35$$

$$(v+1)^2=36$$

$$v+1=\pm 6$$

$$v+1=-6 \qquad v+1=6$$

$$v=-7 \qquad v=5 \qquad \{-7,5\}$$

53) $(a-4)(a+10)=-17$

$$a^2+10a-4a-40=-17$$

$$a^2+6a=23$$

$$a^2+6a+9=23+9$$

$$(a+3)^2=32$$

$$a+3=\pm 4\sqrt{2}$$

$$a=-3\pm 4\sqrt{2}$$

$$\left\{-3-4\sqrt{2},-3+4\sqrt{2}\right\}$$

55) $n+2=3n^2$

$$3n^2-n=2$$

$$n^2-\frac{1}{3}n=\frac{2}{3}$$

$$n-\frac{1}{3}n+\frac{1}{36}=\frac{2}{3}+\frac{1}{36}$$

$$\left(n-\frac{1}{6}\right)^2=\frac{25}{36}$$

$$n-\frac{1}{6}=\pm\frac{5}{6}$$

$$n=\frac{1}{6}\pm\frac{5}{6}$$

$$\left\{-\frac{2}{3},1\right\}$$

57) $(5p+2)(p+4)=1$

$$5p^2+20p+2p+8=1$$

$$5p^2+22p=-7$$

$$p^2+\frac{22}{5}p=-\frac{7}{5}$$

$$p^2+\frac{22}{5}p+\frac{121}{25}=-\frac{7}{5}+\frac{121}{25}$$

$$\left(p+\frac{11}{5}\right)^2=\frac{86}{25}$$

$$p+\frac{11}{5}=\pm\frac{\sqrt{86}}{5}$$

$$p=-\frac{11}{5}\pm\frac{\sqrt{86}}{5}$$

$$\left\{-\frac{11}{5}-\frac{\sqrt{86}}{5},-\frac{11}{5}+\frac{\sqrt{86}}{5}\right\}$$

59) $x =$ width of the Portfolio

$x+7 =$ length of the Portfolio

$170 =$ area of the Portfolio

$$x(x+7)=170$$
$$x^2+7x=170$$
$$x^2+7x+\frac{49}{4}=170+\frac{49}{4}$$
$$\left(x+\frac{7}{2}\right)^2=\frac{729}{4}$$
$$x+\frac{7}{2}=\pm\frac{27}{2}$$
$$x=-\frac{7}{2}-\frac{27}{2} \text{ or } x=-\frac{7}{2}+\frac{27}{2}$$
$$x=\frac{-14}{2}=-17 \quad x=\frac{20}{2}=10$$

length of the potfolio: 10 in

width: 17 in

61) Let h be the height of the triangle

Then base of the triangle is $2h$-1

$$A=\frac{1}{2}bh$$
$$60=\frac{1}{2}(2h-1)h$$
$$120=2h^2-h$$
$$h^2-\frac{h}{2}=60$$
$$h^2-\frac{h}{2}+\frac{1}{16}=\frac{1}{16}+60$$
$$\left(h-\frac{1}{4}\right)^2=\frac{961}{16}$$
$$h-\frac{1}{4}=\pm\frac{31}{4}$$
$$h=\frac{1}{4}\pm\frac{31}{4}$$
$$h=-7.5 \quad h=8$$
$$b=8+7=15 \text{ cm}$$

63) $x^2=(x-16)^2+(x-2)^2$

$$x^2=x^2-32x+256+x^2-4x+4$$
$$x^2=2x^2-36x+260$$
$$x^2-36x=-260$$
$$x^2-36x+324=324-260$$
$$(x-18)^2=8$$
$$x-18=\pm 8$$
$$x=26 \text{ or } x=10$$

the sides are 10, 24, 26

Section 10.3: Exercises

1) $x=\dfrac{-b\pm\sqrt{b^2-4ac}}{2a}$

3) The equation must be written as $3x^2-5x-4=0$ before identifying the values of a, b, and c.

$$x=\frac{-(-5)\pm\sqrt{(-5)^2-4(3)(-4)}}{2(3)}$$

5) $x^2+2x-8=0$

$a=1$, $b=2$ and $c=-8$

$$x=\frac{-2\pm\sqrt{2^2-4(1)(-8)}}{2(1)}$$
$$=\frac{-2\pm\sqrt{4+32}}{2}$$
$$=\frac{-2\pm\sqrt{36}}{2}=\frac{-2\pm 6}{2}$$
$$x\frac{-2+6}{2}=\frac{4}{2}=2, \quad x=\frac{-2-6}{2}=\frac{-8}{2}=-4$$

$\{-4, 2\}$

7) $6z^2 - 7z + 2 = 0$

$a = 6,\ b = -7 \text{ and } c = 2$

$$z = \frac{-(-7) \pm \sqrt{(-7)^2 - 4(6)(2)}}{2(6)}$$

$$= \frac{7 \pm \sqrt{49 - 48}}{12}$$

$$= \frac{7 \pm \sqrt{1}}{12} = \frac{7 \pm 1}{12}$$

$$z = \frac{7+1}{12} = \frac{8}{12} = \frac{2}{3},\quad z = \frac{7-1}{12} = \frac{6}{12} = \frac{1}{2}$$

$$\left\{\frac{1}{2}, \frac{2}{3}\right\}$$

9) $k^2 + 2 = 5k$

$k^2 - 5k + 2 = 0$

$a = 1,\ b = -5 \text{ and } c = 2$

$$k = \frac{-(-5) \pm \sqrt{(-5)^2 - 4(1)(2)}}{2(1)}$$

$$= \frac{5 \pm \sqrt{25 - 8}}{2} = \frac{5 \pm \sqrt{17}}{2}$$

$$\left\{\frac{5 - \sqrt{17}}{2}, \frac{5 + \sqrt{17}}{2}\right\}$$

11) $3w^2 = 2w + 4$

$3w^2 - 2w - 4 = 0$

$a = 3,\ b = -2 \text{ and } c = -4$

$$w = \frac{-(-2) \pm \sqrt{(-2)^2 - 4(3)(-4)}}{2(3)}$$

$$= \frac{2 \pm \sqrt{4 + 48}}{6}$$

$$= \frac{2 \pm \sqrt{52}}{6} = \frac{2 \pm 2\sqrt{13}}{6}$$

$$= \frac{1 \pm \sqrt{13}}{3}$$

$$\left\{\frac{1 - \sqrt{13}}{3}, \frac{1 + \sqrt{13}}{3}\right\}$$

13) $y = 2y^2 + 6$

$2y^2 - y + 6 = 0$

$a = 2,\ b = -1 \text{ and } c = 6$

$$y = \frac{-(-1) \pm \sqrt{(-1)^2 - 4(2)(6)}}{2(1)}$$

$$= \frac{1 \pm \sqrt{1 - 48}}{2} = \frac{1 \pm \sqrt{-47}}{2}$$

There is no real solution. $\varnothing$

15) $m^2 + 11m = 0$

$a = 1,\ b = 11 \text{ and } c = 0$

$$m = \frac{-11 \pm \sqrt{11^2 - 4(1)(0)}}{2(1)}$$

$$= \frac{-11 \pm \sqrt{121}}{2} = \frac{-11 \pm 11}{2}$$

$$= \frac{-11 - 11}{2}, \frac{-11 + 11}{2}$$

$$= -11, 0$$

$\{-11, 0\}$

17) $2p(p-3)=-3$

$2p^2-6p=-3$

$2p^2-6p+3=0$

$a=2,\ b=-6$ and $c=3$

$$p=\frac{-(-6)\pm\sqrt{(-6)^2-4(2)(3)}}{2(2)}$$

$$=\frac{6\pm\sqrt{36-24}}{4}=\frac{6\pm\sqrt{12}}{4}$$

$$=\frac{6\pm2\sqrt{3}}{4}=\frac{2(3\pm\sqrt{3})}{4}=\frac{3\pm\sqrt{3}}{2}$$

$$\left\{\frac{3-\sqrt{3}}{2},\ \frac{3+\sqrt{3}}{2}\right\}$$

19) $(2s+3)(s-1)=s^2-s+6$

$2s^2+s-3=s^2-s+6$

$s^2+2s-9=0$

$a=1,\ b=2$ and $c=-9$

$$s=\frac{-2\pm\sqrt{2^2-4(1)(-9)}}{2(1)}$$

$$=\frac{-2\pm\sqrt{4+36}}{2}=\frac{-2\pm\sqrt{40}}{2}$$

$$=\frac{-2\pm2\sqrt{10}}{2}=\frac{2(-1\pm\sqrt{10})}{2}$$

$$=-1\pm\sqrt{10}$$

$$\{-1-\sqrt{10},\ -1+\sqrt{10}\}$$

21) $k(k+2)=-5$

$k^2+2k+5=0$

$a=1,\ b=2$ and $c=5$

$$k=\frac{-2\pm\sqrt{2^2-4(1)(5)}}{2(1)}$$

$$=\frac{-2\pm\sqrt{4-20}}{2}=\frac{-2\pm\sqrt{-16}}{6}$$

There is no real solution. $\varnothing$

23) $(x-8)(x-3)=3(3-x)$

$x^2-11x+24=9-3x$

$x^2-8x+15=0$

$a=1,\ b=-8$ and $c=15$

$$x=\frac{-(-8)\pm\sqrt{(-8)^2-4(1)(15)}}{2(1)}$$

$$=\frac{8\pm\sqrt{64-60}}{2}=\frac{8\pm\sqrt{4}}{2}=\frac{8\pm2}{2}$$

$$x=\frac{8+2}{2}=\frac{10}{2}=5,$$

$$x=\frac{8-2}{2}=\frac{6}{2}=3 \qquad \{3,\ 5\}$$

25) $8t=1+16t^2$

$16t^2-8t+1=0$

$a=16,\ b=-8$ and $c=1$

$$t=\frac{-(-8)\pm\sqrt{(-8)^2-4(16)(1)}}{2(16)}$$

$$=\frac{8\pm\sqrt{64-64}}{32}=\frac{8\pm\sqrt{0}}{32}$$

$$=\frac{8}{32}=\frac{1}{4} \qquad \left\{\frac{1}{4}\right\}$$

27) $\frac{1}{8}z^2+\frac{3}{4}z+\frac{1}{2}=0$

$$8\left(\frac{1}{8}z^2+\frac{3}{4}z+\frac{1}{2}\right)=8(0)$$
$$z^2+6z+4=0$$

$a=1,\ b=6$ and $c=4$

$$z=\frac{-6\pm\sqrt{6^2-4(1)(4)}}{2(1)}$$
$$=\frac{-6\pm\sqrt{36-16}}{2}=\frac{-6\pm\sqrt{20}}{2}$$
$$=\frac{-6\pm2\sqrt{5}}{2}=\frac{2\left(-3\pm\sqrt{5}\right)}{2}$$
$$=-3\pm\sqrt{5}$$
$$\left\{-3-\sqrt{5},\ -3+\sqrt{5}\right\}$$

29) $\frac{1}{6}k+\frac{1}{2}=\frac{3}{4}k^2$

$$12\left(\frac{1}{6}k+\frac{1}{2}\right)=12\left(\frac{3}{4}k^2\right)$$
$$2k+6=9k^2$$
$$0=9k^2-2k-6$$

$a=9,\ b=-2$ and $c=-6$

$$k=\frac{-(-2)\pm\sqrt{(-2)^2-4(9)(-6)}}{2(9)}$$
$$=\frac{2\pm\sqrt{4+216}}{18}=\frac{2\pm\sqrt{220}}{18}$$
$$=\frac{2\pm2\sqrt{55}}{18}=\frac{2\left(1\pm\sqrt{55}\right)}{18}$$
$$=\frac{1\pm\sqrt{55}}{9}$$
$$\left\{\frac{1-\sqrt{55}}{9},\ \frac{1+\sqrt{55}}{9}\right\}$$

31) $0.8v^2+0.1=0.6v$

$$10\left(0.8v^2+0.1\right)=10(0.6v)$$
$$8v^2+1=6v$$
$$8v^2-6v+1=0$$

$a=8,\ b=-6$ and $c=1$

$$v=\frac{-(-6)\pm\sqrt{(-6)^2-4(8)(1)}}{2(8)}$$
$$=\frac{6\pm\sqrt{36-32}}{16}$$
$$=\frac{6\pm\sqrt{4}}{16}=\frac{6\pm2}{16}$$
$$v=\frac{6-2}{16}=\frac{4}{16}=\frac{1}{4},\quad v=\frac{6+2}{16}=\frac{8}{16}=\frac{1}{2}$$
$$\left\{\frac{1}{4},\frac{1}{2}\right\}$$

33) $16g^2-3=0$

$a=16,\ b=0$ and $c=-3$

$$g=\frac{-0\pm\sqrt{0^2-4(16)(-3)}}{2(16)}$$
$$=\frac{\pm\sqrt{192}}{32}=\frac{\pm8\sqrt{3}}{32}=\pm\frac{\sqrt{3}}{4}$$
$$\left\{-\frac{\sqrt{3}}{4},\ \frac{\sqrt{3}}{4}\right\}$$

35) $9d^2-4=0$

$a=9$, $b=0$ and $c=-4$

$$d=\frac{-0\pm\sqrt{0^2-4(9)(-4)}}{2(9)}$$

$$=\frac{\pm\sqrt{144}}{18}=\frac{\pm 12}{18}=\pm\frac{2}{3}$$

$$\left\{-\frac{2}{3},\frac{2}{3}\right\}$$

37) $3(3-4r)=-4r^2$

$$9-12r=-4r^2$$

$$4r^2-12r+9=0$$

$a=4$, $b=-12$ and $c=9$

$$r=\frac{-(-12)\pm\sqrt{(-12)^2-4(4)(9)}}{2(4)}$$

$$=\frac{12\pm\sqrt{144-144}}{8}=\frac{12\pm\sqrt{0}}{8}$$

$$=\frac{12}{8}=\frac{3}{2} \qquad \left\{\frac{3}{2}\right\}$$

39) $6=7h-3h^2$

$$3h^2-7h+6=0$$

$a=3$, $b=-7$ and $c=6$

$$h=\frac{-(-7)\pm\sqrt{(-7)^2-4(3)(6)}}{2(3)}$$

$$=\frac{7\pm\sqrt{49-72}}{6}=\frac{7\pm\sqrt{-23}}{6}$$

There is no real solution. $\varnothing$

41) $4p^2+6=20p$

$$\frac{4p^2}{2}+\frac{6}{2}=\frac{20p}{2}$$

$$2p^2+3=10p$$

$$2p^2-10p+3=0$$

$a=2$, $b=-10$ and $c=3$

$$p=\frac{-(-10)\pm\sqrt{(-10)^2-4(2)(3)}}{2(2)}$$

$$=\frac{10\pm\sqrt{100-24}}{4}=\frac{10\pm\sqrt{76}}{4}$$

$$=\frac{10\pm 2\sqrt{19}}{4}=\frac{2\left(5\pm\sqrt{19}\right)}{4}$$

$$=\frac{5\pm\sqrt{19}}{2}$$

$$\left\{\frac{5-\sqrt{19}}{2},\frac{5+\sqrt{19}}{2}\right\}$$

43) Let x=length of shorter leg

Then hypotenuse is $2x-1$

$$(\text{hypotenuse})^2=(\text{short leg})^2+(\text{long leg})^2$$

$$(2x-1)^2=x^2+\left(\sqrt{23}\right)^2$$

$$4x^2-4x+1=x^2+23$$

$$3x^2-4x-22=0$$

$a=3$, $b=-4$ and $c=-22$

$$x=\frac{-(-4)\pm\sqrt{(-4)^2-4(3)(-22)}}{2(3)}$$

$$=\frac{4\pm\sqrt{16+264}}{6}=\frac{4\pm\sqrt{280}}{6}$$

$$=\frac{4\pm 2\sqrt{70}}{6}=\frac{2\pm\sqrt{70}}{3}$$

The shorter leg is $\frac{2+\sqrt{70}}{3}\approx 3.46\text{in.}$

45) a) Let $h=8$ and solve for t.

$$8=-16t^2+24t+24$$
$$0=-16t^2+24t+16$$
$$0=2t^2-3t-2$$
$$0=(2t+1)(t-2)$$
$$2t+1=0 \quad \text{or} \quad t-2=0$$
$$2t=-1 \qquad \boxed{t=2}$$
$$t=-\frac{1}{2}$$

The ball reaches 8 feet after 2 sec.

b) Let $h=0$ and solve for t.

$$0=-16t^2+24t+24$$
$$0=2t^2-3t-3$$
$$a=2,\ b=-3 \text{ and } c=-3$$
$$t=\frac{-(-3)\pm\sqrt{(-3)^2-4(2)(-3)}}{2(2)}$$
$$=\frac{3\pm\sqrt{9+24}}{4}=\frac{3\pm\sqrt{33}}{4}$$

Reject $t=\frac{3-\sqrt{33}}{4}$ because it is negative.

The ball will hit the ground after $\frac{3+\sqrt{33}}{4}$ sec ≈ 2.19 sec.

Chapter 10 Putting It All Together

1) $f^2-75=0$

$$f^2=75$$
$$f=\pm\sqrt{75}=\pm 5\sqrt{3}$$
$$\{-5\sqrt{3},\ 5\sqrt{3}\}$$

3)

$$a(a+1)=20$$
$$a^2+a=20$$
$$a^2+a-20=0$$
$$(a+5)(a-4)=0$$
$$a+5=0 \text{ or } a-4=0$$
$$a=-5 \qquad a=4 \qquad \{-5,\ 4\}$$

5) $v^2+6v+7=0$

$$a=1,\ b=6 \text{ and } c=7$$
$$u=\frac{-6\pm\sqrt{6^2-4(1)(7)}}{2(1)}$$
$$=\frac{-6\pm\sqrt{36-28}}{2}=\frac{-6\pm\sqrt{8}}{2}$$
$$=\frac{-6\pm 2\sqrt{2}}{2}=-3\pm\sqrt{2}$$
$$\{-3-\sqrt{2},-3+\sqrt{2}\}$$

7)

$$3x(x+3)=-5(x-1)$$
$$3x^2+9x=-5x+5$$
$$3x^2+14x-5=0$$
$$(3x-1)(x+5)=0$$
$$3x-1=0 \text{ or } x+5=0$$
$$3x=1 \qquad x=-5$$
$$x=\frac{1}{3} \qquad \left\{-5,\frac{1}{3}\right\}$$

9)

$$m^2+12m+42=0$$
$$m^2+12m=-42$$
$$m^2+12m+36=-42+36$$
$$(m+6)^2=-6$$
$$m+6=\pm\sqrt{-6}$$
$$m=-6\pm\sqrt{-6}$$

There is no real solution. $\varnothing$

11) $12+(2k-1)^2=3$

$(2k-1)^2=-9$

There is no real solution. $\varnothing$

12) $1=\frac{x^2}{40}-\frac{3x}{20}$

$40(1)=40\left(\frac{x^2}{40}-\frac{3x}{20}\right)$

$40=x^2-6x$

$0=x^2-6x-40$

$0=(x-10)(x+4)$

$x-10=0$ or $x+4=0$

$x=10$ $\quad x=-4$ $\quad \{-4, 10\}$

14) $b^2-6b=5$

$b^2-6b+9=5+9$

$(b-3)^2=14$

$b-3=\pm\sqrt{14}$

$b=3\pm\sqrt{14}$

$\{3-\sqrt{14}, 3+\sqrt{14}\}$

17) $q(q+12)=3(q^2+5)+q$

$q^2+12q=3q^2+15+q$

$0=2q^2-11q+15$

$0=(2q-5)(q-3)$

$2q-5=0$ or $q-3=0$

$2q=5$ $\quad q=3$

$q=\frac{5}{2}$ $\quad \left\{\frac{5}{2}, 3\right\}$

19) $\frac{9}{c}=1+\frac{18}{c^2}$

$c^2\left(\frac{9}{c}\right)=c^2\left(1+\frac{18}{c^2}\right)$

$9c=c^2+18$

$0=c^2-9c+18$

$0=(c-6)(c-3)$

$c-6=0$ or $c-3=0$

$c=7$ $\quad c=3$ $\quad \{3, 6\}$

21) $(3v+4)(v-2)=-9$

$3v^2-2v-8=-9$

$3v^2-2v+1=0$

$a=3, b=-2$ and $c=1$

$v=\frac{-(-2)\pm\sqrt{(-2)^2-4(3)(1)}}{2(3)}$

$=\frac{2\pm\sqrt{4-12}}{6}=\frac{2\pm\sqrt{-8}}{6}$

There is no real solution. $\varnothing$

23) $2r^2+3r-2=0$

$(2r-1)(r+2)=0$

$2r-1=0$ or $r+2=0$

$r=\frac{1}{2}$ $\quad r=-2$

$\left\{-2, \frac{1}{2}\right\}$

25) $5m=m^2$

$0=m^2-5m$

$0=m(m-5)$

$m-5=0$ or $m=0$

$m=5$ $\quad \{0, 5\}$

27)
$$4m^3 = 9m$$
$$4m^3 - 9m = 0$$
$$m(4m^2 - 9) = 0$$
$$m(2m+3)(2m-3) = 0$$
$$2m+3=0 \text{ or } 2m-3=0 \text{ or } m=0$$
$$2m=-3 \qquad 2m=3$$
$$m=-\frac{3}{2} \qquad m=\frac{3}{2}$$
$$\left\{-\frac{3}{2}, 0, \frac{3}{2}\right\}$$

29)
$$2k^2+3=9k$$
$$2k^2-9k+3=0$$
$$a=2,\ b=-9 \text{ and } c=3$$
$$k=\frac{-(-9)\pm\sqrt{(-9)^2-4(2)(3)}}{2(2)}$$
$$=\frac{9\pm\sqrt{81-24}}{4}=\frac{9\pm\sqrt{57}}{4}$$
$$\left\{\frac{9-\sqrt{57}}{4}, \frac{9+\sqrt{57}}{4}\right\}$$

Section 10.4: Exercises

1) False

3) True

5) $\sqrt{-36}=\sqrt{-1}\cdot\sqrt{36}=i\cdot 6=6i$

7) $\sqrt{-16}=\sqrt{-1}\cdot\sqrt{16}=i\cdot 4=4i$

9) $\sqrt{-5}=\sqrt{-1}\cdot\sqrt{5}=i\cdot\sqrt{5}=i\sqrt{5}$

11) $\sqrt{-27}=\sqrt{-1}\cdot\sqrt{27}=i\cdot 3\sqrt{3}=3i\sqrt{3}$

13) $\sqrt{-45}=\sqrt{-1}\cdot\sqrt{45}=i\cdot 3\sqrt{5}=3i\sqrt{5}$

15) $\sqrt{-32}=\sqrt{-1}\cdot\sqrt{32}=i\cdot 4\sqrt{2}=4i\sqrt{2}$

17) Add the real parts and add the imaginary parts.

19) $(5+2i)+(-8+10i)=-3+12i$

21) $(13-8i)-(9+i)=4-9i$

23)
$$\left(-\frac{5}{8}-\frac{1}{3}i\right)-\left(-\frac{1}{2}+\frac{3}{4}i\right)$$
$$=\left(-\frac{5}{8}+\frac{1}{2}\right)+\left(-\frac{1}{3}i-\frac{3}{4}i\right)$$
$$=\left(-\frac{5}{8}+\frac{4}{8}\right)+\left(-\frac{4}{12}i-\frac{9}{12}i\right)=-\frac{1}{8}-\frac{13}{12}i$$

25)
$$10i-(5+9i)+(5+i)$$
$$=10i-5-9i+5+i=2i$$

27) Both are products of binomials, so we can multiply both using FOIL.

29) $2(7-6i)=14-12i$

31) $-\frac{5}{2}(8+3i)=-20-\frac{15}{2}i$

33)
$$3i(-2+7i)=-6i+21i^2$$
$$=-6i+21(-1)$$
$$=-21-6i$$

35) $5i(4-3i)=20i-15i^2$
$=20i-15(-1)$
$=15+20i$

37) $(3+i)(2+8i)=6+24i+2i+8i^2$
$=6+26i+8(-1)$
$=6+26i-8$
$=-2+26i$

39) $(-1+3i)(4-6i)=-4+6i+12i-18i^2$
$=-4+18i-18(-1)$
$=-4+18i+18$
$=14+18i$

41) $(6-2i)(8-7i)=48-42i-16i+14i^2$
$=48-58i+14(-1)$
$=48-58i-14$
$=34-58i$

43) $\left(\frac{2}{3}+\frac{2}{3}i\right)\left(\frac{1}{5}+\frac{2}{5}i\right)$
$=\frac{2}{15}+\frac{4}{15}i+\frac{2}{15}i+\frac{4}{15}i^2$
$=\frac{2}{15}+\frac{6}{15}i+\frac{4}{15}(-1)$
$=\frac{2}{15}+\frac{6}{15}i-\frac{4}{15}=-\frac{2}{15}+\frac{2}{5}i$

45) $(3+5i)$ conjugate $(3-5i)$
Product $=(3+5i)(3-5i)$
$=9-15i+15i-25i^2$
$=9-25(-1)=9+25=34$

47) $(-10-2i)$ conjugate $(-10+2i)$
Product $(-10-2i)(-10+2i)$
$=100-20i+20i-4i^2$
$=100-4(-1)=100+4=104$

49) $(-7+9i)$ conjugate $(-7-9i)$
Product $(-7+9i)(-7-9i)$
$=49+63i-63i-81i^2$
$=49-81(-1)=49+81=130$

51) Multiply the numerator and denominator by the conjugate of the denominator. Write the result in the form $a+bi$.

53) $\frac{2}{1-6i}=\frac{2}{1-6i}\cdot\frac{1+6i}{1+6i}=\frac{2+12i}{1^2+6^2}$
$=\frac{2+12i}{1+36}=\frac{2+12i}{37}=\frac{2}{37}+\frac{12}{37}i$

55) $\frac{8}{-3+5i}=\frac{8}{-3+5i}\cdot\frac{-3-5i}{-3-5i}=\frac{-24-40i}{(-3)^2+5^2}$
$=\frac{-24-40i}{9+25}=\frac{-24-40i}{9+25}$
$=-\frac{24}{34}-\frac{40}{34}i=-\frac{12}{17}-\frac{20}{17}i$

57) $\dfrac{-10i}{-6+3i}=\dfrac{-10i}{-6+3i}\cdot\dfrac{-6-3i}{-6-3i}$
$=\dfrac{60i+30i^2}{(-6)^2+3^2}=\dfrac{60i+30(-1)}{36+9}$
$=\dfrac{60i-30}{45}=-\dfrac{30}{45}+\dfrac{60}{45}i$
$=-\dfrac{2}{3}+\dfrac{4}{3}i$

59) $\dfrac{3-8i}{-6+7i}=\dfrac{3-8i}{-6+7i}\cdot\dfrac{-6-7i}{-6-7i}$
$=\dfrac{-18-21i+48i+56i^2}{(-6)^2+7^2}$
$=\dfrac{-18+27i+56(-1)}{36+49}$
$=\dfrac{-74+27i}{85}=-\dfrac{74}{85}+\dfrac{27}{85}i$

61) $\dfrac{1+8i}{2-5i}=\dfrac{1+8i}{2-5i}\cdot\dfrac{2+5i}{2+5i}$
$=\dfrac{2+5i+16i+40i^2}{2^2+5^2}$
$=\dfrac{2+21i+40(-1)}{4+25}$
$=\dfrac{-38+21i}{29}=-\dfrac{38}{29}+\dfrac{21}{29}i$

63) $\dfrac{7}{i}=\dfrac{7}{i}\cdot\dfrac{-i}{-i}=\dfrac{-7i}{1^2}=-7i$

65) $t^2=-9$
$t=\pm\sqrt{-9}$
$d=\pm 3i$ $\quad\{-3i,3i\}$

67) $n^2+11=0$
$n^2=-11$
$n=\pm\sqrt{-11}$
$n=\pm i\sqrt{11}$ $\quad\{-i\sqrt{11},i\sqrt{11}\}$

69) $z^2-5=-33$
$z^2=5-33$
$z^2=-28$
$z=\pm\sqrt{-28}$
$z=\pm 2i\sqrt{7}$ $\quad\{-2i\sqrt{7},2i\sqrt{7}\}$

71) $-23=3p^2+52$
$3p^2=-23-52$
$3p^2=-75$
$p^2=-25$
$p=\pm\sqrt{-25}$
$p=\pm 5i$ $\quad\{-5i,5i\}$

73) $(c+3)^2-4=-29$
$(c+3)^2=-25$
$c+3=\pm\sqrt{-25}$
$c+3=\pm 5i$
$c=-3\pm 5i$ $\quad\{-3-5i,-3+5i\}$

75) $3=23+(d-8)^2$
$(d-8)^2=-20$
$d-8=\pm\sqrt{-20}$
$d-8=\pm 2i\sqrt{5}$
$d=8\pm 2i\sqrt{5}$
$\{8-2i\sqrt{5},8+2i\sqrt{5}\}$

77) $5+(4b+1)^2=2$

$$(4b+1)^2=-3$$
$$4b+1=\pm\sqrt{-3}$$
$$4b+1=\pm i\sqrt{3}$$
$$4b=-1\pm i\sqrt{3}$$
$$b=\frac{-1\pm i\sqrt{3}}{4}$$
$$\left\{\frac{-1-i\sqrt{3}}{4},\frac{-1+i\sqrt{3}}{4}\right\}$$

79) $3x^2+x+2=0$

$a=3$, $b=1$ and $c=2$

$$x=\frac{-1\pm\sqrt{1^2-4(3)(2)}}{2(3)}$$
$$=\frac{-1\pm\sqrt{1-24}}{6}$$
$$=\frac{-1\pm\sqrt{-23}}{6}=\frac{-1\pm i\sqrt{23}}{6}$$
$$\left\{\frac{-1-i\sqrt{23}}{6},\frac{-1+i\sqrt{23}}{6}\right\}$$

81) $h^2+4=-2h$

$$h^2+2h+4=0$$

$a=1$, $b=2$ and $c=4$

$$h=\frac{-2\pm\sqrt{2^2-4(1)(4)}}{2(1)}$$
$$=\frac{-2\pm\sqrt{4-16}}{2}=\frac{-2\pm\sqrt{-12}}{2}$$
$$=\frac{-2\pm 2i\sqrt{3}}{2}=-1\pm i\sqrt{3}$$
$$\left\{-1+i\sqrt{3},-1-i\sqrt{3}\right\}$$

83) $4k=k^2+29$

$$k^2-4k+29=0$$

$a=1$, $b=-4$ and $c=29$

$$v=\frac{-(-4)\pm\sqrt{(-4)^2-4(1)(29)}}{2(1)}$$
$$=\frac{4\pm\sqrt{16-116}}{2}=\frac{4\pm\sqrt{-100}}{2}$$
$$=\frac{4\pm 10i}{2}=2\pm 5i$$
$$\{2+5i,2-5i\}$$

85) $2(n^2+3n)=7$

$$2n^2+6n-7=0$$

$a=2$, $b=6$ and $c=-7$

$$v=\frac{-6\pm\sqrt{6^2-4(2)(-7)}}{2(2)}$$
$$=\frac{-6\pm\sqrt{36+56}}{4}=\frac{-6\pm\sqrt{92}}{4}$$
$$=\frac{-6\pm 2\sqrt{23}}{4}=\frac{-3\pm\sqrt{23}}{2}$$
$$\left\{\frac{-3-\sqrt{23}}{2},\frac{-3+\sqrt{23}}{2}\right\}$$

87) $4q^2=2q-3$

$$4q^2-2q+3=0$$

$a=4$, $b=-2$ and $c=3$

$$q=\frac{-(-2)\pm\sqrt{(-2)^2-4(4)(3)}}{2(4)}$$
$$=\frac{2\pm\sqrt{4-48}}{8}=\frac{2\pm\sqrt{-44}}{8}$$
$$=\frac{2\pm 2i\sqrt{11}}{8}=\frac{1\pm i\sqrt{11}}{4}$$
$$\left\{\frac{1-i\sqrt{11}}{4},\frac{1+i\sqrt{11}}{4}\right\}$$

89) $3y(y+2)=-5$

$3y^2+6y+5=0$

$a=3,\ b=6$ and $c=5$

$$y=\frac{-6\pm\sqrt{6^2-4(3)(5)}}{2(3)}$$

$$=\frac{-6\pm\sqrt{36-60}}{6}=\frac{-6\pm\sqrt{-24}}{6}$$

$$=\frac{-6\pm 2i\sqrt{6}}{6}=\frac{-3\pm i\sqrt{6}}{3}$$

$$\left\{\frac{-3-i\sqrt{6}}{3},\frac{-3+i\sqrt{6}}{3}\right\}$$

91) $5g^2+9=12g$

$5g^2-12g+9=0$

$a=5,\ b=-12$ and $c=9$

$$g=\frac{-(-12)\pm\sqrt{(-12)^2-4(5)(9)}}{2(5)}$$

$$=\frac{12\pm\sqrt{144-180}}{10}=\frac{12\pm\sqrt{-36}}{10}$$

$$=\frac{12\pm 6i}{10}=\frac{6\pm 3i}{5}$$

$$\left\{\frac{6-3i}{5},\frac{6+3i}{5}\right\}$$

93) $n^2-4n+5=0$

$n^2-4n=-5$

$n^2-4n+4=4-5$

$(n-2)^2=-1$

$n-2=\pm\sqrt{-1}$

$n=2\pm i$

$\{2-i,2+i\}$

95) $z^2+8z+29=0$

$z^2+8z=-29$

$z^2+8z+16=16-29$

$(z+4)^2=-13$

$z+4=\pm\sqrt{-13}$

$z=-4\pm i\sqrt{13}$

$\{-4-i\sqrt{13},-4+i\sqrt{13}\}$

97) $2p^2-12p+54=0$

$p^2-6p=-27$

$p^2-6p+9=9-27$

$(p-3)^2=-18$

$p-3=\pm\sqrt{-18}$

$p=3\pm 3i\sqrt{2}$

$\{3-3i\sqrt{2},3+3i\sqrt{2}\}$

99) $3k^2+9k+12=0$

$k^2+3k=-4$

$$k^2+3k+\frac{9}{4}=\frac{9}{4}-4$$

$$\left(k+\frac{3}{2}\right)^2=-\frac{7}{4}$$

$$k+\frac{3}{2}=\pm\sqrt{\frac{-7}{4}}$$

$$k=-\frac{3}{2}\pm i\frac{\sqrt{7}}{2}$$

$$\left\{-\frac{3}{2}-i\frac{\sqrt{7}}{2},-\frac{3}{2}+i\frac{\sqrt{7}}{2}\right\}$$

Section 10.5: Exercises

1) The x-coordinate of the vertex is $-\dfrac{b}{2a}$. Substitute that value into the equation to find the y-coordinate of the vertex.

3) Vertex: $(0, -4)$

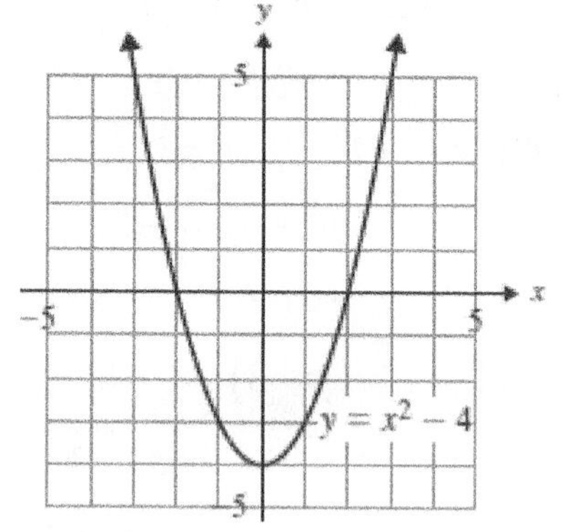

5) Vertex: $(0, -1)$

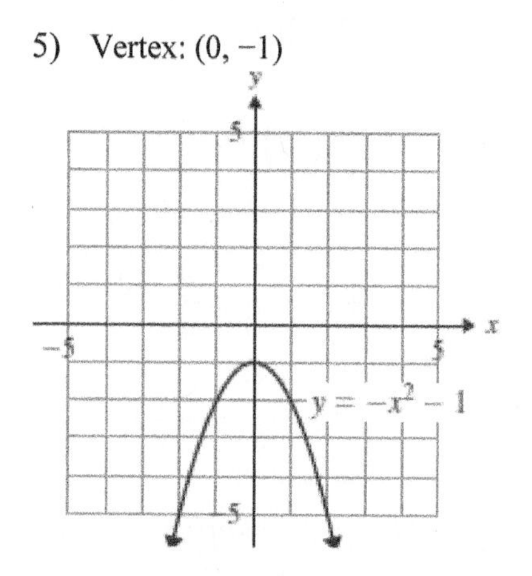

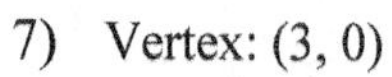

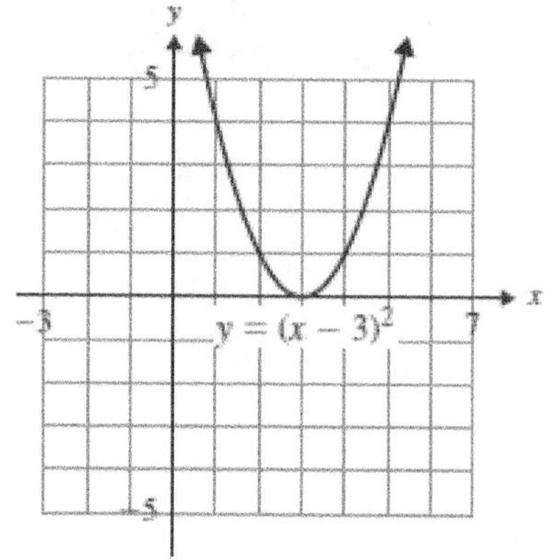

9) Vertex: $(2, -1)$

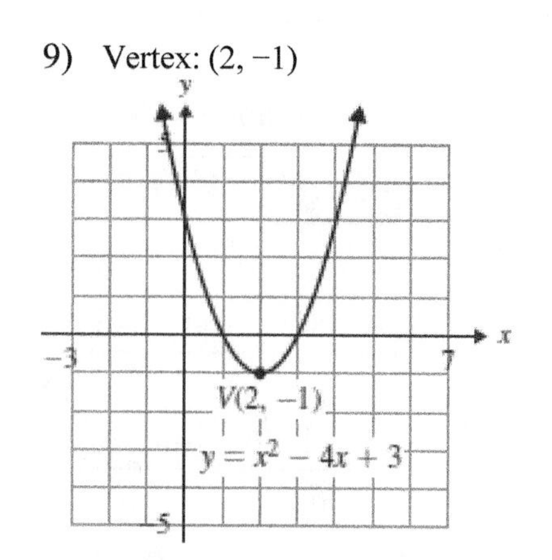

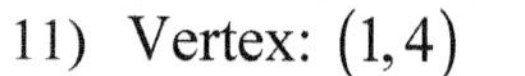

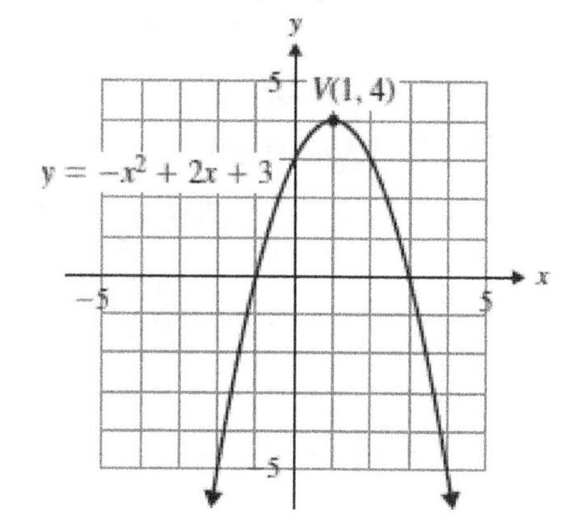

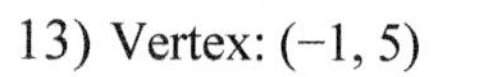

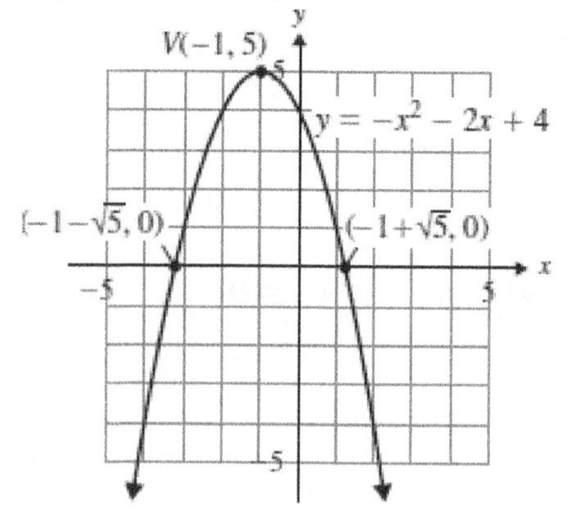

15) Vertex: $(-3, 2)$

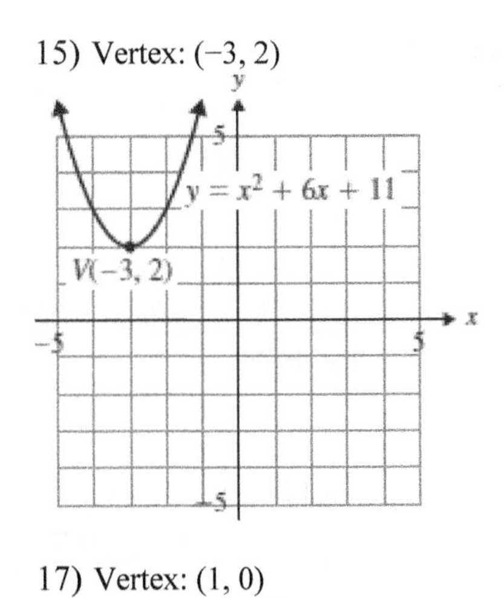

17) Vertex: $(1, 0)$

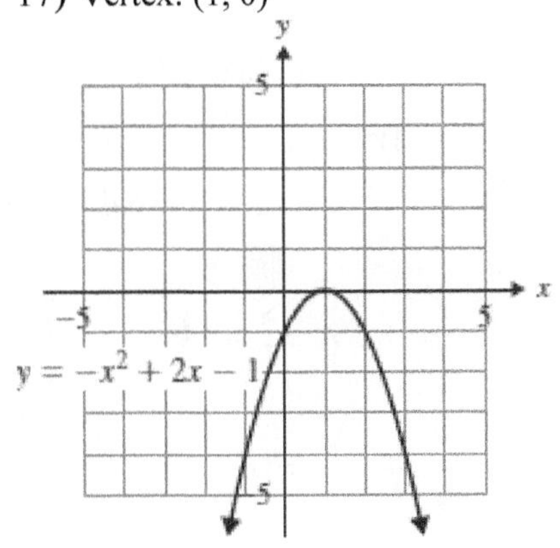

19) Vertex: $(1, -2)$

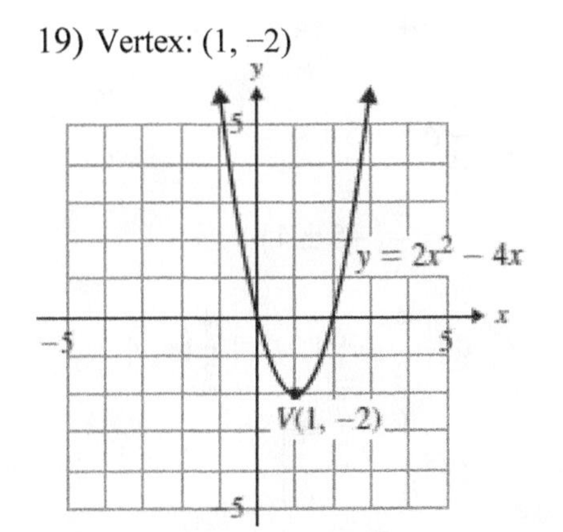

21)
$$x^2 - 2x - 3 = 0$$
$$(x-3)(x+1) = 0$$
$$x = 3,1$$
$$\{-1,3\}$$

23)
$$x^2 + 2x + 1 = 0$$
$$(x+1)(x+1) = 0$$
$$x = -1$$
$$\{-1\}$$

25)
$$x^2 - 4x + 5 = 0$$
$$x^2 - 4x = -5$$
$$x^2 - 4x + 4 = -5 + 4$$
$$(x-2)^2 = -1$$
$$x - 2 = \pm\sqrt{-1}$$
No real soltuion. $\varnothing$

27) $(0.4, 3)$

29) $(-4.5, -0.25)$

31) $(-3.65, 1.65)$

33) Answers may vary. One equation is
$(2,0)$ and $(4,0)$
$$x = 2 \qquad x = 4$$
$$x - 2 = 0 \qquad x - 4 = 0$$
$$(x-2)(x-4) = 0$$
$$x^2 - 6x + 8 = 0$$
$$y = x^2 - 6x + 8$$

35) Answers may vary. One equation is
$(-3,0)$ and $(1,0)$
$$x = -3 \qquad x = 1$$
$$x + 3 = 0 \qquad x - 1 = 0$$
$$(x+3)(x-1) = 0$$
$$x^2 + 2x - 3 = 0$$
$$y = x^2 + 2x - 3$$

37) $(-\infty,\infty)$

39) $[-2,\infty)$

41) $(-\infty,0]$

43) $(-\infty,3]$

45) $[-1,\infty)$

47) a) $x=-\frac{b}{2a}=-\frac{16}{2(-2)}=-\frac{16}{-4}$

$=-(-4)=4$

b) $f(4)=-2x^2+16x-25$

$f(4)=-2(4)^2+16(4)-25$

$f(4)=-2(16)+64-25$

$f(4)=-32+64-25$

$f(4)=7$

$g(2)=-3(2)^2+12(2)-2$

$g(2)=-3(4)+24-2$

$g(2)=-12+24-2$

$g(2)=10$

49) a) $t=-\frac{b}{2a}=-\frac{288}{2(-16)}=-\frac{288}{-32}$

$=-(-9)=9$

b) $h(t)=-16t^2+288t$

$h(9)=-16(9)^2+288(9)$

$h(9)=-16(81)+2592$

$h(9)=-1296+2592$

$h(9)=1296$

1296 ft

Chapter 10: Review Exercises

1) $x^2=81$

$x=\pm\sqrt{81}$

$x=9 \quad x=-9 \qquad \{-9,9\}$

3) $3a^2+7=40$

$3a^2=33$

$a^2=11$

$a=\pm\sqrt{11} \qquad (-\sqrt{11},\sqrt{11})$

5) $5=(k-4)^2+18$

$(k-4)^2=-13$

There is no real solution. $\varnothing$

7) $\left(\frac{3}{4}c+5\right)^2-2=14$

$\left(\frac{3}{4}c+5\right)^2=16$

$\frac{3}{4}c+5=\pm4$

$\frac{3}{4}c=-5\pm4$

$c=\frac{4}{3}(-5\pm4)$

$c=\frac{4}{3}(-9);\ p=\frac{4}{3}(-1)$

$c=\left\{-12,\frac{4}{3}\right\}$

9) $x^2+18x+81=(x+9)^2$

11) $z^2-3z+\frac{9}{4}=\left(z-\frac{3}{2}\right)^2$

13) $d^2+10d+9=0$

$$d^2+10d=-9$$
$$d^2+10d+25=-9+25$$
$$(d+5)^2=16$$
$$d+5=\pm\sqrt{16}$$
$$d+5=\pm4$$
$$d+5=4 \text{ or } d+5=-4$$
$$z=-1 \qquad z=-9$$
$$\{-9,-1\}$$

15) $r^2+15=4r$

$$r^2-4r=-15$$
$$r^2-4r+4=-15+4$$
$$(r-2)^2=-11$$

There is no real solution. $\varnothing$

17) $a^2+5a+2=0$

$$a^2+5a=-2$$
$$a^2+5a+\frac{25}{4}=-2+\frac{25}{4}$$
$$\left(a+\frac{5}{2}\right)^2=\frac{17}{4}$$
$$a+\frac{5}{2}=\pm\frac{\sqrt{17}}{2}$$
$$a=-\frac{5}{2}\pm\frac{\sqrt{17}}{2}$$
$$\left\{-\frac{5}{2}-\frac{\sqrt{17}}{2},-\frac{5}{2}+\frac{\sqrt{17}}{2}\right\}$$

19) $4n^2-8n=21$

$$n^2-2n=\frac{21}{4}$$
$$n^2-2n+1=\frac{21}{4}+1$$
$$(n-1)^2=\frac{25}{4}$$
$$n-1=\pm\frac{5}{2}$$
$$n=1\pm\frac{5}{2}$$
$$n=-\frac{3}{2};\quad n=\frac{7}{2}$$
$$\left\{-\frac{3}{2},\frac{7}{2}\right\}$$

21) $x=\dfrac{-b\pm\sqrt{b^2-4ac}}{2a}$

23) $x^2+7x+12=0$

$a=1$, $b=7$ and $c=12$

$$x=\frac{7\pm\sqrt{7^2-4(1)(12)}}{2(1)}$$
$$=\frac{7\pm\sqrt{49-48}}{2}$$
$$=\frac{7\pm\sqrt{1}}{2}=\frac{7\pm1}{2}$$
$$x=\frac{-7-1}{2}=\frac{-8}{2}=-4,\quad x=\frac{-7+1}{2}=\frac{-6}{2}=-3$$
$$\{-4,-3\}$$

25) $2r^2+3r=-6$

$$2r^2+3r+6=0$$

$a=2$, $b=3$ and $c=6$

$$r=\frac{-3\pm\sqrt{3^2-4(2)(6)}}{2(2)}$$

$$=\frac{-3\pm\sqrt{9-48}}{4}=\frac{-3\pm\sqrt{-39}}{4}$$

There is no real solution. $\varnothing$

27) $t(t-10)=-7$

$$t^2-10t+7=0$$

$a=1$, $b=-10$ and $c=7$

$$t=\frac{-(-10)\pm\sqrt{(-10)^2-4(1)(7)}}{2(1)}$$

$$=\frac{10\pm\sqrt{100-28}}{2}=\frac{10\pm\sqrt{72}}{2}$$

$$=\frac{10\pm6\sqrt{2}}{2}=5\pm3\sqrt{2}$$

$$\{5-3\sqrt{2},5+3\sqrt{2}\}$$

29) $(2w+3)(w+4)=6(w+1)$

$$2w^2+8w+3w+12=6w+6$$

$$2w^2+11w+12-6w-6=0$$

$$2w^2+5w+6=0$$

$a=2$, $b=5$ and $c=6$

$$w=\frac{5\pm\sqrt{5^2-4(2)(6)}}{2(2)}$$

$$=\frac{5\pm\sqrt{25-48}}{4}=\frac{5\pm\sqrt{-23}}{4}$$

There is no real solution. $\varnothing$

31) a) Let $h=6$ and solve for t.

$$6=-16t^2+22t+26$$

$$0=-16t^2+22t+20$$

$$0=8t^2-11t-10$$

$$0=(8t+5)(t-2)$$

$8t+5=0$ or $t-2=0$

$8t=-5$ $\quad \boxed{t=2}$

$$t=-\frac{5}{8}$$

The ball reaches 8 feet after 2 sec.

b) Let $h=0$ and solve for t.

$$0=-16t^2+22t+26$$

$$0=8t^2-11t-13$$

$a=8$, $b=-11$ and $c=-13$

$$t=\frac{-(-11)\pm\sqrt{(-11)^2-4(8)(-13)}}{2(8)}$$

$$=\frac{11\pm\sqrt{121+416}}{16}=\frac{11\pm\sqrt{537}}{16}$$

Reject $t=\dfrac{11-\sqrt{537}}{16}$ because it is negative.

The ball will hit the ground after $\dfrac{11+\sqrt{537}}{16}$ sec ≈ 2.14 sec.

33) $2y(y-3)=-1$

$$2y^2-6y=-1$$
$$2y^2-6y+1=0$$

$a=2,\ b=-6$ and $c=1$

$$y=\frac{-(-6)\pm\sqrt{(-6)^2-4(2)(1)}}{2(2)}$$
$$=\frac{6\pm\sqrt{36-8}}{4}=\frac{6\pm\sqrt{28}}{4}$$
$$=\frac{6\pm2\sqrt{7}}{4}=\frac{2\left(3\pm\sqrt{7}\right)}{4}=\frac{3\pm\sqrt{7}}{2}$$
$$\left\{\frac{3-\sqrt{7}}{2},\frac{3+\sqrt{7}}{2}\right\}$$

35) $\frac{1}{18}p^2+\frac{7}{9}p+\frac{8}{3}=0$

$$18\left(\frac{1}{18}p^2+\frac{7}{9}p+\frac{8}{3}\right)=18(0)$$
$$p^2+14p+48=0$$
$$(p+6)(p+8)=0$$
$$p+6=0,\quad p+8=0$$
$$p=-6,\quad p=-8$$
$$\{-8,\ -6\}$$

37) $(3a-4)^2+15=11$

$$(3a-4)^2=-4$$
$$3a-4=\pm\sqrt{-4}$$

There is no real solution. $\varnothing$

39) $4m^2-25=0$

$$4m^2=25$$
$$m^2=\frac{25}{4}$$
$$m=\pm\frac{5}{2}\qquad\left\{-\frac{5}{2},\frac{5}{2}\right\}$$

41) $w^2+6w+4=0$

$a=1,\ b=6$ and $c=4$

$$w=\frac{-6\pm\sqrt{6^2-4(1)(4)}}{2(1)}$$
$$=\frac{-6\pm\sqrt{36-16}}{2}=\frac{-6\pm\sqrt{20}}{2}$$
$$=\frac{-6\pm2\sqrt{5}}{2}=-3\pm\sqrt{5}$$
$$\left\{-3-\sqrt{5},-3+\sqrt{5}\right\}$$

43) $0.01x^2+0.09x+0.12=0$

$$100\left(0.01x^2+0.09x+0.12\right)=100(0)$$
$$x^2+9x+12=0$$

$a=1,\ b=9$ and $c=12$

$$x=\frac{-9\pm\sqrt{9^2-4(1)(12)}}{2(1)}$$
$$=\frac{-9\pm\sqrt{81-48}}{2}$$
$$=\frac{-9\pm\sqrt{33}}{2}$$
$$\left\{\frac{-9-\sqrt{33}}{2},\frac{-9+\sqrt{33}}{2}\right\}$$

45) $\sqrt{-6}=\sqrt{-1}\cdot\sqrt{6}=i\sqrt{6}$

47) $\sqrt{-125}=\sqrt{-1}\cdot\sqrt{125}=i\cdot 5\sqrt{5}=5i\sqrt{5}$

49) $(4+i)-(1-7i)$
$=4+i-1+7i=3+8i$

51) $(-2-11i)-(9+4i)+(3-6i)$
$=-2-11i-9-4i+3-6i=-8-21i$

53) $\left(\frac{2}{3}-\frac{1}{8}i\right)+\left(\frac{5}{6}+\frac{2}{5}i\right)$
$=\left(\frac{2}{3}+\frac{5}{6}\right)+\left(-\frac{1}{8}i+\frac{2}{5}i\right)$
$=\left(\frac{4}{6}+\frac{5}{6}\right)+\left(-\frac{5}{40}i+\frac{16}{40}i\right)=\frac{3}{2}+\frac{11}{40}i$

55) $-4(5-9i)=-20+36i$

57) $2i(7+i)=14i+2i^2$
$=14i+2(-1)$
$=-2+14i$

59) $(3+2i)(7-6i)=21-18i+14i-12i^2$
$=21-4i-12(-1)$
$=21-4i+12$
$=33-4i$

61) Multiply the numerator and denominator by the conjugate of the denominator.

63) $\frac{2i}{5-6i}=\frac{2i}{5-6i}\cdot\frac{5+6i}{5+6i}$
$=\frac{10i+12i^2}{5^2+6^2}$
$=\frac{10i+12(-1)}{25+36}=\frac{-12+10i}{61}$
$=-\frac{12}{61}+\frac{10}{61}i$

65) $\frac{6-6i}{1+7i}=\frac{6-6i}{1+7i}\cdot\frac{1-7i}{1-7i}$
$=\frac{6-42i-6i+42i^2}{1^2+7^2}$
$=\frac{6-48i+42(-1)}{1+49}$
$=\frac{-36-48i}{50}=\frac{-36}{50}-\frac{48}{50}i$
$=-\frac{18}{25}-\frac{24}{25}i$

67) a) $(5+2i)+(3-8i)$
$=5+3+2i-8i=8-6i$

b) $(5+2i)(3-8i)$
$=15-40i+6i-16i^2$
$=15-34i-16(-1)$
$=15-34i+16$
$=31-34i$

69) $\frac{5}{2}(-6+7i)=-15+\frac{35}{2}i$

71) $(9-i)(2+3i)$
$=18+27i-2i-3i^2$
$=18+25i-3(-1)$
$=18+25i+3$
$=21+25i$

73) $(10+3i)-(2-11i)$
$=10-2+3i+11i=8+14i$

75) $\dfrac{6-8i}{3-2i}=\dfrac{6-8i}{3-2i}\cdot\dfrac{3+2i}{3+2i}$

$=\dfrac{18+12i-24i-16i^2}{3^2+2^2}$

$=\dfrac{18-12i-16(-1)}{9+4}$

$=\dfrac{34-12i}{13}=\dfrac{34}{13}-\dfrac{12}{13}i$

77) $4y^2+19=6$

$4y^2=6-19$

$4y^2=-13$

$y^2=-\dfrac{13}{4}$

$y=\pm\dfrac{\sqrt{-17}}{2}$

$y=\pm i\dfrac{\sqrt{-13}}{2}$ $\quad\left\{-i\dfrac{\sqrt{13}}{2},i\dfrac{\sqrt{13}}{2}\right\}$

79) $3x^2+2x+5=0$

$a=3$, $b=2$ and $c=5$

$x=\dfrac{-2\pm\sqrt{2^2-4(3)(5)}}{2(3)}$

$=\dfrac{-2\pm\sqrt{4-60}}{6}=\dfrac{-2\pm\sqrt{-56}}{6}$

$=\dfrac{-2\pm 2i\sqrt{14}}{6}=\dfrac{2\left(-1\pm i\sqrt{14}\right)}{6}$

$=\dfrac{-1\pm i\sqrt{14}}{3}$

$\left\{\dfrac{-1-i\sqrt{14}}{3},\dfrac{-1+i\sqrt{14}}{3}\right\}$

81) $k(k+10)=-34$

$k^2+10k=-34$

$k^2+10k+34=0$

$a=1$, $b=10$ and $c=34$

$k=\dfrac{-10\pm\sqrt{10^2-4(1)(34)}}{2(1)}$

$=\dfrac{-10\pm\sqrt{100-136}}{2}$

$=\dfrac{-10\pm\sqrt{-36}}{2}$

$=\dfrac{-10\pm 6i}{2}$

$=-5\pm 3i$

$\{-5-3i,-5+3i\}$

83) $(2a-1)^2+11=9$

$(2a-1)^2=-2$

$2a-1=\pm\sqrt{-2}$

$2a-1=\pm i\sqrt{2}$

$2a=1\pm i\sqrt{2}$

$a=\dfrac{1\pm i\sqrt{2}}{2}$

$\left\{\dfrac{1-i\sqrt{2}}{2},\dfrac{1+i\sqrt{2}}{2}\right\}$

85) The x-coordinate of the vertex is $-\dfrac{b}{2a}$. Substitute this value into the equation to find the y-ccordinate of the vertex.

87) $y=(x+2)^2=x^2+4x+4$

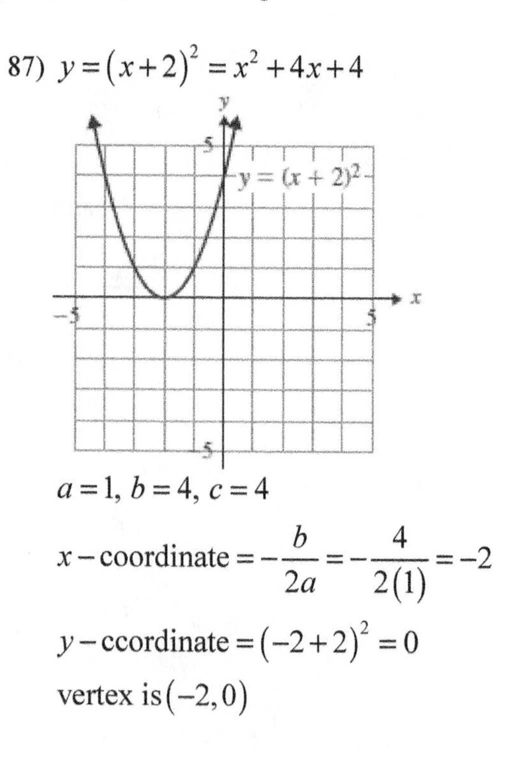

$a=1,\ b=4,\ c=4$

$x-\text{coordinate}=-\dfrac{b}{2a}=-\dfrac{4}{2(1)}=-2$

$y-\text{ccordinate}=(-2+2)^2=0$

vertex is $(-2,0)$

89) $y=-x^2+2x+2$

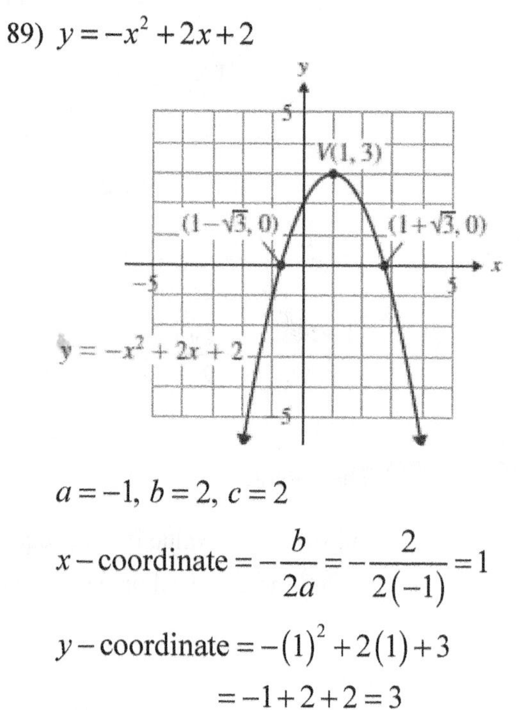

$a=-1,\ b=2,\ c=2$

$x-\text{coordinate}=-\dfrac{b}{2a}=-\dfrac{2}{2(-1)}=1$

$y-\text{coordinate}=-(1)^2+2(1)+3$

$=-1+2+2=3$

vertex is $(1,3)$

91) $y=x^2+6x+8$

For $x^2+6x+8=0$, the solutions and $x-$intercepts are: $x=-4,\quad x=-2$

93) vertex is $(0.59,3.41)$

95) Domain $(-\infty,\infty)$ Range $(-\infty,5]$

97) Domain $(-\infty,\infty)$ Range $[-3,\infty)$

Chapter 10 Test

1) $3a^2+4=22$

$3a^2=18$

$a^2=6$

$a=\pm\sqrt{6}$

$\{-\sqrt{6},\sqrt{6}\}$

3) $m^2-12m+26=0$

$m^2-12m=-26$

$m^2-12m+36=-26+36$

$(m-6)^2=10$

$m-6=\pm\sqrt{10}$

$m=6\pm\sqrt{10}$

$\{6-\sqrt{10},6+\sqrt{10}\}$

5) $\dfrac{12\pm8\sqrt{5}}{4}=\dfrac{12}{4}\pm\dfrac{8\sqrt{5}}{4}=3\pm2\sqrt{5}$

7) $(p-7)^2+1=13$

$(p-7)^2=12$

$p-7=\pm\sqrt{12}$

$p=7\pm2\sqrt{3}$

$\{7-2\sqrt{3},7+2\sqrt{3}\}$

9) $(2k-3)(2k+5)=6(k-2)$

$(2k-3)(2k+5)=6(k-2)$

$4k^2+10k-6k-15=6k-12$

$4k^2+4k-15-6k+12=0$

$4k^2-2k-3=0$

$a=4$, $b=-2$ and $c=-3$

$$k=\frac{-(-2)\pm\sqrt{(-2)^2-4(4)(-3)}}{2(4)}$$

$$=\frac{2\pm\sqrt{4+48}}{8}=\frac{2\pm\sqrt{52}}{8}=\frac{2\pm2\sqrt{13}}{8}$$

$$=\frac{2(1\pm\sqrt{13})}{8}=\frac{1\pm\sqrt{13}}{4}$$

$$\left\{\frac{1-\sqrt{13}}{4},\frac{1+\sqrt{13}}{4}\right\}$$

11) a) Let $h=8$ and solve for t.

$8=-16t^2+40t+32$

$0=-16t^2+40t+24$

$0=2t^2-5t-3$

$0=(2t+1)(t-3)$

$2t+1=0$ or $t-3=0$

$2t=-1$ $\boxed{t=3}$

$t=-\frac{1}{2}$

The ball reaches 8 feet after 3 sec.

b) Let $h=0$ and solve for t.

$0=-16t^2+40t+32$

$0=2t^2-5t-4$

$a=2$, $b=-5$ and $c=-4$

$$t=\frac{-(-5)\pm\sqrt{(-5)^2-4(2)(-4)}}{2(2)}$$

$$=\frac{5\pm\sqrt{25+32}}{4}=\frac{5\pm\sqrt{57}}{4}$$

Reject $t=\frac{5-\sqrt{57}}{4}$ because it is negative.

The ball will hit the ground after $\frac{5+\sqrt{57}}{4}$ sec ≈ 3.14 sec.

13) $\sqrt{-48}=\sqrt{-1}\sqrt{48}=i\sqrt{16}\sqrt{3}=4i\sqrt{3}$

15) $(2-9i)-(6-10i)=2-6-9i+10i=-4+i$

17) $\dfrac{10}{9-2i}=\dfrac{10}{9-2i}\cdot\dfrac{9+2i}{9+2i}=\dfrac{90+20i}{9^2+2^2}$

$=\dfrac{90+20i}{81+4}=\dfrac{90}{85}+\dfrac{20}{85}i$

$=\dfrac{18}{17}+\dfrac{4}{17}i$

19) $(r+8)^2+15=12$

$(r+8)^2=-3$

$r+8=\pm\sqrt{-3}$

$r+8=\pm i\sqrt{3}$

$r=-8\pm 3i\sqrt{3}$

$\{-8-i\sqrt{3},-8+i\sqrt{3}\}$

21) $y=-(x-1)^2=-x^2+2x-1$

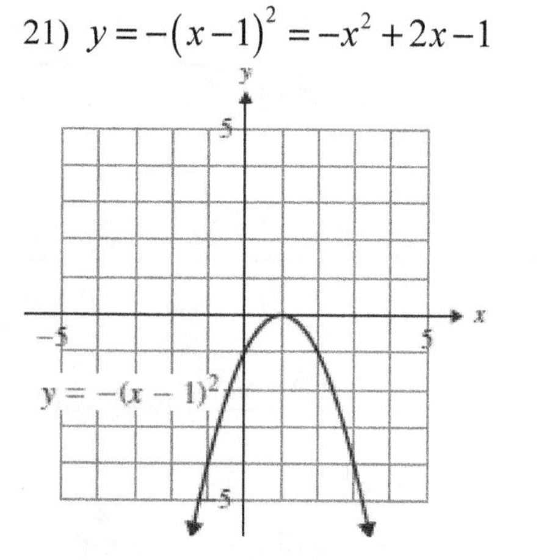

$a=-1,\ b=2,\ c=-1$

$x-\text{coordinate}=-\dfrac{b}{2a}=-\dfrac{2}{2(-1)}=1$

$y-\text{coordinate}=-(1)^2+2(1)-1=-1+2-1=0$

vertex is $(1,0)$

23) $y=-x^2+4x-1$

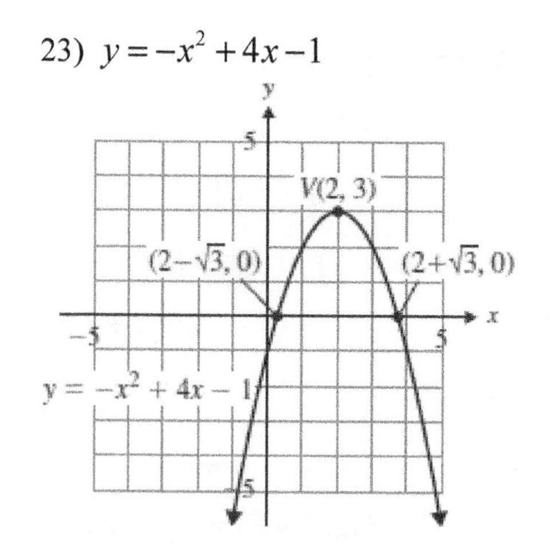

$a=-1,\ b=4,\ c=-1.$

x-coordinate $=-\dfrac{b}{2a}=-\dfrac{4}{2(-1)}=2$

y-ccordinate $=-(2)^2+4(2)-1=-4+8-1=3$

vertex is $(2,3)$

25) Domain $(-\infty,\infty)$ Range $[-4,\infty)$

Cumulative Review: Chapters 1-10

1) $\dfrac{5}{8}-\dfrac{2}{7}=\dfrac{35}{56}-\dfrac{16}{56}=-\dfrac{19}{56}$

3) Perimeter $=4+8+4+8=24$ mm

Area $=(\text{base})(\text{height})=(8)(3.7)=29.6\,\text{mm}^2$

5) $-66+49=-17^\circ$

7) a) $(-12)^0=1$

b) $(5)^0+(4)^0=1+1=2$

c) $-6^{-2}=-\dfrac{1}{6^2}=-\dfrac{1}{36}$

d) $2^{-4}=\dfrac{1}{2^4}=\dfrac{1}{16}$

9) 5.75×10^{-5}

11)
$$\begin{aligned} 8b-7&=57 \\ 8b-7+7&=57+7 \\ 8b&=64 \\ \frac{8b}{8}&=\frac{64}{8} \\ b&=8 \end{aligned}$$
$\{8\}$

13) Let x = a number
$$\begin{aligned} x+9&=2x-1 \\ x+9+1&=2x-1+1 \\ x+10&=2x \\ x+10-x&=2x-x \\ 10&=x \end{aligned}$$
The number is 10.

15)
$$\begin{aligned} -15<4p-7&\le 5 \\ -15+7<4p-7+7&\le 5+7 \\ -8<4p&\le 12 \\ \frac{-8}{4}<\frac{4p}{4}&\le\frac{12}{4} \\ -2<p&\le 3 \end{aligned}$$

$(-2, 3]$

17) $x-2y=6$

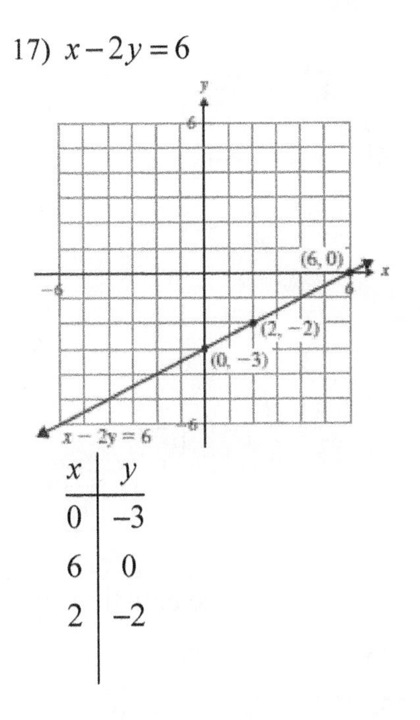

x	y
0	-3
6	0
2	-2

19) $(x_1, y_1)=(6, 5);\ m_{\text{perp}}=2$

$$\begin{aligned} y-y_1&=m(x-x_1) \\ y-5&=2(x-6) \\ y-5&=2x-12 \\ y&=2x-12+5 \\ y&=2x-7 \end{aligned}$$

21)

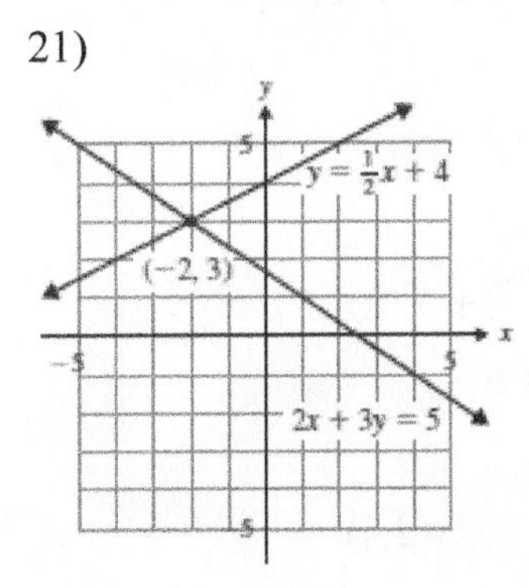

23) $\frac{3}{4}x - y = \frac{1}{2}$

$4\left(\frac{3}{4}x - y\right) = 4 \cdot \frac{1}{2}$

$3x - 4y = 2$ $\boxed{\text{I}}$

$-\frac{x}{3} + \frac{y}{2} = -\frac{1}{6}$

$6\left(-\frac{x}{3} + \frac{y}{2}\right) = 6 \cdot \left(-\frac{1}{6}\right)$

$-2x + 3y = -1$ $\boxed{\text{II}}$

Multiply the equation $\boxed{\text{I}}$ by 2

$3x - 4y = 2$

$6x - 8y = 4$

Multiply the equation $\boxed{\text{II}}$ by 3

$-2x + 3y = -1$

$-6x + 9y = -3$

Add the equations

$6x - 8y = 4$

$-6x + 9y = -3$

$y = 1$

Substitute $y = 1$ into first equation.

$3x - 4y = 2$

$3x - 4(1) = 2$

$3x - 4 = 2$

$3x = 2 + 4$

$3x = 6$

$x = 2$

$(2,1)$

25)

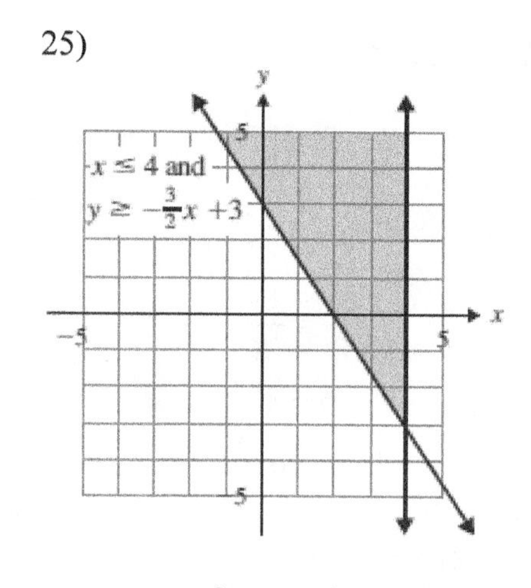

27)
$$\begin{array}{r} 5.8p^3 - 1.2p^2 + \quad p - 7.5 \\ +\ 2.1p^3 + 6.3p^2 + 3.8p + 3.9 \\ \hline 7.9p^3 + 5.1p^2 + 4.8p - 3.6 \end{array}$$

29) $-5(7w - 12)(w + 3)$

$= (-35w + 60)(w + 3)$

$= -35w^2 - 105w + 60w + 180$

$= -35w^2 - 45w + 180$

31) $t^2 - 2tu - 63u^2 = (t + 7u)(t - 9u)$

33) $a^2 + 16a + 64 = (a + 8)^2$

35) $x + 4 = \text{Length}$

$x + 1 = \text{Width}$

$28 = \text{Area}$

$(x + 4)(x + 1) = 60$

$x^2 + x + 4x + 4 = 28$

$x^2 + 5x - 24 = 0$

$(x - 3)(x + 8) = 0$

$x = 3$ or $x = -8$

Length: 7 cm; Width: 4 cm

37) $\frac{10p+3}{4p+4}-\frac{8}{p^2-6p-7}$

$=\frac{10p+3}{4(p+1)}-\frac{8}{(p-7)(p+1)}$

$=\frac{(10p+3)(p-7)}{4(p+1)(p-7)}-\frac{32}{4(p-7)(p+1)}$

$=\frac{10p^2-70p+3p-21-32}{4(p-7)(p+1)}=\frac{10p^2-67p-53}{4(p-7)(p+1)}$

39) $\dfrac{\frac{c}{c+2}+\frac{1}{c^2-4}}{1-\frac{3}{c+2}}$

$=\dfrac{(c+2)(c-2)\left(\frac{c}{c+2}+\frac{1}{(c+2)(c-2)}\right)}{(c+2)(c-2)\left(1-\frac{3}{c+2}\right)}$

$=\frac{c(c-2)+1}{(c+2)(c-2)-3(c-2)}$

$=\frac{c^2-2c+1}{c^2-4-3c+6}$

$=\frac{(c-1)^2}{c^2-3c+2}$

$=\frac{(c-1)(c-1)}{(c-1)(c-2)}$

$=\frac{c-1}{c-2}$

41) $\frac{\sqrt{200k^{21}}}{\sqrt{2k^5}}=\sqrt{100k^{16}}=10k^8$

43) $\frac{z-4}{\sqrt{z}+2}\cdot\frac{\sqrt{z}-2}{\sqrt{z}-2}=\frac{z-4}{z-4}\cdot\frac{\sqrt{z}-2}{1}=\sqrt{z}-2$

45) $\sqrt{6x+9}-\sqrt{2x+1}=4$

$\sqrt{6x+9}=4+\sqrt{2x+1}$

$\left(\sqrt{6x+9}\right)^2=\left(4+\sqrt{2x+1}\right)^2$

$6x+9=16+8\sqrt{2x+1}+2x+1$

$6x+9=2x+17+8\sqrt{2x+1}$

$4x-8=8\sqrt{2x+1}$

$x-2=2\sqrt{2x+1}$

$(x-2)^2=\left(2\sqrt{2x+1}\right)^2$

$x^2-4x+4=4(2x+1)$

$x^2-4x+4=8x+4$

$x^2-4x+4-8x-4=0$

$x^2-12x=0$

$x(x-12)=0$

Possible Solutions: $x=0 \quad x=12$

Check: $x=0$

$\sqrt{6(0)+9}-\sqrt{2(0)+1}=4$

$\sqrt{9}-\sqrt{1}=4$

$2=4$ False

Check: $x=12$

$\sqrt{6(12)+9}-\sqrt{2(12)+1}=4$

$\sqrt{81}-\sqrt{25}=4$

$9-5=4$

$4=4$ True

Solution $\{12\}$

47) $t^2+9=-4t$

$t^2+4t=-9$

$t^2+4t+4=-9+4$

$(t+2)^2=-5$

$t+2=\pm\sqrt{-5}$

$t=-2\pm i\sqrt{5}$

$\{-2-i\sqrt{5},-2+i\sqrt{5}\}$

49) $(4-6i)(3-6i)=12-24i-18i+36i^2$

$=12-42i+36(-1)$

$=12-42i-36$

$=-24-42i$

NOTES

NOTES

NOTES

NOTES

NOTES

NOTES

NOTES

NOTES

NOTES

NOTES

NOTES

NOTES

NOTES

NOTES